T0189165

Foundations of Quantitative Finance

Chapman & Hall/CRC Financial Mathematics Series

Series Editors

M.A.H. Dempster
Centre for Financial Research
Department of Pure Mathematics and Statistics
University of Cambridge, UK

Dilip B. Madan
Robert H. Smith School of Business
University of Maryland, USA

Rama Cont
Department of Mathematics
Imperial College, UK

Robert A. Jarrow
Lynch Professor of Investment Management
Johnson Graduate School of Management
Cornell University, USA

Recently Published Titles

Machine Learning for Factor Investing: Python Version
Guillaume Coqueret and Tony Guida

Introduction to Stochastic Finance with Market Examples, Second Edition
Nicolas Privault

Commodities: Fundamental Theory of Futures, Forwards, and Derivatives Pricing, Second Edition
Edited by M.A.H. Dempster and Ke Tang

Introducing Financial Mathematics: Theory, Binomial Models, and Applications
Mladen Victor Wickerhauser

Financial Mathematics: From Discrete to Continuous Time
Kevin J. Hastings

Financial Mathematics: A Comprehensive Treatment in Discrete Time
Giuseppe Campolieti and Roman N. Makarov

Introduction to Financial Derivatives with Python
Elisa Alòs and Raúl Merino

The Handbook of Price Impact Modeling
Kevin T. Webster

Sustainable Life Insurance: Managing Risk Appetite for Insurance Savings & Retirement Products
Aymeric Kalife with Saad Mouti, Ludovic Goudenege, Xiaolu Tan, and Mounir Bellmane

Geometry of Derivation with Applications
Norman L. Johnson

For more information about this series please visit: https://www.crcpress.com/Chapman-and-HallCRC-Financial-Mathematics-Series/book series/CHFINANCMTH

Foundations of Quantitative Finance

Book IV: Distribution Functions and Expectations

Robert R. Reitano
Brandeis International Business School
Waltham, MA

CRC Press
Taylor & Francis Group
Boca Raton London New York

CRC Press is an imprint of the
Taylor & Francis Group, an **informa** business

A CHAPMAN & HALL BOOK

First edition published 2023
by CRC Press
6000 Broken Sound Parkway NW, Suite 300, Boca Raton, FL 33487-2742

and by CRC Press
4 Park Square, Milton Park, Abingdon, Oxon, OX14 4RN

CRC Press is an imprint of Taylor & Francis Group, LLC

ISBN: 978-1-032-20653-0 (hbk)
ISBN: 978-1-032-20652-3 (pbk)
ISBN: 978-1-003-26458-3 (ebk)

DOI: 10.1201/9781003264583

Typeset in CMR10
by KnowledgeWorks Global Ltd.

Publisher's note: This book has been prepared from camera-ready copy provided by the authors.

to Dorothy and Domenic

Contents

Preface   xi

Author   xiii

Introduction   xv

1 Distribution and Density Functions   **1**
  1.1   Summary of Book II Results 1
      1.1.1   Distribution Functions on $\mathbb{R}$ 1
      1.1.2   Distribution Functions on $\mathbb{R}^n$ 3
  1.2   Decomposition of Distribution Functions on $\mathbb{R}$ 6
  1.3   Density Functions on $\mathbb{R}$ 10
      1.3.1   The Lebesgue Approach 10
      1.3.2   Riemann Approach 13
      1.3.3   Riemann-Stieltjes Framework 14
  1.4   Examples of Distribution Functions on $\mathbb{R}$ 16
      1.4.1   Discrete Distribution Functions 16
      1.4.2   Continuous Distribution Functions 19
      1.4.3   Mixed Distribution Functions 25

2 Transformed Random Variables   **29**
  2.1   Monotonic Transformations 29
  2.2   Sums of Independent Random Variables 32
      2.2.1   Distribution Functions of Sums 32
      2.2.2   Density Functions of Sums 39
  2.3   Ratios of Random Variables 43
      2.3.1   Independent Random Variables 44
      2.3.2   Example without Independence 53

3 Order Statistics   **57**
  3.1 M-Samples and Order Statistics 57
  3.2   Distribution Functions of Order Statistics 59
  3.3   Density Functions of Order Statistics 60
  3.4   Joint Distribution of All Order Statistics 62
  3.5   Density Functions on $\mathbb{R}^n$ 67
  3.6   Multivariate Order Functions 69
      3.6.1   Joint Density of All Order Statistics 69
      3.6.2   Marginal Densities and Distributions 70
      3.6.3   Conditional Densities and Distributions 73
  3.7   The Rényi Representation Theorem 75

4 Expectations of Random Variables 1   **81**

  4.1 General Definitions 81
      4.1.1 Is Expectation Well Defined? 83
      4.1.2 Formal Resolution of Well-Definedness 85
  4.2 Moments of Distributions 88
      4.2.1 Common Types of Moments 88
      4.2.2 Moment Generating Function 89
      4.2.3 Moments of Sums – Theory 91
      4.2.4 Moments of Sums – Applications 94
      4.2.5 Properties of Moments 98
      4.2.6 Examples–Discrete Distributions 102
      4.2.7 Examples–Continuous Distributions 104
  4.3 Moment Inequalities 108
      4.3.1 Chebyshev's Inequality 109
      4.3.2 Jensen's Inequality 111
      4.3.3 Kolmogorov's Inequality 113
      4.3.4 Cauchy-Schwarz Inequality 114
      4.3.5 Hölder and Lyapunov Inequalities 117
  4.4 Uniqueness of Moments 119
      4.4.1 Applications of Moment Uniqueness 123
  4.5 Weak Convergence and Moment Limits 125

5 Simulating Samples of RVs – Examples   **135**

  5.1 Random Samples 135
      5.1.1 Discrete Distributions 135
      5.1.2 Simpler Continuous Distributions 140
      5.1.3 Normal and Lognormal Distributions 141
      5.1.4 Student T Distribution 144
  5.2 Ordered Random Samples 147
      5.2.1 Direct Approaches 147
      5.2.2 The Rényi Representation 149

6 Limit Theorems   **153**

  6.1 Introduction 153
  6.2 Weak Convergence of Distributions 156
      6.2.1 Student T to Normal 156
      6.2.2 Poisson Limit Theorem 158
      6.2.3 "Weak Law of Small Numbers" 159
      6.2.4 De Moivre-Laplace Theorem 161
      6.2.5 The Central Limit Theorem 1 165
      6.2.6 Smirnov's Theorem on Uniform Order Statistics 168
      6.2.7 A Limit Theorem on General Quantiles 172
      6.2.8 A Limit Theorem on Exponential Order Statistics 174
  6.3 Laws of Large Numbers 175
      6.3.1 Tail Events and Kolmogorov's 0-1 Law 176
      6.3.2 Weak Laws of Large Numbers 178
      6.3.3 Strong Laws of Large Numbers 183
      6.3.4 A Limit Theorem in EVT 187
  6.4 Convergence of Empirical Distributions 189
      6.4.1 Definition and Basic Properties 190
      6.4.2 The Glivenko-Cantelli Theorem 195
      6.4.3 Distributional Estimates for $D_n(s)$ 197

7 Estimating Tail Events 2   **201**

     7.1   Large Deviation Theory 2 201

           7.1.1   Chernoff Bound 202

           7.1.2   Cramér-Chernoff Theorem 208

     7.2   Extreme Value Theory 2   210

           7.2.1   Fisher-Tippett-Gnedenko Theorem 211

           7.2.2   The Hill Estimator, $\gamma > 0$ 212

           7.2.3 $F \in D(G_\gamma)$ is Asymptotically Pareto for $\gamma > 0$ 217

           7.2.4 $F \in D(G_\gamma)$, $\gamma > 0$, then $\gamma_H \approx \gamma$ 226

           7.2.5 $F \in D(G_\gamma)$, $\gamma > 0$, then $\gamma_H \to_1 \gamma$ 232

           7.2.6   Asymptotic Normality of the Hill Estimator 236

           7.2.7   The Pickands-Balkema-de Haan Theorem: $\gamma > 0$ 237

Bibliography   **243**

Index   **247**

Preface

The idea for a reference book on the mathematical foundations of quantitative finance has been with me throughout my professional and academic careers in this field, but the commitment to finally write it didn't materialize until completing my first "introductory" book in 2010.

My original academic studies were in "pure" mathematics in a field of mathematical analysis, and neither applications generally nor finance in particular were then even on my mind. But on completion of my degree, I decided to temporarily investigate a career in applied math, becoming an actuary, and in short order became enamored with mathematical applications in finance.

One of my first inquiries was into better understanding yield curve risk management, ultimately introducing the notion of partial durations and related immunization strategies. This experience led me to recognize the power of greater precision in the mathematical specification and solution of even an age-old problem. From there my commitment to mathematical finance was complete, and my temporary investigation into this field became permanent.

In my personal studies, I found that there were a great many books in finance that focused on markets, instruments, models and strategies, and which typically provided an informal acknowledgement of the background mathematics. There were also many books in mathematical finance focusing on more advanced mathematical models and methods, and typically written at a level of mathematical sophistication requiring a reader to have significant formal training and the time and motivation to derive omitted details.

The challenge of acquiring expertise is compounded by the fact that the field of quantitative finance utilizes advanced mathematical theories and models from a number of fields. While there are many good references on any of these topics, most are again written at a level beyond many students, practitioners and even researchers of quantitative finance. Such books develop materials with an eye to comprehensiveness in the given subject matter, rather than with an eye toward efficiently curating and developing the theories needed for applications in quantitative finance.

Thus the overriding goal I have for this collection of books is to provide a complete and detailed development of the many foundational mathematical theories and results one finds referenced in popular resources in finance and quantitative finance. The included topics have been curated from a vast mathematics and finance literature for the express purpose of supporting applications in quantitative finance.

I originally budgeted 700 pages per book, in two volumes. It soon became obvious this was too limiting, and two volumes ultimately turned into ten. In the end, each book was dedicated to a specific area of mathematics or probability theory, with a variety of applications to finance that are relevant to the needs of financial mathematicians.

My target readers are students, practitioners and researchers in finance who are quantitatively literate, and recognize the need for the materials and formal developments presented. My hope is that the approach taken in these books will motivate readers to navigate these details and master these materials.

Most importantly for a reference work, all ten volumes are extensively self-referenced. The reader can enter the collection at any point of interest, and then using the references cited, work backwards to prior books to fill in needed details. This approach also works for a course on a given volume's subject matter, with earlier books used for reference, and for both course-based and self-study approaches to sequential studies.

The reader will find that the developments herein are presented at a much greater level of detail than most advanced quantitative finance books. Such developments are of necessity typically longer, more meticulously reasoned, and therefore can be more demanding on the reader. Thus before committing to a detailed line-by-line study of a given result, it is always more efficient to first scan the derivation once or twice to better understand the overall logic flow.

I hope the additional details presented will support your journey to better understanding.

I am grateful for the support of my family: Lisa, Michael, David, and Jeffrey, as well as the support of friends and colleagues at Brandeis International Business School.

Robert R. Reitano
Brandeis International Business School

Author

Robert R. Reitano is Professor of the Practice of Finance at the Brandeis International Business School where he specializes in risk management and quantitative finance. He previously served as MSF Program Director, and Senior Academic Director. He has a PhD in mathematics from MIT, is a fellow of the Society of Actuaries, and a Chartered Enterprise Risk Analyst. Dr. Reitano consults in investment strategy and asset/liability risk management, and previously had a 29-year career at John Hancock/Manulife in investment strategy and asset/liability management, advancing to Executive Vice President & Chief Investment Strategist. His research papers have appeared in a number of journals and have won an Annual Prize of the Society of Actuaries and two F.M. Redington Prizes of the Investment Section of the Society of the Actuaries. Dr. Reitano serves on various not-for-profit boards and investment committees.

Introduction

Foundations of Quantitative Finance is structured as follows:
  Book I: *Measure Spaces and Measurable Functions*
  Book II: *Probability Spaces and Random Variables*
  Book III: *The Integrals of Riemann, Lebesgue, and (Riemann-)Stieltjes*
  Book IV: *Distribution Functions and Expectations*
  Book V: *General Measure and Integration Theory*
  Book VI: *Densities, Transformed Distributions, and Limit Theorems*
  Book VII: *Brownian Motion and Other Stochastic Processes*
  Book VIII: *Itô Integration and Stochastic Calculus 1*
  Book IX: *Stochastic Calculus 2 and Stochastic Differential Equations*
  Book X: *Classical Models and Applications in Finance*

The series is logically sequential. Books I, III, and V develop foundational mathematical results needed for the probability theory and finance applications of Books II, IV, and VI, respectively. Then Books VII, VIII, and IX develop results in the theory of stochastic processes. While these latter three books introduce ideas from finance as appropriate, the final realization of the applications of these stochastic models to finance is deferred to Book X.

This Book IV, *Distribution Functions and Expectations,* extends the investigations of Book II using the formidable tools afforded by the Riemann, Lebesgue, and Riemann-Stieltjes integration theories of Book III.

To set the stage, Chapter 1 opens with a short review of the key results on distribution functions from Books I and II. The focus here is on the connections between distribution functions of random variables and random vectors and distribution functions induced by Borel measures on $\mathbb{R}$ and $\mathbb{R}^n$. A complete functional characterization of distribution functions on $\mathbb{R}$ is then derived, providing a natural link between general probability theory and the discrete and continuous theories commonly encountered. This leads to an investigation into the existence of density functions associated with various distribution functions. Here the integration theories from Book III are recalled to frame this investigation, and the general results to be seen in Book VI using the Book V integration theory are introduced. The chapter ends with a catalog of many common distribution and density functions from the discrete and continuous probability theories.

Chapter 2 investigates transformations of random variables. For example, given a random variable X and associated distribution/density function, what is the distribution/density function of the random variable $g(X)$ given a Borel measurable function $g(x)$? More generally, what are the distribution functions and densities of sums and ratios of random variables, where now $g(x)$ is a multivariate function? The first section addresses the distribution function question for strictly monotonic $g(x)$ and the density question when such $g(x)$ is differentiable. More general transformations are deferred to Book VI using the change of variable results from the integration theory of Book V. A number of results are then derived for the distribution and density functions of sums of independent random variables using the integration theories of Book III. The various forms of such distribution functions then reflect the assumptions made on the underlying distribution functions and/or density functions. Examples and exercises connect the theory with the Chapter 1 catalog of distribution functions. The chapter ends with an investigation into ratios of independent random variables, as well as an example using dependent random variables.

A special example of an order statistic was introduced in Chapter 9 of Book II on extreme value theory, where this random variable was defined as the maximum of a collection of independent, identically distributed random variables. Order statistics, the subject of Chapter 3, generalize this notion, converting such a collection into ordered random variables. Distribution and density functions of such variates are first derived, before turning to the joint distribution function of all order statistics. This latter derivation introduces needed combinatorial ideas as well as results on multivariate integration from Book III and a more general result from Book V. Various density functions of order statistics are then derived, beginning with the joint density and then proceeding to the various marginal and conditional densities of these random variables. The final investigation is into the Rényi representation theorem for the order statistics of an exponential distribution. While seemingly of narrow applicability as a result of exponential variables, this theorem will be seen to be more widely applicable.

Expectations of random variables and transformed random variables are introduced in Chapter 4 in the general context of a Riemann-Stieltjes integral. In the special case of discrete or continuous probability theory, this definition reduces to the familiar notions from these theories using Book III results. But this definition also raises existence and consistency questions. The roadmap to a final solution is outlined, foretelling needed results from the integration theory of Book V and the final detailed resolution in Book VI. Various moments, the moment generating function, and properties of such are then developed, as well as examples from the distribution functions introduced earlier. Moment inequalities of Chebyshev, Jensen, Kolmogorov, Cauchy-Schwarz, Hölder, and Lyapunov are derived, before turning to the question of uniqueness of moments and the moment generating function. The chapter ends with an investigation of weak convergence of distributions and moment limits, developing a number of results underlying the "method of moments."

Given a random variable X defined on a probability space, Chapter 4 of Book II derived the theoretical basis for, and several constructions of, a probability space on which could be defined a countable collection of independent random variables, identically distributed with X. Such spaces provide a rigorous framework for the laws of large numbers of that book, and the limit theorems of this book's Chapter 6. This framework is in the background for Chapter 5, but the focus here is on the actual generation of random sample collections using the previous theory and the various distribution functions introduced in previous chapters. The various sections then exemplify simulation approaches for discrete distributions, and then continuous distributions, using the left-continuous inverse function $F^*(y)$ and independent, continuous uniform variates commonly provided by various mathematical software. For generating normal, lognormal, and Student's T variates, these constructions are, at best, approximate, and the chapter derives the exact constructions underlying the Box-Muller transform and the Bailey transform, respectively. The final section turns to the simulation of order statistics, both directly and with the aid of the Rényi representation theorem.

Chapter 6 begins with a more formal short review of the theoretical framework of Book II for the construction of a probability space on which a countable collection of independent, identically distributed random variables can be defined, and thus on which limit theorems of various types can be addressed. The first section then addresses weak convergence of various distribution function sequences. Among those studied are the Student's T, Poisson, DeMoivre-Laplace, and a first version of the central limit theorem, as well as Smirnov's result on uniform order statistics, a general result on exponential order statistics, and finally a limit theorem on quantiles. The next section generalizes the study of laws of large numbers of Book II using moment defined limits, and proves a limit theorem on extreme value theory identified in that book. The final section studies empirical distribution functions, and in particular, derives the Glivenko-Cantelli theorem on convergence of empirical distributions to the underlying distribution function. Kolmogorov's theorem on the limiting distribution of the maximum error in an empirical distribution is also discussed, as are related results.

Continuing the study initiated in Chapter 9 of Book II, Chapter 7 again has two main themes. The first topic is large deviation theory. Following a summary of the main result and open questions of Book II, the section introduces and exemplifies the Chernoff bound, which requires the existence of the moment generating function. Following an analysis of properties of this bound, and introducing tilted distributions and their relevant properties, the section concludes with the Cramér-Chernoff theorem, which conclusively settles the open questions of Book II. The second major section is on extreme value theory and focuses on two matters. The first is a study of the Hill estimator for the extreme value index γ for $\gamma > 0$, the index values most commonly encountered in finance applications. This estimator is introduced and exemplified in the context of Pareto distributions, and the Hill result of convergence with probability 1 derived, along with a variety of related results. For this, earlier developments in order statistics will play a prominent role, as does a representation theorem of Karamata. The second major investigation is into the Pickands-Balkema-de Haan theorem, a result that identifies the limiting distribution of certain conditional tail distributions. This final result was approximated in the Book II development, but here it can be derived in detail with another representation theorem of Karamata. Using an example, it is then shown that the convergence promised by this result need not be fast.

I hope this book and the other books in the collection serve you well.

Notation 0.1 (Referencing within FQF Series) *To simplify the referencing of results from other books in this series, we use the following convention.*

A reference to "Proposition I.3.33" is a reference to Proposition 3.33 of Book I, while "Chapter III.4" is a reference to Chapter 4 of Book III, and "II.(8.5)" is a reference to formula (8.5) of Book II, and so forth.

1

Distribution and Density Functions

1.1 Summary of Book II Results

As the goal of this book is to study distribution functions, it is perhaps worthwhile to summarize some of the key results of earlier books.

Notation 1.1 ($\mu \to \lambda$) *In Book II, probability spaces were generally denoted by $(\mathcal{S}, \mathcal{E}, \mu)$, where $\mathcal{S}$ is the measure space, here also called a "sample" space, $\mathcal{E}$ is the sigma algebra of measurable sets, here also called the collection of "events," and μ is the probability measure defined on all sets in $\mathcal{E}$. In this book we retain most of this notational convention. However, because μ will often be called upon in later chapters to represent the "mean" of a given distribution as is conventional, we will represent the probability measure herein by λ or by another Greek letter.*

1.1.1 Distribution Functions on $\mathbb{R}$

Book II introduced definitions and basic properties. Beginning with Definition II.3.1:

Definition 1.2 (Random variable) *Given a probability space $(\mathcal{S}, \mathcal{E}, \lambda)$, a **random variable (r.v.)** is a real-valued function:*

$$X : \mathcal{S} \longrightarrow \mathbb{R},$$

such that for any bounded or unbounded interval, $(a, b) \subset \mathbb{R}$:

$$X^{-1}(a, b) \in \mathcal{E}.$$

*The **distribution function (d.f.)**, or **cumulative distribution function (c.d.f.)**, associated with X, denoted by F or F_X, is defined on $\mathbb{R}$ by*

$$F(x) = \lambda[X^{-1}(-\infty, x]]. \tag{1.1}$$

Properties of such functions were summarized in Proposition II.6.1.

Proposition 1.3 (Properties of a d.f. $F(x)$) *Given a random variable X on a probability space $(\mathcal{S}, \mathcal{E}, \lambda)$, the distribution function $F(x)$ associated with X has the following properties:*

1. *$F(x)$ is a nonnegative, increasing function on $\mathbb{R}$ which is Borel, and hence, Lebesgue measurable.*

2. *For all x :*

$$\lim_{y \to x+} F(y) = F(x), \tag{1.2}$$

DOI: 10.1201/9781003264583-1

and thus $F(x)$ is right continuous, and:

$$\lim_{y \to x-} F(y) = F(x) - \lambda(\{X(s) = x\}), \tag{1.3}$$

so $F(x)$ has left limits.

3. *$F(x)$ has, at most, countably many discontinuities.*

4. *The limits of $F(x)$ exists as $x \to \pm\infty$:*

$$\lim_{y \to -\infty} F(y) = 0. \tag{1.4}$$

$$\lim_{y \to \infty} F(y) = 1. \tag{1.5}$$

This result also reflects the characterizing properties of such distribution functions. A function $F : \mathbb{R} \to \mathbb{R}$ can be identified as a distribution function by Proposition II.3.6:

Proposition 1.4 (Identifying a distribution function) *Let $F(x)$ be an increasing function, which is right continuous and satisfies $F(-\infty) = 0$ and $F(\infty) = 1$, defined as limits.*

Then there exists a probability space $(\mathcal{S}, \mathcal{E}, \lambda)$ and random variable X so that $F(x) = \lambda[X^{-1}(-\infty, x]]$. In other words, every such function is the distribution function of a random variable.

Distribution functions are also intimately linked with Borel measures by Proposition II.6.3. Recall that $\mathcal{A}'$ denoted the semi-algebra of right semi-closed intervals $(a, b]$, $\mathcal{A}$ the associated algebra of finite disjoint unions of such sets, and $\mathcal{B}(\mathbb{R})$ the Borel sigma algebra of Definition I.2.13. By definition, $\mathcal{A} \subset \mathcal{B}(\mathbb{R})$.

The outer measure $\lambda_{\mathcal{A}}^*$ is defined in I.(5.8) and induced by the set function $\lambda_{\mathcal{A}}$ defined on $\mathcal{A}'$ by (1.6) and extended to $\mathcal{A}$ by finite additivity. A set A is said to be $\lambda_{\mathcal{A}}^*$-measurable, or **Carathéodory measurable** after **Constantin Carathéodory** (1873–1950), if for every set $E \subset \mathbb{R}$:

$$\lambda_{\mathcal{A}}^*(E) = \lambda_{\mathcal{A}}^* \left(A \bigcap E \right) + \lambda_{\mathcal{A}}^* \left(\tilde{A} \bigcap E \right).$$

Proposition 1.5 (The Borel measure λ_F induced by d.f. $F : \mathbb{R} \to \mathbb{R}$) *Given a probability space $(\mathcal{S}, \mathcal{E}, \lambda)$ and random variable $X : \mathcal{S} \longrightarrow \mathbb{R}$, the associated distribution function F defined on $\mathbb{R}$ by (1.1), induces a **unique probability measure** λ_F on the Borel sigma algebra $\mathcal{B}(\mathbb{R})$ that is defined on $\mathcal{A}'$ by:*

$$\lambda_F [(a, b]] = F(b) - F(a). \tag{1.6}$$

*In detail, with $\mathcal{M}_F(\mathbb{R})$ defined as the **collection of $\lambda_{\mathcal{A}}^*$-measurable sets**:*

1. *$\mathcal{A} \subset \mathcal{M}_F(\mathbb{R})$, and for all $A \in \mathcal{A}$:*

$$\lambda_{\mathcal{A}}^*(A) = \lambda_{\mathcal{A}}(A).$$

2. *$\mathcal{M}_F(\mathbb{R})$ is a complete sigma algebra and thus contains every set $A \subset \mathbb{R}$ with $\lambda_{\mathcal{A}}^*(A) = 0$.*

3. *$\mathcal{M}_F(\mathbb{R})$ contains the Borel sigma algebra, $\mathcal{B}(\mathbb{R}) \subset \mathcal{M}_F(\mathbb{R})$.*

4. *If λ_F denotes the restriction of $\lambda_{\mathcal{A}}^*$ to $\mathcal{M}_F(\mathbb{R})$, then λ_F is a probability measure and hence $(\mathbb{R}, \mathcal{M}_F(\mathbb{R}), \lambda_F)$ is a complete probability space.*

5. *The probability measure λ_F is the unique extension of $\lambda_{\mathcal{A}}$ from $\mathcal{A}$ to the smallest sigma algebra generated by $\mathcal{A}$, which is $\mathcal{B}(\mathbb{R})$ by Proposition I.8.1.*

6. *For all $A \in \mathcal{B}(\mathbb{R})$:*

$$\lambda_F(A) = \mu \left[X^{-1}(A) \right]. \tag{1.7}$$

1.1.2 Distribution Functions on $\mathbb{R}^n$

Definitions II.3.28 and II.3.30 provided equivalent formulations for the notion of a joint distribution of a random vector. This equivalence was proved in Proposition II.3.32.

Definition 1.6 (Random vector 1) *If $X_j : \mathcal{S} \longrightarrow \mathbb{R}$ are random variables on a probability space $(\mathcal{S}, \mathcal{E}, \lambda)$, $j = 1, 2, ..., n$, define the **random vector** $X = (X_1, X_2, ..., X_n)$ as the vector-valued function:*

$$X : \mathcal{S} \longrightarrow \mathbb{R}^n,$$

defined on $s \in \mathcal{S}$ by:

$$X(s) = (X_1(s), X_2(s), ..., X_n(s)).$$

*The **joint distribution function (d.f.)**, or **joint cumulative distribution function (c.d.f.)**, associated with X, denoted by F or F_X, is then defined on $(x_1, x_2, ..., x_n) \in \mathbb{R}^n$ by*

$$F(x_1, x_2, ..., x_n) = \lambda \left[\bigcap_{j=1}^{n} X_j^{-1}(-\infty, x_j] \right]. \tag{1.8}$$

Definition 1.7 (Random vector 2) *Given a probability space $(\mathcal{S}, \mathcal{E}, \lambda)$, a **random vector** is a mapping $X : \mathcal{S} \longrightarrow \mathbb{R}^n$, so that for all $A \in \mathcal{B}(\mathbb{R}^n)$, the sigma algebra of Borel measurable sets on $\mathbb{R}^n$:*

$$X^{-1}(A) \in \mathcal{E}. \tag{1.9}$$

*The **joint distribution function (d.f.)**, or **joint cumulative distribution function (c.d.f.)**, associated with X, denoted by F or F_X, is then defined on $(x_1, x_2, ..., x_n) \in \mathbb{R}^n$ by*

$$F(x_1, x_2, ..., x_n) = \lambda \left[X^{-1} \left(\prod_{j=1}^{n}(-\infty, x_j] \right) \right]. \tag{1.10}$$

Properties of such functions, and the link to Borel measures on $\mathbb{R}^n$, were summarized in Proposition II.6.9. But first, recall the definitions of **continuous from above** and *n*-**increasing**.

Definition 1.8 (Continuous from above; n-increasing) *A function $F : \mathbb{R}^n \to \mathbb{R}$ is said to be **continuous from above** if given $x = (x_1, ..., x_n)$ and a sequence $x^{(m)} = (x_1^{(m)}, ..., x_n^{(m)})$ with $x_i^{(m)} \geq x_i$ for all i and m, and $x^{(m)} \to x$ as $m \to \infty$, then:*

$$F(x) = \lim_{m \to \infty} F(x^{(m)}). \tag{1.11}$$

*Further, F is said to be n-**increasing** or to satisfy the n-**increasing condition,** if for any bounded right semi-closed rectangle $\prod_{i=1}^{n}(a_i, b_i]$:*

$$\sum_x sgn(x)F(x) \geq 0. \tag{1.12}$$

Each $x = (x_1, ..., x_n)$ in the summation is one of the 2^n vertices of this rectangle, so $x_i = a_i$ or $x_i = b_i$, and $sgn(x)$ is defined as -1 if the number of a_i-components of x is odd, and $+1$ otherwise.

These properties are common to all joint distribution functions, and are exactly the properties needed to generate Borel measures, by Proposition II.6.9. Below we quote this result and add properties from Propositions I.8.15 and I.8.16.

Proposition 1.9 (The Borel measure λ_F induced by d.f. $F : \mathbb{R}^n \to \mathbb{R}$) *Let $X : \mathcal{S} \longrightarrow \mathbb{R}^n$ be a random vector defined on $(\mathcal{S}, \mathcal{E}, \lambda)$, and $F : \mathbb{R}^n \to \mathbb{R}$ the joint distribution function defined in (1.8) or (1.10). Then F is continuous from above and satisfies the n-increasing condition.*

In addition, $F(-\infty) = 0$ and $F(\infty) = 1$ defined as limits, meaning that for any $\epsilon > 0$, there exists N so that $F(x_1, x_2, ..., x_n) \geq 1 - \epsilon$ if all $x_i \geq N$, and $F(x_1, x_2, ..., x_n) \leq \epsilon$ if all $x_i \leq -N$.

Thus by Propositions I.8.15 and I.8.16, F induces a unique probability measure λ_F on the Borel sigma algebra $\mathcal{B}(\mathbb{R}^n)$, so that for all bounded right semi-closed rectangles $\prod_{i=1}^{n}(a_i, b_i] \subset \mathbb{R}^n$:

$$\lambda_F \left(\prod_{i=1}^{n}(a_i, b_i] \right) = \sum_x sgn(x)F(x). \tag{1.13}$$

In addition, for all $A \in \mathcal{B}(\mathbb{R}^n)$:

$$\lambda_F(A) = \lambda \left[X^{-1}(A) \right], \tag{1.14}$$

and thus by (1.10), for $A = \prod_{j=1}^{n}(-\infty, x_j]$:

$$\lambda_F \left(\prod_{j=1}^{n}(-\infty, x_j] \right) = F(x).$$

We can identify distribution functions by Proposition II.6.10.

Proposition 1.10 ($(\mathcal{S}, \mathcal{E}, \lambda)$ and X induced by d.f. $F : \mathbb{R}^n \to \mathbb{R}$) *Given a function $F : \mathbb{R}^n \to \mathbb{R}$ that is n-increasing and continuous from above, with $F(-\infty) = 0$ and $F(\infty) = 1$ defined as limits, there exists a probability space $(\mathcal{S}, \mathcal{E}, \lambda)$ and a random vector X so that F is the joint distribution function of X.*

Joint distribution functions induce both **marginal distribution functions** and **conditional distribution functions**. We recall Definitions II.3.34 and II.3.39.

Definition 1.11 (Marginal distribution function) *Let $X : \mathcal{S} \longrightarrow \mathbb{R}^n$ be the random vector $X = (X_1, X_2, ..., X_n)$ defined on $(\mathcal{S}, \mathcal{E}, \lambda)$, where $X_j : \mathcal{S} \longrightarrow \mathbb{R}$ are random variables on $(\mathcal{S}, \mathcal{E}, \lambda)$ for $j = 1, 2, ..., n$.*

1. **Special Case** $n = 2$: *Given the joint distribution function $F(x_1, x_2)$ defined on $\mathbb{R}^2$, the two **marginal distribution functions** on $\mathbb{R}$, $F(x_1)$ and $F(x_2)$, are defined by:*

$$F_1(x_1) \equiv \lim_{x_2 \to \infty} F(x_1, x_2), \quad F_2(x_2) \equiv \lim_{x_1 \to \infty} F(x_1, x_2). \tag{1.15}$$

2. **General Case:** *Given $F(x_1, x_2, ..., x_n)$ and $I = \{i_1, ..., i_m\} \subset \{1, 2, ..., n\}$, let $x_J \equiv (x_{j_1}, x_{j_2}, ..., x_{j_{n-m}})$ for $j_k \in J \equiv \tilde{I}$. The **marginal distribution function** $F_I(x_I) \equiv F_I(x_{i_1}, x_{i_2}, ..., x_{i_m})$ is defined on $\mathbb{R}^m$ by:*

$$F_I(x_I) \equiv \lim_{x_J \to \infty} F(x_1, x_2, ..., x_n). \tag{1.16}$$

Definition 1.12 (Conditional distribution function) *Let $X : \mathcal{S} \longrightarrow \mathbb{R}^n$ be the random vector $X = (X_1, X_2, ..., X_n)$ defined on $(\mathcal{S}, \mathcal{E}, \lambda)$, $J \equiv \{j_1, ..., j_m\} \subset \{1, 2, ..., n\}$ and $X_J \equiv (X_{j_1}, X_{j_2}, ..., X_{j_m})$. Given a Borel set $B \in \mathcal{B}(\mathbb{R}^m)$ with $\lambda \left[X_J^{-1}(B) \right] \neq 0$, define the **conditional distribution function** of X given $X_J \in B$, denoted by $F(x|X_J \in B) \equiv F(x_1, x_2, ..., x_n|X_J \in B)$, in terms of the **conditional probability measure**:*

$$F(x|X_J \in B) \equiv \lambda \left[X^{-1} \left(\prod_{i=1}^{n}(-\infty, x_i] \right) \middle| X_J^{-1}(B) \right].$$

Thus by Definition II.1.31,

$$F(x|X_J \in B) = \lambda \left[X^{-1} \left(\prod_{i=1}^n (-\infty, x_i] \right) \bigcap X_J^{-1}(B) \right] \Big/ \lambda \left[X_J^{-1}(B) \right]. \tag{1.17}$$

This distribution function is sometimes denoted by $F_{J|B}(x)$.

Finally, recall the notion of **independent random variables** from Definition II.3.47, and the characterization of the joint distribution function of such variables by Proposition II.3.53.

Definition 1.13 (Independent random variables/vectors) *If $X_j : S \longrightarrow \mathbb{R}$ are random variables on $(S, \mathcal{E}, \lambda)$, $j = 1, 2, ..., n$, we say that $\{X_j\}_{j=1}^n$ are **independent random variables** if $\{\sigma(X_j)\}_{j=1}^n$ are independent sigma algebras in the sense of Definition II.1.15. That is, given $\{B_j\}_{j=1}^n$ with $B_j \in \sigma(X_j)$:*

$$\lambda \left(\bigcap_{j=1}^n B_j \right) = \prod_{j=1}^n \lambda(B_j). \tag{1.18}$$

Equivalently, given $\{A_j\}_{j=1}^n \subset \mathcal{B}(\mathbb{R})$:

$$\lambda \left(\bigcap_{j=1}^n X_j^{-1}(A_j) \right) = \prod_{j=1}^n \lambda \left(X_j^{-1}(A_j) \right). \tag{1.19}$$

*If $X_j : S \longrightarrow \mathbb{R}^{n_j}$ are random vectors on $(S, \mathcal{E}, \lambda)$, $j = 1, 2, ..., n$, we say that $\{X_j\}_{j=1}^n$ are **independent random vectors** if $\{\sigma(X_j)\}_{j=1}^n$ are independent sigma algebras. That is, given $\{B_j\}_{j=1}^n$ with $B_j \in \sigma(X_j)$:*

$$\lambda \left(\bigcap_{j=1}^n B_j \right) = \prod_{j=1}^n \lambda(B_j). \tag{1.20}$$

Equivalently, given $\{A_j\}_{j=1}^n$ with $A_j \in \mathcal{B}(\mathbb{R}^{n_j})$:

$$\lambda \left(\bigcap_{j=1}^n X_j^{-1}(A_j) \right) = \prod_{j=1}^n \lambda \left(X_j^{-1}(A_j) \right). \tag{1.21}$$

*A countable collection of random variables $\{X_j\}_{j=1}^\infty$ defined on $(S, \mathcal{E}, \lambda)$ are said to be **independent random variables** if given any **finite** index subcollection, $J = (j(1), j(2), ..., j(n))$, $\{X_{j(i)}\}_{i=1}^n$ are independent random variables. The analogous definition of independence applies to a countable collection of **random vectors**.*

Proposition 1.14 (Characterization of the joint D.F.) *Let $X_j : S \longrightarrow \mathbb{R}$, $j = 1, 2, ..., n$, be random variables on a probability space $(S, \mathcal{E}, \lambda)$ with distribution functions $F_j(x)$. Let the random vector $X : S \longrightarrow \mathbb{R}^n$ be defined on this space by $X(s) = (X_1(s), X_2(s), ..., X_n(s))$ and with joint distribution function $F(x)$. Then $\{X_j\}_{j=1}^n$ are independent random variables if and only if:*

$$F(x_1, x_2, ..., x_n) = \prod_{j=1}^n F_j(x_j). \tag{1.22}$$

Countably many random variables $\{X_j\}_{j=1}^\infty$ defined on $(S, \mathcal{E}, \lambda)$ are independent random variables if and only if for every finite index subcollection, $J = (j(1), j(2), ..., j(n))$:

$$F(x_{j(1)}, x_{j(2)}, ..., x_{j(n)}) = \prod_{i=1}^n F_{j(i)}(x_{j(i)}). \tag{1.23}$$

These results are valid for random vectors $X_j : S \longrightarrow \mathbb{R}^{n_j}$, noting that $F_j(x_j)$ are joint distribution functions on $\mathbb{R}^{n_j}$ as given in (1.8) or (1.10).

1.2   Decomposition of Distribution Functions on $\mathbb{R}$

In this section, we utilize some of the results from Chapter III.3 to derive what is sometimes called the **canonical decomposition** of the distribution function of a random variable. This result identifies three components in the decomposition, two of which will be recognized within the discrete and continuous probability theories. The third component, while fascinating to contemplate, has not found compelling applications in finance so far.

To formally state this proposition requires some terminology. The following definition introduces **saltus functions,** which are generalized step functions with potentially countably many steps defined by a collection $\{x_n\}_{n=1}^{\infty}$. The word saltus is from Latin, meaning "leap" or "jump," and reflects a break in continuity. The general definition also allows two jumps at each domain point x_n, one of size u_n reflecting a discontinuity from the left, and one of size v_n, reflecting the discontinuity from the right, where u_n and v_n can in this definition be positive or negative.

Since distribution functions are right continuous and increasing by Proposition 1.3, for the application below it will always be the case that $u_n > 0$ and $v_n = 0$.

Definition 1.15 (Saltus function) *Given* $\{x_n\}_{n=1}^{\infty} \subset \mathbb{R}$ *and real sequences* $\{u_n\}_{n=1}^{\infty}$, $\{v_n\}_{n=1}^{\infty}$ *which are absolutely convergent:*

$$\sum_{n=1}^{\infty} |u_n| < \infty, \quad \sum_{n=1}^{\infty} |v_n| < \infty,$$

a ***saltus function*** $f(x)$ *is defined as:*

$$f(x) = \sum_{n=1}^{\infty} f_n(x),$$

where:

$$f_n(x) = \begin{cases} 0, & x < x_n, \\ u_n, & x = x_n, \\ u_n + v_n & x > x_n. \end{cases}$$

In other words,

$$f(x) = \sum_{x_n \leq x} u_n + \sum_{x_n < x} v_n.$$

In addition to saltus functions, recall the following functions from Definitions III.3.49 and III.3.54.

Definition 1.16 (Singular function; absolutely continuous function) *A   function* $f(x)$ *is* ***singular*** *on the interval* $[a, b]$ *if* $f(x)$ *is continuous, monotonically increasing with* $f(b) > f(a)$, *and* $f'(x) = 0$ *almost everywhere.*

A function $f(x)$ *is* ***absolutely continuous*** *on the interval* $[a, b]$ *if for any* $\epsilon > 0$ *there is a* δ *so that:*

$$\sum_{i=1}^{n} |f(x_i) - f(x_i')| < \epsilon,$$

for any finite collection of disjoint subintervals, $\{(x_i', x_i)\}_{i=1}^{n} \subset [a, b]$, *with:*

$$\sum_{i=1}^{n} |x_i - x_i'| < \delta.$$

Remark 1.17 *The relevant facts from Book III on such functions are:*

- *Proposition III.3.53: Singular functions exist, and this proposition illustrates this with the Cantor function of Definition III.3.51, named for Georg Cantor (1845–1918).*

- *Proposition III.3.62: Absolutely continuous functions are characterized as follows:*

  A function $f(x)$ is absolutely continuous on $[a, b]$ if and only if $f(x)$ equals the Lebesgue integral of its derivative on this interval:

$$f(x) = f(a) + (\mathcal{L}) \int_a^x f'(y) dy.$$

  Implicit in this result is that $f'(x)$ exists almost everywhere and is Lebesgue integrable.

  Any continuously differentiable function $f(x)$ is absolutely continuous by the mean value theorem (Remark III.3.56), and then the above representation is also valid as a Riemann integral.

We are now ready for the main result.

Proposition 1.18 (Decomposition of $F(x)$) *Let X be a random variable defined on a probability space $(\mathcal{S}, \mathcal{E}, \lambda)$ with distribution function $F(x) \equiv \lambda\left[X^{-1}(-\infty, x]\right]$. Then $F(x)$ is differentiable almost everywhere and:*

$$F(x) = \alpha F_{SLT}(x) + \beta F_{AC}(x) + \gamma F_{SN}(x), \tag{1.24}$$

where:

- *α, β, γ are nonnegative, with $\alpha + \beta + \gamma = 1$.*

- *$F_{SLT}(x)$ is a saltus distribution function.*

- *$F_{AC}(x)$ is an absolutely continuous distribution function.*

- *$F_{SN}(x)$ is a singular distribution function.*

Proof. *By Proposition 1.3, $F(x)$ is increasing and thus differentiable almost everywhere by Proposition III.3.12. In addition, $F(x)$ is continuous from the right, has left limits, and has at most countably many points of discontinuity which we denote $\{x_n\}_{n=1}^\infty$. At such points define:*

$$u_n = F(x_n) - F(x_n^-),$$

where $F(x^-) \equiv \lim_{y \to x-} F(y)$. Hence from (1.3):

$$u_n = \lambda(X^{-1}(x_n)) > 0,$$

and so $0 \le \sum_{n=1}^\infty u_n \le 1$. Let $\alpha = 0$ if this sum is 0, and otherwise define $\alpha = \sum_{n=1}^\infty u_n$, and:

$$F_{SLT}(x) \equiv \sum_{x_n \le x} u_n \Big/ \alpha. \tag{1.25}$$

Then $F_{SLT}(x)$ is increasing, and by definition $F_{SLT}(-\infty) = 0$ and $F_{SLT}(\infty) = 1$ defined as limits. To prove right continuity, let x and $\epsilon > 0$ be given. Then for $x \le y$:

$$F_{SLT}(y) - F_{SLT}(x) = \sum_{x < x_n \le y} u_n \Big/ \alpha$$

where this summation is finite or countable. In the first case, there is δ so that for $y \leq x + \delta$ this summation is zero and this difference is then bounded by ϵ. If countable, then since $\sum_{n=1}^{\infty} u_n \leq 1$, the summation is convergent and can be made arbitrarily small by eliminating finitely many terms. Reducing y to eliminate this finite set, the above difference can again be made arbitrarily small. Thus $F_{SLT}(x)$ is a saltus distribution function by Definition 1.15 and Proposition 1.4.

Next, let $G(x) \equiv F(x) - \alpha F_{SLT}(x)$. If $y > x$ then $F(y^-) \geq F(x^-)$ and:

$$\alpha F_{SLT}(y) - \alpha F_{SLT}(x) \equiv \sum_{x < x_n \leq y} \left[F(x_n) - F(x_n^-) \right]$$
$$\leq F(y) - F(x) - \left[F(y^-) - F(x^-) \right]$$
$$\leq F(y) - F(x).$$

This obtains $G(y) - G(x) \geq 0$ and so $G(x)$ is increasing. Also by consideration of the component functions, $G(-\infty) = 0$ and $G(\infty) = 1 - \alpha$ defined as limits.

Further, $G(x)$ is continuous. First, right continuity follows from right continuity of the components. If x is a left discontinuity of $G(x)$, so $G(x) - G(x^-) > 0$, then this implies:

$$F(x) - \alpha F_{SLT}(x) - \left[F(x^-) - \alpha F_{SLT}(x^-) \right]$$
$$= F(x) - F(x^-) - \lim_{y \to x-} \sum_{y < x_n \leq x} u_n$$
$$> 0.$$

But this is a contradiction. Either x is a continuity point of F so $F(x) = F(x^-)$ and this sum converges to 0, or, $x = x_n$ is a left discontinuity of F and this sum converges to $u_n = F(x_n) - F(x_n^-)$ by construction.

By Proposition III.3.12, increasing $G(x)$ is differentiable almost everywhere, $G'(x) \geq 0$ by Corollary III.3.13, and $G'(x)$ is Lebesgue integrable on every interval $[a, b]$ by Proposition III.3.19 with:

$$(\mathcal{L}) \int_a^b G'(y) dy \leq G(b) - G(a). \tag{1}$$

It follows that the value of this integral increases as $a \to -\infty$ and/or $b \to \infty$, and that the limit exists since bounded by:

$$0 \leq (\mathcal{L}) \int_{-\infty}^{\infty} G'(y) dy \leq 1 - \alpha.$$

If this improper integral equals 0 then define $\beta = 0$, and otherwise, let $\beta \equiv (\mathcal{L}) \int_{-\infty}^{\infty} G'(y) dy$.
Define:

$$\tilde{F}_{AC}(x) \equiv (\mathcal{L}) \int_{-\infty}^{x} G'(y) dy. \tag{2}$$

By Proposition III.3.58, for any a :

$$\tilde{F}_{AC}(x) - \tilde{F}_{AC}(a) = (\mathcal{L}) \int_a^x G'(y) dy,$$

is absolutely continuous for $x \in [a, b]$ for any b.

It then follows that $F_{AC}(x) - F_{AC}(a)$ equals the Lebesgue integral of its derivative almost everywhere by Proposition III.3.62:

$$\tilde{F}_{AC}(x) - \tilde{F}_{AC}(a) = (\mathcal{L}) \int_a^x \tilde{F}'_{AC}(y) dy, \quad a.e.,$$

and $\tilde{F}'_{AC}(y) = G'(y)$ *a.e. by the proof of this result. As this integral identity is true for all a and* $\tilde{F}_{AC}(-\infty) = 0$ *by definition:*

$$\tilde{F}_{AC}(x) = (\mathcal{L})\int_{-\infty}^{x} \tilde{F}'_{AC}(y)dy. \tag{3}$$

By (2) and (3), the integrals of $G'(y)$ *and* $\tilde{F}'_{AC}(y)$ *agree over* $(-\infty, x]$ *for all x, and so also* $\beta \equiv (\mathcal{L})\int_{-\infty}^{\infty} \tilde{F}'_{AC}dy.$
  Now define:

$$F_{AC}(x) \equiv (\mathcal{L})\int_{-\infty}^{x} \tilde{F}'_{AC}(y)dy \Big/ \beta. \tag{1.26}$$

Then $F_{AC}(x) = \tilde{F}_{AC}(x)/\beta$ *is absolutely continuous, and is a distribution function since* $F_{AC}(-\infty) = 0$ *and* $F_{AC}(\infty) = 1$.
  Finally consider $H(x) \equiv G(x) - \beta F_{AC}(x)$. *Note that* $H(-\infty) = 0$ *and* $H(\infty) = 1 - \alpha - \beta$. *As a difference of continuous increasing functions,* $H(x)$ *is continuous and of bounded variation (Proposition III.3.29), and differentiable almost everywhere by Corollary III.3.32. But* $H'(x) \equiv G'(x) - \beta F'_{AC}(x) = 0$ *a.e., and this obtains by item 3 of Proposition III.2.49 and (1):*

$$\beta F_{AC}(b) - \beta F_{AC}(a) = \beta \int_{a}^{b} F'_{AC}(y)dy = \int_{a}^{b} G'(y)dy \leq G(b) - G(a).$$

Hence $H(a) \leq H(b)$ *and* $H(x)$ *is increasing.*
  If $H(x)$ *is constant it must be identically 0 since* $H(x) \to 0$ *as* $x \to -\infty$, *and, in this case define* $\gamma = 0$. *Otherwise, there exists an interval* $[a, b]$ *for which* $H(a) < H(b)$, *and then by definition* $H(x)$ *is a singular function. In this case let* $\gamma = H(\infty) - H(-\infty)$ *defined as limits, then* $\gamma = 1 - \alpha - \beta$ *and we set:*

$$F_{SN}(x) \equiv H(x)/\gamma. \tag{1.27}$$

∎

Remark 1.19 (Decomposition of joint distributions on $\mathbb{R}^n$**)** *There does not appear to be an n-dimensional analog to Proposition 1.18 for joint distribution functions* $F(x_1, x_2, ..., x_n)$. *By **Sklar's theorem** of Proposition II.7.41, named for **Abe Sklar** (1925–2020), who published these results in 1959, any such distribution function can be represented by:*

$$F(x_1, x_2, ..., x_n) = C(F_1(x_1), F_2(x_2), ..., F_n(x_n)). \tag{1.28}$$

Here, $\{F_j(x)\}_{j=1}^{n}$ *are the marginal distribution functions of* $F(x_1, x_2, ..., x_n)$, *and* $C(u_1, u_2, ..., u_n)$ *is a distribution function defined on* $[0, 1]^n$ *with **uniform marginal distributions**.*
  By uniform marginals is meant that given $I = \{i\} \subset \{1, 2, ..., n\}$, *and* $u_J \equiv (u_{j_1}, u_{j_2}, ..., u_{j_{n-1}})$ *for* $j_k \in J \equiv \tilde{I}$, *the **marginal distribution function** $C_i(u_i)$ defined on* $\mathbb{R}$ *by Definition 1.11:*

$$C_i(u_i) \equiv \lim_{u_J \to \infty} C(u_1, u_2, ..., u_n),$$

satisfies:

$$C_i(u_i) = u_i.$$

The joint distribution functions $C(u_1, u_2, ..., u_n)$ *are called **copulas**.*
  While the above proposition applies to the marginals $\{F_j(x)\}_{j=1}^{n}$ *of* $F(x_1, x_2, ..., x_n)$, *there does not appear to be a feasible way to use this to decompose* $F(x_1, x_2, ..., x_n)$.

1.3   Density Functions on $\mathbb{R}$

In this section, we investigate the density functions associated with the distribution functions of Proposition 1.3, when these densities exist. There are several approaches for identifying these functions, which we call the "Lebesgue approach," the "Riemann approach," and the "Riemann-Stieltjes framework," for reasons that will become apparent.

The Lebesgue approach is introduced first because it will be the more enduring model that will see various generalizations in Section 4.2.3 and in Book V. This approach leads to a precise answer to the question of existence of a density function, which is that, a density exists if and only if the distribution function is absolutely continuous.

The Riemann approach is then seen to obtain a specialized subset of the Lebesgue approach, which provides the basis for continuous probability theory. This approach, identifies densities only for the subset of absolutely continuous distribution functions that are continuously differentiable.

The Riemann-Stieltjes framework expands the final Riemann result. Representing distribution functions as such integrals connect the continuous and discrete probability theories, and obtains densities for both continuously differentiable and discrete distribution functions using Book III results.

See Sections 3.5 and 4.2.3 for a discussion of density functions on $\mathbb{R}^n$.

1.3.1   The Lebesgue Approach

Let X be a random variable on a probability space $(\mathcal{S}, \mathcal{E}, \lambda)$ with distribution function $F(x) \equiv \lambda \left[X^{-1}(-\infty, x] \right]$, and let $f(x)$ be a nonnegative Lebesgue measurable function defined on $\mathbb{R}$ with:

$$(\mathcal{L}) \int_{-\infty}^{\infty} f(y)dy = 1.$$

We say that $f(x)$ is **a density function associated with $F(x)$ in the Lebesgue sense** if for all x:

$$F(x) = (\mathcal{L}) \int_{-\infty}^{x} f(y)dy. \tag{1.29}$$

A density $f(x)$ cannot be unique since if $g(x) = f(x)$ a.e., then $g(x)$ is also a density for $F(x)$ by item 3 of Proposition III.2.31.

This then leads to the question:

Which distribution functions, or equivalently, which random variables X on $(\mathcal{S}, \mathcal{E}, \lambda)$, have density functions in the Lebesgue sense?

A density function can also be defined relative to the measure μ_F induced by $F(x)$ of Proposition 1.5, and this will prove useful for generalizations in Book V. We say $f(x)$ is **a density function associated with λ_F** if for all x and $A_x \equiv (-\infty, x]$:

$$\lambda_F(A_x) = (\mathcal{L}) \int_{A_x} f(y)dy \equiv (\mathcal{L}) \int_{-\infty}^{x} f(y)dy.$$

By definition such a function satisfies the integrability condition above since $\lambda_F(\mathbb{R}) = 1$.

By finite additivity of measures and item 7 of Proposition III.2.31, it then follows that for any right, semi-closed interval $A_{(a,b]} = (a, b]$:

$$\lambda_F((a,b]) = (\mathcal{L}) \int_{A_{(a,b]}} f(y)dy \equiv (\mathcal{L}) \int_{a}^{b} f(y)dy,$$

and this raises the question:

What is the connection between $\lambda_F(A)$ and the density $f(x)$ for other Borel sets $A \in \mathcal{B}(\mathbb{R})$?

A very good start to an answer would be to investigate the set function λ_f defined on the Borel sigma algebra $\mathcal{B}(\mathbb{R})$ by:

$$\lambda_f(A) = (\mathcal{L}) \int_A f(y) dy. \tag{1.30}$$

The set function λ_f is well-defined since such Lebesgue integrals are well-defined by Definition III.2.9.

By definition, the set function λ_f and the Borel measure λ_F induced by F agree on $\mathcal{A}'$, the semi-algebra of right, semi-closed intervals. By extension they also agree on the associated algebra $\mathcal{A}$ of finite disjoint unions of such sets. The uniqueness of extensions theorem of Proposition I.6.14 then assures that for all $A \in \mathcal{B}(\mathbb{R})$:

$$\lambda_f(A) = \lambda_F(A),$$

since $\mathcal{B}(\mathbb{R})$ is the smallest sigma algebra that contains $\mathcal{A}$ by Proposition I.8.1.

Generalizing, it will be proved in Book V that given **any** Lebesgue integrable function $f(x)$, the set function λ_f of (1.30) is always a measure on $\mathcal{B}(\mathbb{R})$.

Returning to the existence question on a density function associated with the distribution function $F(x)$, there are two ways to frame an answer.

Summary 1.20 (Existence of a Lebesgue density) *When does a distribution function $F(x)$, or the induced measure μ_F, have a density function $f(x)$ in the above sense?*

1. *By Proposition III.3.62, a Lebesgue measurable density $f(x)$ that satisfies (1.29) exists if and only if $F(x)$ is an* **absolutely continuous function** *on every interval $[a, b]$. This follows because (1.29) and item 7 of Proposition III.2.31 obtain for all a:*

$$F(x) = F(a) + (\mathcal{L}) \int_a^x f(y) dy.$$

   This proposition also states that (1.29) is satisfied with $f(x) \equiv F'(x)$, and then by item 8 of Proposition III.2.31, it is also satisfied by any measurable function $f(x)$ such that $f(x) = F'(x)$ a.e.

   Thus $F(x)$ has a density function $f(x)$ in the Lebesgue sense **if and only if** *$F(x)$ is* **absolutely continuous.**

2. *By item 3 of Proposition III.2.31, if the induced measure λ_F has a density function, then $\lambda_F(A) = 0$ for every measurable set with $m(A) = 0$, where m denotes Lebesgue measure. This follows from (1.30) and $\mu_F = \mu_f$, since by Definition III.2.9:*

$$\int_A f(y) dy \equiv \int \chi_A(x) f(y) dy.$$

   Here $\chi_A(x) = 1$ on A and is 0 otherwise and thus $\lambda_F(A) = 0$ since $\chi_A(x) f(y) = 0$ a.e.

   Thus if the measure λ_F induced by $F(x)$ has a density function $f(x)$, then $\lambda_F(A) = 0$ **for every set for which** *$m(A) = 0$.*

Remark 1.21 (Radon-Nikodým theorem) *At the moment, item 2 provides a weaker "if" characterization than does item 1, which obtains "if and only if." But it will be seen in Book V that this characterization is again "if and only if," in that if $\lambda_F(A) = 0$ for every set for which $m(A) = 0$, then λ_F will have a density function in the above sense.*

This will follow from a deep and general result known as the **Radon-Nikodým theorem**, *named for* **Johann Radon** *(1887–1956) who proved this result on* $\mathbb{R}^n$, *and* **Otto Nikodým** *(1887–1974) who generalized Radon's result to all* σ-*finite measure spaces.*

A consequence of this result, which will see generalizations in many ways, is then:

The measure λ_F *induced by* $F(x)$ *has a density function* $f(x)$ *if and only if* $\lambda_F(A) = 0$ *for every set for which* $m(A) = 0$.

The following proposition summarizes the implications of these observations in terms of the decomposition of Proposition 1.18.

Proposition 1.22 (Decomposition of $F(x)$: Lebesgue density functions) *Let X be a random variable defined on a probability space $(\mathcal{S}, \mathcal{E}, \lambda)$ with distribution function $F(x) \equiv \lambda\left[X^{-1}(-\infty, x]\right]$, and decomposition in (1.24):*

$$F(x) = \alpha F_{SLT}(x) + \beta F_{AC}(x) + \gamma F_{SN}(x).$$

Then only $F_{AC}(x)$ has a density function in the Lebesgue sense, with:

$$f_{AC}(x) = F'_{AC}(x) \ a.e.$$

Proof. *This result is Proposition III.3.62, which states that $F(x)$ has a density if and only if $F(x)$ is absolutely continuous.* ∎

Remark 1.23 (Lebesgue decomposition theorem, Book V) *As implied by Remark 1.21, the current discussion touches some very general ideas to be developed in Book V, there in the form of the* **Radon-Nikodým theorem**. *The decomposition of Proposition 1.18 also has a measure-theoretic interpretation, which we introduce here.*

To simplify notation, assume that the constants α, β, and γ have been incorporated into the distribution functions F_{SLT}, F_{AC}, and F_{SN}, and thus these are now increasing, right continuous functions with all $F_{\#}(-\infty) = 0$ and $F_{\#}(\infty)$ equal to α, β, and γ, respectively.

In Book V, the conclusion of the second item of Summary 1.20, that $\lambda_{F_{AC}}(A) = 0$ for every set for which $m(A) = 0$, will be denoted:

$$\lambda_{F_{AC}} \ll m.$$

This notation is read: The Borel measure $\lambda_{F_{AC}}$ is **absolutely continuous** *with respect to Lebesgue measure m. While this notation is suggestive, that $m(A) = 0$ forces $\lambda_{F_{AC}}(A) = 0$, it is not intended to imply any other relationship between these measures on other sets.*

For F_{SLT}, let $E_1 \equiv \{x_n\}_{n=1}^{\infty}$, which in the notation of the proof of Proposition 1.18 are the discontinuities of $F(x)$. Then $m(E_1) = \lambda_{F_{SLT}}(\widetilde{E}_1) = 0$, where $\widetilde{E}_1 \equiv \mathbb{R} - E_1$ denotes the complement of E_1. This follows because in the notation of Proposition 1.3,

$$\lambda_{F_{SLT}}(x_n) \equiv F_{SLT}(x_n) - F_{SLT}(x_n^-) = u_n,$$

and this measure is 0 on every set that is outside E_1. For F_{SN}, let E_2 be defined as the set of measure 0 on which $F'_{SN}(x) \neq 0$. Then by definition $m(E_2) = 0$, while $\lambda_{F_{SN}}(\widetilde{E}_2) = 0$ by Proposition I.5.30.

Thus, the sets on which m and $\lambda_{F_{SLT}}$ are "supported," meaning on which they have nonzero measure, are complementary. The same is true for m and $\lambda_{F_{SN}}$. In Book V this relationship will be denoted:

$$m \perp \lambda_{F_{SLT}}, \quad m \perp \lambda_{F_{SN}}.$$

This reads, Lebesgue measure m and the Borel measure $\lambda_{F_{SLT}}$ (respectively $\lambda_{F_{SN}}$) are **mutually singular**.

Now define $E = E_1 \bigcup E_2$. *Then* $m(E) = \lambda_{F_{SLT}}(\widetilde{E}) = \lambda_{F_{SN}}(\widetilde{E}) = 0$, *and thus:*

$$m \perp (\lambda_{F_{SLT}} + \lambda_{F_{SN}}).$$

By the uniqueness theorem of Proposition I.6.14 it follows that:

$$\lambda_F = \lambda_{F_{AC}} + \lambda_{F_{SLT}} + \lambda_{F_{SN}},$$

and we have derived a special case of **Lebesgue's decomposition theorem**, *named for* **Henri Lebesgue (1875–1941)**.

- **Lebesgue's decomposition theorem:** *Given a σ-finite measure space (Definition I.5.34), here* $(\mathbb{R}, \mathcal{B}(\mathbb{R}), m)$, *and a σ-finite measure defined on* $\mathcal{B}(\mathbb{R})$, *here* λ_F, *there exists measures* ν_{ac} *and* ν_s *defined on* $\mathcal{B}(\mathbb{R})$ *so that* $\nu_{ac} \ll m$, $\nu_s \perp m$, *and:*

$$\lambda_F = \nu_{ac} + \nu_s.$$

By the discussion above, when λ_F is the Borel measure induced by a distribution function $F(x)$, then:

$$\nu_{ac} = \lambda_{F_{AC}}, \quad \nu_s = \lambda_{F_{SLT}} + \lambda_{F_{SN}},$$

is such a decomposition.

1.3.2 Riemann Approach

As noted in the introduction, the Riemann approach is a subset of the Lebesgue approach. Let X be a random variable on a probability space $(\mathcal{S}, \mathcal{E}, \lambda)$ with distribution function $F(x) \equiv \lambda \left[X^{-1}(-\infty, x] \right]$, and let $f(x)$ be a nonnegative Riemann integrable function defined on $\mathbb{R}$ with:

$$(\mathcal{R}) \int_{-\infty}^{\infty} f(y) dy = 1.$$

We say that $f(x)$ is **a density function associated with $F(x)$ in the Riemann sense** if for all x:

$$F(x) = (\mathcal{R}) \int_{-\infty}^{x} f(y) dy. \tag{1.31}$$

Again in this context, such a density $f(x)$ cannot be unique, at least if bounded. By the Lebesgue existence theorem of Proposition III.1.22, if the above $f(x)$ is bounded, then it must be continuous almost everywhere on every interval $[a, b]$, and thus continuous almost everywhere. If $g(x)$ is bounded and continuous almost everywhere, and $g(x) = f(x)$ a.e., then $g(x)$ is also a density for $F(x)$. This follows because by the proof of Proposition III.1.22, the value of the Riemann integral is determined by the continuity points of the integrand. Proposition III.1.33 then states that such a distribution function $F(x)$ is differentiable almost everywhere, and $F'(x) = f(x)$ at each continuity point of $f(x)$.

For bounded densities, it follows from Proposition III.2.56 that the Lebesgue integral of $f(x)$ over $\mathbb{R}$ is 1, and that $F(x)$ is also definable as a Lebesgue integral. Thus every density in the Riemann sense is a density in the Lebesgue sense.

Continuous probability theory further specializes the above discussion to continuous densities $f(x)$. Such densities are unique within the class of continuous functions, and now Proposition III.1.33 assures that $F'(x) = f(x)$ for all x. This result also obtains that the only distribution functions that have continuous densities are the continuously differentiable distribution functions.

Exercise 1.24 (Continuous densities are unique) *Prove that if a distribution func-tion $F(x)$ has two continuous density functions $f(x)$ and $\tilde{f}(x)$ in the Riemann sense, then $f(x) = \tilde{f}(x)$ for all x. Hint: If $f(x_0) > \tilde{f}(x_0)$ then by continuity this inequality applies on $(x_0 - \epsilon, x_0 + \epsilon)$. Calculate $F(x_0 + \epsilon) - F(x_0 - \epsilon)$.*

1.3.3 Riemann-Stieltjes Framework

While the above Lebesgue approach provides the framework that will support significant generalizations in Book V, the conclusion of Proposition 1.22 is a bit disappointing. The reader likely already knows that the saltus distribution functions of discrete probability theory do have well-defined density functions, and may wonder if singular distributions do too. Certainly, neither type of distribution function can have a density function in the Lebesgue sense, but perhaps in some other sense. The Riemann-Stieltjes framework provides an approach to connecting the density functions of continuous and discrete probability theory.

Let X be a random variable on a probability space $(\mathcal{S}, \mathcal{E}, \lambda)$ with distribution function $F(x) \equiv \lambda\left[X^{-1}(-\infty, x]\right]$. Then since F is increasing and bounded:

$$G(x) \equiv \int_{-\infty}^{x} dF,$$

exists as a Riemann-Stieltjes integral for all x by Proposition III.4.21. Further by item 4 of Proposition III.4.24:

$$G(b) - G(a) = \int_{a}^{b} dF.$$

But an evaluation with Riemann-Stieltjes sums of Definition III.4.3 obtains for all $[a, b]$:

$$G(b) - G(a) = F(b) - F(a).$$

Letting $a \to -\infty$, then $G(-\infty) = F(-\infty) = 0$ obtains that $G(b) = F(b)$ for all b and thus:

$$F(x) \equiv \int_{-\infty}^{x} dF. \tag{1.32}$$

By Proposition 1.18:

$$F(x) = \alpha F_{SLT}(x) + \beta F_{AC}(x) + \gamma F_{SN}(x),$$

and each of these distribution functions can be represented as in (1.32). Thus by item 5 of Proposition III.4.24:

$$F(x) = \alpha \int_{-\infty}^{x} dF_{SLT} + \beta \int_{-\infty}^{x} dF_{AC} + \gamma \int_{-\infty}^{x} dF_{SN}. \tag{1.33}$$

We now can obtain density functions for two of these distribution functions by Propo-sition III.4.28. Further, the density functions in (1.34) and (1.35) are the density functions of **discrete probability theory**, and **continuous probability theory**, respectively.

Proposition 1.25 (Decomposition of $F(x)$; R-S density functions) *Let X be a ran-dom variable defined on a probability space $(\mathcal{S}, \mathcal{E}, \lambda)$ with distribution function $F(x) \equiv \lambda\left[X^{-1}(-\infty, x]\right]$, and decomposition in (1.33). Then:*

1. If $F_{SLT}(x)$ is defined as in (1.25), and $\{x_n\}_{n=1}^{\infty}$ have no accumulation points, then:

$$\int_{-\infty}^{x} dF_{SLT} = \sum_{x_n \leq x} u_n \Big/ \alpha.$$

In this case, we define the density function f_{SLT} of F_{SLT} by:

$$f_{SLT}(x_n) = u_n/\alpha,$$

and thus:

$$F_{SLT}(x) = \sum_{x_n \leq x} f_{SLT}(x_n). \tag{1.34}$$

2. If $F_{AC}(x)$ is continuously differentiable, then:

$$\int_{-\infty}^{x} dF_{AC} = (\mathcal{R}) \int_{-\infty}^{x} F'_{AC}(y)dy,$$

defined as a Riemann integral. In this case, we define the unique continuous density function f_{AC} of F_{AC} by:

$$f_{AC}(x) = F'_{AC}(x),$$

and thus:

$$F_{AC}(x) = (\mathcal{R}) \int_{-\infty}^{x} f_{AC}(y)dy. \tag{1.35}$$

Proof. *Proposition III.4.28.* ∎

Remark 1.26 (On Proposition 1.25) *A few comments on the above result.*

1. *If $F_{SLT}(x)$ is defined as in (1.25) and $\{x_n\}_{n=1}^{\infty}$ has accumulation points, it is conventional to extend (1.34) to this case even though this does not follow directly from Proposition III.4.28. Instead, we can use right continuity of distribution functions.*

   For any x_n, it follows by definition that:

$$F_{SLT}(x_n) = \sum_{x'_n \leq x_n} f_{SLT}(x'_n). \tag{1}$$

   Given an accumulation point $x \neq x_n$ for any n, consider $\{x_n > x\}$. If $x_n \to x$, then by right continuity of F_{SLT} and (1) :

$$F_{SLT}(x) = \lim_{n \to \infty} F_{SLT}(x_n) = \sum_{x'_n \leq x} f_{SLT}(x'_n).$$

2. *If $F_{AC}(x)$ is continuously differentiable then it is absolutely continuous by the mean value theorem (Remark III.3.56), and thus item 2 of Proposition 1.25 is a special case of the general Lebesgue model above. And by Proposition III.2.56, the Riemann and Lebesgue integrals of $f_{AC}(y) = F'_{AC}(y)$ agree over every interval $(-\infty, x]$.*

3. *Finally, there is little hope of defining a density function for a singular distribution function $F_{SN}(x)$. Since $F'_{SN}(x) = 0$ almost everywhere, the Lebesgue integral of this function is zero, and (1.35) cannot be valid. Similarly, a representation as in (1.34) would in general not be feasible, summing over all $x_\alpha \leq x$ for which $F'_{SN}(x_\alpha) \neq 0$. For the Cantor function of Definition III.3.51, for example, such points are uncountable.*

1.4   Examples of Distribution Functions on $\mathbb{R}$

The above sections provide a framework for thinking about the distribution functions and density functions of a random variable. In this section, we exemplify a variety of popular examples of such functions within this framework, adding to the discussion in Section II.1.3.

1.4.1   Discrete Distribution Functions

Discrete probability theory studies random variables for which $\beta = \gamma = 0$ in (1.24), and so $\alpha = \sum_{n=1}^{\infty} u_n = 1$ and:

$$F(x) = F_{SLT}(x). \tag{1.36}$$

Since $u_n = \mu(X^{-1}(x_n)) > 0$, it is conventional as noted in Proposition 1.25 to define the **probability density function (p.d.f.)** associated with a discrete random variable by:

$$f(x) = \begin{cases} u_n, & x = x_n, \\ 0, & \text{otherwise.} \end{cases}$$

The **distribution function (d.f.)** is then given by:

$$F(x) = \sum_{x_n \leq x} f(x_n). \tag{1.37}$$

In virtually all cases in discrete probability theory, it is the probability density functions that are explicitly defined in a given application. Frequently encountered examples of discrete distribution functions are the discrete rectangular, binomial, geometric, negative binomial, and Poisson distribution functions.

Example 1.27   *1.* **Discrete Rectangular Distribution:** *The defining collection $\{x_j\}_{j=1}^{n}$ for this distribution is finite and can otherwise be arbitrary. However, this collection is conventionally taken as $\{j/n\}_{j=1}^{n}$, and so $\{x_j\}_{j=1}^{n} \subset [0,1]$, and the discrete rectangular random variable is modeled by $X^R : \mathcal{S} \longrightarrow [0,1]$.*

For given n, the probability density function of the **discrete rectangular distribution,** *also called the* **discrete uniform distribution,** *is defined on $\{j/n\}_{j=1}^{n}$ by:*

$$f_R(j/n) = 1/n, \quad j = 1, 2, .., n. \tag{1.38}$$

In effect, this random variable splits $\mathcal{S} = \bigcup_{j=1}^{n} S_j$ into n sets with $S_j \equiv X^{-1}(j/n)$ and:

$$\lambda(S_j) = 1/n.$$

By rescaling, this distribution can be translated to any interval $[a, b]$, defining

$$Y^R = (b - a)X^R + a.$$

2. **Binomial Distribution:** *For given p, $0 < p < 1$, the* **standard binomial** *random variable is defined: $X_1^B : \mathcal{S} \longrightarrow \{0, 1\}$, where the associated p.d.f. is defined by:*

$$f(1) = p, \quad f(0) = p' \equiv 1 - p.$$

This is often expressed:

$$X_1^B = \begin{cases} 1, & \Pr = p, \\ 0, & \Pr = p', \end{cases}$$

or to emphasize the associated p.d.f.:

$$f_{B_1}(j) = \left\{ \begin{array}{ll} p, & j = 1, \\ p', & j = 0. \end{array} \right. \tag{1.39}$$

A simple application for this random variable is as a model for a single coin flip. So $\mathcal{S} = \{H, T\}$, *a probability measure is defined on* $\mathcal{S}$ *by:* $\lambda(H) = p$, *and* $\lambda(T) = p'$, *and the random variable defined by* $X_1^B(H) \equiv 1$ *and* $X_1^B(T) \equiv 0$. *This random variable is sometimes referred to as a* **Bernoulli trial,** *and the associated distribution function as the* **Bernoulli distribution** *after* **Jakob Bernoulli** *(1654–1705).*

This standard formulation is then translated to a **shifted standard binomial** *random variable:* $Y_1^B = b + (a - b)X_1^B$, *which is defined:*

$$Y_1^B = \left\{ \begin{array}{ll} a, & \mathrm{Pr} = p, \\ b, & \mathrm{Pr} = p', \end{array} \right.$$

where the example of $b = -a$ *is common in discrete time asset price modeling for example.*

3. **General Binomial:** *The binomial model can also be extended to accommodate sample spaces of n-coin flips, producing the* **general binomial** *random variable with two parameters,* p *and* $n \in \mathbb{N}$. *That is,* $\mathcal{S} = \{(F_1, F_2, ..., F_n) \mid F_j \in \{H, T\}\}$, *and* X_n^B *is defined as the "head counting" random variable:*

$$X_n^B(F_1, F_2, ..., F_n) = \sum_{j=1}^{n} X_1^B(F_j).$$

It is apparent that X_n^B *assumes values* $0, 1, 2, ..., n$, *and using a standard combinatorial analysis that the associated probabilities are given for* $j = 0, 1, .., n$ *by:*

$$X_n^B = \{ \ j, \quad \mathrm{Pr} = \binom{n}{j} p^j (1-p)^{n-j}.$$

To emphasize the associated p.d.f.:

$$f_{B_n}(j) = \binom{n}{j} p^j (1-p)^{n-j}, \quad j = 0, 1, 2, ..., n. \tag{1.40}$$

Recall that $\binom{n}{j}$ *denotes the* **binomial coefficient** *defined by:*

$$\binom{n}{j} = \frac{n!}{(n-j)!j!}, \tag{1.41}$$

where $0! = 1$ *by convention. This expression is sometimes denoted* $_nC_j$ *and read, "n choose j."*

The name "binomial coefficient" follows from the expansion of a binomial $a + b$ *raised to the power* n, *producing the* **binomial theorem:**

$$(a + b)^n = \sum_{j=0}^{n} \binom{n}{j} a^j b^{n-j}. \tag{1.42}$$

4. **Geometric Distribution:** *For given* p, $0 < p < 1$, *the* **geometric distribution** *is defined on the nonnegative integers, and its p.d.f. is given by:*

$$f_G(j) = p(1-p)^j, \quad j = 0, 1, 2, .. \tag{1.43}$$

and thus by summation:

$$F_G(j) = 1 - (1-p)^{j+1}, \quad j = 0, 1, 2, ..$$ (1.44)

This distribution is related to the binomial distribution in a natural way.

The underlying sample space can be envisioned as the collection of all coin-flip sequences which terminate on the first H. So:

$$S = \{H, TH, TTH, TTTH, ...\},$$

and the random variable X^G defined as the number of flips before the first H. Consequently, $f_G(j)$ above is the probability in S of the sequence of j-Ts and then 1-H. That is, the probability that the first H occurs after j-Ts. Of course, $f_G(j)$ is indeed a p.d.f. in that $\sum_{j=0}^{\infty} p(1-p)^j = 1$ as follows from (1.44), letting $j \to \infty$.

The geometric distribution is sometimes parametrized as:

$$f_{G'}(j) = p(1-p)^{j-1}, \quad j = 1, 2, ..,$$

and then represents the probability of the first head in a coin flip sequence appearing on flip j. These representations are conceptually equivalent, but mathematically distinct due to the shift in domain.

*One way of generalizing the geometric distribution is to allow the probability of a head to vary with the sequential number of the coin flip. This is the basic model in all financial calculations relating to payments **contingent on death or survival**, as well as to various other vitality-based outcomes. Specifically, if*

$$\Pr[H \mid kth\ flip] = p_k,$$

*then with a simplifying change in notation to exclude the case $j = 0$, a **generalized geometric distribution** can be defined by:*

$$f_{GG}(j) = p_j \prod_{k=1}^{j-1} (1-p_k), \quad j = 1, 2, 3, ...,$$ (1.45)

where $f_{GG}(j)$ is the probability of the first head appearing on flip j. By convention, $\prod_{k=1}^{0}(1-p_k) \equiv 1$ when $j = 1$.

*If $p_k = p > 0$ for all k, then $f_G(j)$ is a p.d.f. as noted above. With nonconstant probabilities, this conclusion is also true with some restrictions to assure that the distribution function is bounded. For example, if $0 < a \le p_k \le b < 1$ for all j, then the summation is finite since $f_{GG}(j) < b(1-a)^{j-1}$ and thus $\sum_{j=1}^{\infty} f_{GG}(j)(j) < b/a$ by a geometric series summation. Additional restrictions are required to make this sum equal to 1. Details are left to the interested reader, or see **Reitano (2010)** pp. 314–319 for additional discussion.*

5. ***Negative Binomial Distribution:*** *The name of this distribution calls out yet another connection to the binomial distribution, and here we generalize the idea behind the geometric distribution. There, $f_G(j)$ was defined as the probability of j-Ts before the first H. The negative binomial, $f_{NB}(j)$ introduces another parameter k, and is defined as the probability of j-Ts before the kth-H. So when $k = 1$, the negative binomial is the same as the geometric.*

The p.d.f. is then defined with parameters p with $0 < p < 1$ and $k \in \mathbb{N}$ by:

$$f_{NB}(j) = \binom{j+k-1}{k-1} p^k (1-p)^j, \quad j = 0, 1, 2, ..$$ (1.46)

This formula can be derived analogously to that for the geometric by considering the sample space of coin flip sequences, each terminated on the occurrence of the kth-H. The probability of a sequence with j-Ts and k-Hs is then $p^k(1-p)^j$. Next, we must determine the number of such sequences in the sample space. Since every such sequence terminates with an H, there are only the first $j+k-1$ positions that need to be addressed. Each such sequence is then determined by the placement of the first $(k-1)$-Hs, and so the total count of these sequences is $\binom{j+k-1}{k-1}$. Multiplying the probability and the count, we have (1.46).

6. **Poisson Distribution:** *The Poisson distribution is named for* **Siméon-Denis Poisson (1781–1840)** *who discovered this distribution and studied its properties. This distribution is characterized by a single parameter $\lambda > 0$, and is defined on the nonnegative integers by:*

$$f_P(j) = e^{-\lambda}\frac{\lambda^j}{j!}, \quad j = 0, 1, 2, \ldots. \tag{1.47}$$

That $\sum_{j=0}^{\infty} f_P(j) = 1$ is an application of the Taylor series expansion for e^{λ} :

$$e^{\lambda} = \sum_{j=0}^{\infty} \lambda^j/j!$$

One important application of the Poisson distribution is provided by the **Poisson Limit theorem** *of Proposition II.1.11. See also Proposition 6.5. This result states that with $\lambda = np$ fixed:*

$$\lim_{n \to \infty} \binom{n}{j}p^j(1-p)^{n-j} = e^{-\lambda}\frac{\lambda^j}{j!}.$$

Thus when the binomial parameter p is small, and n is large, the binomial probabilities in (1.40) can be approximated:

$$\binom{n}{j}p^j(1-p)^{n-j} \simeq e^{-np}\frac{(np)^j}{j!}. \tag{1.48}$$

By p small and n large is usually taken to mean $n \geq 100$ and $np \leq 10$.

Another important property of the Poisson distribution is that it characterizes "arrivals" during a given period of time under reasonable and frequently encountered assumptions. For example, the model might be one of automobile arrivals at a stop light; or telephone calls to a switchboard; or internet searches to a server; or radio-active particles to a Geiger counter; or insurance claims of any type (injuries, deaths, automobile accidents, etc.) from a large group of policyholders; or defaults from a large portfolio of loans and bonds; etc. For this result, see pp. 300–301 in **Reitano (2010)**.

1.4.2   Continuous Distribution Functions

Continuous probability theory studies a subset of the absolutely continuous distribution functions, so $\alpha = \gamma = 0$ and $\beta = 1$ in (1.24), and for which:

$$F(x) \equiv F_{AC}(x),$$

is also assumed to be continuously differentiable. Then $f(x) \equiv F'(x)$ exists everywhere, is continuous, and by the fundamental theorem of calculus of Proposition III.1.32:

$$F(x) = (\mathcal{R})\int_{-\infty}^{x} f(y)dy. \tag{1.49}$$

A random variable $X : S \to \mathbb{R}$ with such a distribution function is said to have a **continuous density function,** and the function $f(x)$ is called the **probability density function (p.d.f.) associated with the distribution function** F. A continuous density function is unique among Riemann integrable functions that satisfy (1.49) for given $F(x)$, since the fundamental theorem of calculus of Proposition III.1.33 obtains that for all x :

$$F'(x) = f(x). \tag{1.50}$$

There are many distributions with continuous density functions used in finance and in other applications, but some of the most common are the uniform, exponential, gamma (including chi-squared), beta, normal, lognormal, and Cauchy. In addition, other examples are found with the **extreme value distributions** discussed in Chapter II.9 and Chapter 7, as well as Student's T (also called Student T) and F distributions introduced in Example 2.28.

It should be noted that it is not required that a continuous density function $f(x)$ be continuous on $\mathbb{R}$, but only on its domain of definition. Then by Proposition III.1.33, $F'(x) = f(x)$ at every such continuity point.

Example 1.28 *1. **Continuous Uniform Distribution***: *Perhaps the simplest continuous probability density that can be imagined is one that is constant on its domain. The domain of this distribution is arbitrary, though of necessity bounded, and is conventionally denoted as the interval* $[a, b]$. *The p.d.f. of the **continuous uniform distribution,** sometimes called the **continuous rectangular distribution,** is defined on* $[a, b]$ *by the density function:*

$$f_U(x) = \frac{1}{(b-a)}, \quad x \in [a, b], \tag{1.51}$$

and $f_U(x) = 0$ *otherwise.*

This distribution is called "uniform" because if $a \leq s < t \leq b$ *and* $X_U : S \to \mathbb{R}$ *denotes the underlying random variable, then:*

$$\lambda\{X_U^{-1}[s, t]\} = F_U(t) - F_U(s) = \frac{t-s}{b-a}.$$

*This probability is thus **translation invariant** within* $[a, b]$, *meaning that* $[s, t]$ *and* $[s+\epsilon, t+\epsilon]$ *have the same measure when* $[s+\epsilon, t+\epsilon] \subset [a, b]$, *and this justifies the uniform label. It is then also the case that the range of* X_U *under these induced probabilities can be identified with the probability space* $((a, b), \mathcal{B}(a, b), m)$, *where* m *is Lebesgue measure and* $\mathcal{B}(a, b)$ *is the Borel sigma algebra, noting that nothing is lost in omitting the endpoints which have probability 0.*

Defined on $[0, 1]$, *this distribution is more important than it may first appear due to the results in Chapter II.4. To summarize these results below, we first recall Definition II.3.12 on the left-continuous inverse of a distribution function, or any increasing function. This inverse function was proved to be increasing and left continuous in Proposition II.3.16.*

Definition 1.29 (Left-continuous inverse) *Let* $F : \mathbb{R} \to \mathbb{R}$ *be a distribution or other increasing function. The **left-continuous inverse of** F, denoted by* F^*, *is defined:*

$$F^*(y) = \inf\{x | y \leq F(x)\}. \tag{1.52}$$

By convention, if $\{x | F(x) \geq y\} = \emptyset$ *then we define* $F^*(y) = \infty$, *while if* $\{x | F(x) \geq y\} = \mathbb{R}$, *then* $F^*(y) = -\infty$ *by definition.*

For the following Chapter II.4 results, let $(\mathcal{S}, \mathcal{E}, \lambda)$ be given and $X : (\mathcal{S}, \mathcal{E}, \lambda) \to (\mathbb{R}, \mathcal{B}(\mathbb{R}), m)$ a random variable with distribution function $F(x)$ and left-continuous inverse $F^*(y)$. Let $(\mathcal{S}', \mathcal{E}', \lambda')$ be given and $X_U : (\mathcal{S}', \mathcal{E}', \lambda') \to ((0,1), \mathcal{B}((0,1)), m)$ a random variable with continuous uniform distribution function. Then:

Proposition II.4.5: *If* $F(x)$ *is continuous and strictly increasing, so* $F^* = F^{-1}$ *by Proposition II.3.22, then:*

(a) $F(X) : (\mathcal{S}, \mathcal{E}, \lambda) \to ((0,1), \mathcal{B}((0,1)), m)$ *defined by* $F(X)(s) = F(X(s))$, *is a random variable on* $\mathcal{S}$ *which has a continuous uniform distribution on* $(0,1)$, *meaning* $F_{F(X)}(x) = x$.

(b) $F^{-1}(X_U) : (\mathcal{S}', \mathcal{E}', \lambda') \to (\mathbb{R}, \mathcal{B}(\mathbb{R}), m)$ *defined by* $F^{-1}(X_U)(s') = F^{-1}(X_U(s'))$, *is a random variable on* $\mathcal{S}'$ *with distribution function* $F(x)$.

Proposition II.4.8: *For general increasing* $F(x)$:

(a) $F(X) : (\mathcal{S}, \mathcal{E}, \lambda) \to ((0,1), \mathcal{B}((0,1)), m)$ *defined as above has distribution function* $F_{F(X)}(x) \leq x$, *and this function is continuous uniform with* $F_{F(X)}(x) = x$ *if and only if* $F(x)$ *is continuous;*

(b) $F^*(X_U) : (\mathcal{S}', \mathcal{E}', \lambda') \to (\mathbb{R}, \mathcal{B}(\mathbb{R}), m)$ *defined as above is a random variable on* $\mathcal{S}'$ *with distribution function* $F(x)$.

Proposition II.4.9:

(a) *If* $\{U_j\}_{j=1}^n$ *are independent, continuous uniformly distributed random variables, then* $\{X_j\}_{j=1}^n \equiv \{F^*(U_j)\}_{j=1}^n$ *are independent random variables with distribution function* $F(x)$.

(b) *If* $F(x)$ *is continuous and* $\{X_j\}_{j=1}^n$ *are independent random variables with distribution function* $F(x)$, *then* $\{U_j\}_{j=1}^n \equiv \{F(X_j)\}_{j=1}^n$ *are independent, continuous uniformly distributed random variables.*

The significance of these Book II results is that one can convert a uniformly distributed random sample $\{U_j\}_{j=1}^n$ into a sample of the $\{X_j\}_{j=1}^n$ variables by defining:

$$X_j = F^*(U_j),$$

with $F^*(U_j)$ defined as in (1.52). And note that despite being defined as an infimum, the value of $F^*(U_j)$ is truly in the domain of the random variable X because $F(x)$ is right continuous.

We will return to an application of these results in Chapter 5.

2. **Exponential Distribution:** The **exponential density function** is defined with parameter $\lambda > 0$:

$$f_E(x) = \lambda e^{-\lambda x}, \quad x \geq 0, \tag{1.53}$$

and $f_E(x) = 0$ for $x < 0$. The associated distribution function is then:

$$F_E(x) = 1 - \exp(-\lambda x), \quad x \geq 0, \tag{1.54}$$

and $F_E(x) = 0$ for $x < 0$. When $\lambda = 1$, this is often called the **standard exponential distribution.**

The exponential distribution is important in the generation of **ordered random samples.** See Chapter 3.

3. **Gamma Distribution:** *The **gamma distribution** is a two parameter generalization of the exponential, with $\alpha > 0$ and $\lambda > 0$, and density function $f_\Gamma(x)$ defined by:*

$$f_\Gamma(x) = \frac{1}{\Gamma(\alpha)} \lambda^\alpha x^{\alpha-1} e^{-\lambda x}, \quad x \geq 0, \tag{1.55}$$

*and $f_\Gamma(x) = 0$ for $x < 0$. When $0 < \alpha < 1$, $f_\Gamma(x)$ is unbounded at $x = 0$ but is integrable. The **gamma function**, $\Gamma(\alpha)$, is defined by:*

$$\Gamma(\alpha) = \int_0^\infty x^{\alpha-1} e^{-x} dx, \tag{1.56}$$

and thus $\int_0^\infty f_\Gamma(x) dx = 1$ with a change of variables. The gamma function is continuous in $\alpha > 0$, meaning that $\Gamma(\alpha) \to \Gamma(\alpha_0)$ if $\alpha \to \alpha_0$. Since this integral equals the Lebesgue counterpart by Proposition III.2.56, this can be proved using Lebesgue's dominated convergence theorem of Proposition III.2.52. Details are left as an exercise.

An integration by parts shows that for $\alpha > 0$, the gamma function satisfies:

$$\Gamma(\alpha + 1) = \alpha \Gamma(\alpha), \tag{1.57}$$

and since $\Gamma(1) = 1$, this function generalizes the factorial function in that for any integer n:

$$\Gamma(n) = (n-1)! \tag{1.58}$$

This identity also provides the logical foundation for defining $0! = 1$, as noted in the discussion of the general binomial distribution.

Another interesting identity for the gamma function is:

$$\Gamma(1/2) = \sqrt{\pi}. \tag{1.59}$$

From (1.56) and the substitution $x = y^2/2$:

$$\begin{aligned} \Gamma(1/2) &= \int_0^\infty x^{-1/2} e^{-x} dx \\ &= \sqrt{2} \int_0^\infty e^{-y^2/2} dy. \end{aligned}$$

This integral equals $\sqrt{2\pi}/2$ by symmetry and the discussion on the normal distribution in item 5.

When the parameter $\alpha = k$ is a positive integer, the probability density function becomes by (1.58):

$$f_\Gamma(x) = \frac{1}{(k-1)!} \lambda^k x^{k-1} e^{-\lambda x}, \quad x \geq 0.$$

The associated distribution function is:

$$F_\Gamma(x) = e^{-\lambda x} \sum_{j=k}^\infty \frac{(\lambda x)^j}{j!}, \tag{1.60}$$

and it can be verified by differentiation that $F_\Gamma'(x) = f_\Gamma(x)$.

Remark 1.30 (Chi-squared) *When a random variable Y has a gamma distribution with $\lambda = 1/2$ and $\alpha = n/2$, it is said to have a **chi-squared distribution with n degrees of freedom**, which is sometimes denoted $\chi^2_{n \, d.f.}$. See Examples 2.4 and 2.28.*

4. **Beta Distribution:** *The* **beta distribution** *contains two shape parameters,* $v > 0$ *and* $w > 0$*, and is defined on the interval* $[0, 1]$ *by the density function:*

$$f_\beta(x) = \frac{1}{B(v, w)} x^{v-1} (1 - x)^{w-1}. \tag{1.61}$$

The **beta function** $B(v, w)$ *is defined by a definite integral which in general requires numerical evaluation:*

$$B(v, w) = \int_0^1 y^{v-1} (1 - y)^{w-1} dy. \tag{1.62}$$

By definition, therefore, $\int_0^1 f_\beta(x) dx = 1$.

If v *or* w *or both parameters are less than* 1*, the beta density and integrand of* $B(v, w)$ *are unbounded at* $x = 0$ *or* $x = 1$ *or both. However the integral converges in the limit as an improper integral:*

$$B(v, w) = \lim_{\substack{a \to 0^+ \\ b \to 1^-}} \int_a^b y^{v-1} (1 - y)^{w-1} dy,$$

because the exponents of the y *and* $1 - y$ *terms exceed* -1.

If both parameters are greater than 1*, this density function is* 0 *at the interval endpoints, and has a unique maximum at* $x = (v - 1) / (v + w - 2)$. *When both parameters equal* 1*, this distribution reduces to the continuous uniform distribution on* $[0, 1]$.

The beta function is closely related to the gamma function above. Substituting $z = y/(1 - y)$ *into the integral in (1.62):*

$$B(v, w) = \int_0^\infty \frac{z^{v-1}}{(1 + z)^{v+w}} dz. \tag{1}$$

Letting $\alpha = v + w$ *in (1.56) obtains by substitution* $x = (1 + z)y$:

$$\frac{1}{(1 + z)^{v+w}} = \frac{1}{\Gamma(v + w)} \int_0^\infty y^{v+w-1} e^{-(1+z)y} dy.$$

Multiplying by z^{v-1}*, integrating, and applying (1) obtains:*

$$B(v, w) = \frac{1}{\Gamma(v + w)} \int_0^\infty y^{v+w-1} e^{-y} \left(\int_0^\infty e^{-zy} z^{v-1} dz \right) dy.$$

A final substitution $x = zy$ *in the inner integral, and two applications of (1.56) produces the identity:*

$$B(v, w) = \frac{\Gamma(v)\Gamma(w)}{\Gamma(v + w)}. \tag{1.63}$$

Thus for integer n *and* m*, it follows from (1.58) that:*

$$B(n, m) = \frac{(n - 1)! \, (m - 1)!}{(n + m - 1)!}. \tag{1.64}$$

5. **Normal Distribution:** *The* **normal distribution** *is defined on* $(-\infty, \infty)$*, depends on a location parameter* $\mu \in \mathbb{R}$ *and a scale parameter* $\sigma > 0$*, and is given by the probability density function:*

$$f_N(x) = \frac{1}{\sigma\sqrt{2\pi}} \exp\left(-(x - \mu)^2 / 2\sigma^2 \right), \tag{1.65}$$

where $\exp A \equiv e^A$ *to simplify notation. In many books, the parametrization is defined in terms of μ and σ^2, where σ is taken as the positive square root in (1.65). We use these specifications interchangeably.*

*When $\mu = 0$ and $\sigma = 1$, this is known as the **standard normal distribution** or **unit normal distribution** and denoted $\phi(x)$:*

$$\phi(x) = \frac{1}{\sqrt{2\pi}} \exp\left(-x^2/2\right). \tag{1.66}$$

A change of variables in the associated integral shows that if a random variable X is normally distributed with parameters μ and σ, then $(X - \mu)/\sigma$ has a standard normal distribution. Conversely, if X is standard normal, then $\sigma X + \mu$ is normally distributed with parameters μ and σ^2.

Perhaps the greatest significance of the normal distribution is that it can be used as an approximating distribution to the distribution of sums and averages of a sample of "scaled" random variables under relatively mild assumptions.

(a) *When applied to approximate the binomial distribution, this result is called the **De Moivre-Laplace theorem**, named for **Abraham de Moivre** (1667–1754), who demonstrated the special case of $p = 1/2$, and **Pierre-Simon Laplace** (1749– 1827), who years later generalized to all p, $0 < p < 1$. See Proposition 6.11.*

(b) *In the most general cases, this result is known as the **central limit theorem**. See Proposition 6.13 for one example, and Book VI for others.*

Remark 1.31 ($\int_{-\infty}^{\infty} f_N(x)dx = 1$) *By change of variable, $\int_{-\infty}^{\infty} f_N(x)dx = 1$ if and only if $\int_{-\infty}^{\infty} \phi(x)dx = 1$. While there are no elementary proofs of the latter identity, there is clever derivation that involves embedding the integral $\int_{-\infty}^{\infty} \phi(x)dx$ into a 2-dimensional Riemann integral, and the ability to move back and forth between 2-dimensional and iterated Riemann integrals. This is largely justified in Corollary III.1.77, but the earlier result requires a generalization to improper integrals. The derivation below also requires a change of variables to polar coordinates which allows direct calculation, a step that is perhaps familiar to the reader but one that will not be formally justified until Book V.*

In detail:

$$\int_{-\infty}^{\infty} \phi(x)dx \int_{-\infty}^{\infty} \phi(y)dy = \frac{1}{2\pi} \int_{-\infty}^{\infty} \int_{-\infty}^{\infty} \exp\left(-\left[x^2 + y^2\right]/2\right) dxdy,$$

which is an iterated integral that equals a 2-dimensional Riemann integral over $\mathbb{R}^2$.

The change of variables $x = r\cos\theta$ and $y = r\sin\theta$ for $r \in [0, \infty)$ and $\theta \in [0, 2\pi)$ then obtains $dxdy = rd\theta dr$, and a 2-dimensional Riemann integral over the rectangle $R = [0, 2\pi) \times [0, \infty)$. Thus:

$$\int_{-\infty}^{\infty} \phi(x)dx \int_{-\infty}^{\infty} \phi(y)dy = \frac{1}{2\pi} \int_R re^{-r^2} d\theta dr.$$

Returning to iterated integrals obtains:

$$\int_{-\infty}^{\infty} \phi(x)dx \int_{-\infty}^{\infty} \phi(y)dy = \frac{1}{2\pi} \int_0^{\infty} re^{-r^2} dr \int_0^{2\pi} d\theta = 1.$$

6. **Lognormal Distribution:** *The* **lognormal distribution** *is defined on* $[0, \infty)$, *depends on a location parameter* $\mu \in \mathbb{R}$ *and a shape parameter* $\sigma > 0$, *and unsurprisingly is intimately related to the normal distribution above. However, to some, the name "lognormal" appears to be opposite to the relationship that exists.*

Stated one way, a random variable X *is lognormal with parameters* (μ, σ^2) *if* $X = e^Z$ *where* Z *is normal with the same parameters. So* X *can be understood as an exponentiated normal. Stated another way, a random variable* X *is lognormal with parameters* (μ, σ^2) *if* $\ln X$ *is normal with the same parameters. Thus the name comes from the second interpretation, that the log of a lognormal variate is a normal variate.*

The probability density function of the lognormal is defined as follows, again using $\exp A \equiv e^A$ *to simplify notation:*

$$f_L(x) = \frac{1}{\sigma x \sqrt{2\pi}} \exp\left(-(\ln x - \mu)^2 / 2\sigma^2\right). \tag{1.67}$$

By change of variable, $\int_0^\infty f_L(x)dx = 1$ *if and only if* $\int_{-\infty}^\infty f_N(x)dx = 1$.

7. **Cauchy Distribution:** *The* **Cauchy distribution** *is named for* **Augustin Louis Cauchy** *(1789–1857). It is of interest as an example of a distribution function that has no finite moments as will be seen in item 4 of Section 4.2.7, and has another surprising property to be illustrated in Example 2.18.*

The associated density function is defined on $\mathbb{R}$ *as a function of a location parameter* $x_0 \in \mathbb{R}$, *and a scale parameter* $\gamma > 0$, *by:*

$$f_C(x) = \frac{1}{\pi \gamma} \frac{1}{1 + ([x - x_0]/\gamma)^2}. \tag{1.68}$$

The **standard Cauchy distribution** *is parameterized with* $x_0 = 0$ *and* $\gamma = 1$:

$$f_C(x) = \frac{1}{\pi} \frac{1}{1 + x^2}. \tag{1.69}$$

A substitution into the integral of $\tan z = (x - x_0)/\gamma$ *shows that this probability function integrates to 1, as expected.*

1.4.3   Mixed Distribution Functions

Mixed distribution functions, for which $\gamma = 0$ in (1.24), are commonly encountered in finance and defined:

$$F(x) \equiv \alpha F_{SLT}(x) + (1 - \alpha)F_{AC}(x), \tag{1.70}$$

where again it is assumed that $F'_{AC}(x) = f(x)$ is continuous. In other words, such a distribution function represents a random variable with both continuously distributed and discrete parts.

Example 1.32 (Example II.1.7) *Example II.1.7 examined default experience on a loan portfolio, and of particular interest was the modeling of default losses. For a portfolio of* n *loans, if* X *denotes the number of defaults, then* $X : \mathcal{S} \to \{0, 1, 2, ..., n\}$ *is discrete, while if* $Y : \mathcal{S} \to \mathbb{R}$ *denotes the dollars of loss, then the model for* Y *will typically be "mixed."*

This follows because with fixed default probability p say, the probability of no defaults assuming independence is given by:

$$\Pr\{Y = 0\} = \Pr\{X = 0\} = (1 - p)^n.$$

This probability will typically be quite large compared to $\Pr\{0 < Y \leq \epsilon\}$, which is the probability of one or several defaults but very low loss amounts.

For example, assume that each bond's loss given default is uniformly distributed on $[0, L_j]$, where L_j denotes the loan amount on the jth bond. Then $F(y)$ will have a discrete part at $y = 0$, and then a continuous or mixed distribution for $y > 0$. In the former case, which we confirm below:

$$F(y) \equiv (1 - p)^n \chi_{[0,\infty)}(y) + (1 - (1 - p)^n) F_{AC}(y),$$

where $\chi_{[0,\infty)}(y) = 1$ for $y \in [0, \infty)$ and is 0 otherwise, and $F_{AC}(y)$ is absolutely continuous with $F_{AC}(0) = 0$.

*As $F(y)$ is continuous from the right as a distribution, and certainly not left continuous at $y = 0$, to investigate left continuity for $y > 0$, recall the **law of total probability** in Proposition II.1.35. If Y denotes total losses and $0 < \epsilon < y$, then noting that $\Pr\{y - \epsilon < Y \leq y \mid 0 \text{ defaults}\} = 0$:*

$$\Pr\{y - \epsilon < Y \leq y\} = \sum_{k=1}^{n} \Pr\{y - \epsilon < Y \leq y \mid k \text{ def.}\} \Pr\{k \text{ def.}\}, \tag{1.71}$$

Since:

$$\Pr\{k \text{ defaults}\} = \binom{n}{k} p^k (i - p)^{n-k},$$

this can be stated in terms of $F(y)$:

$$F(y) - F(y - \epsilon) = \sum_{k=1}^{n} \binom{n}{k} p^k (i - p)^{n-k} [F_k(y) - F_k(y - \epsilon)] \tag{1}$$

where $F_k(y)$ is the conditional distribution function of Y given k losses.

Put another way, $F_k(y)$ is the distribution function of the sum of k random variables, defined as the losses on the k defaulted bonds. If loss given default is uniformly distributed on $[0, L_j]$, where L_j denotes the loan amount on the jth bond, then this distribution function again reflects which subset of k bonds actually defaulted, assuming $\{L_j\}_{j=1}^n$ are not constant. Again by the law of total probability, assuming all $N_k \equiv \binom{n}{k}$ collections of bonds are equally likely to default:

$$F_k(y) - F_k(y - \epsilon) = \frac{1}{N_k} \sum_{i=1}^{N_k} \left[F_k^{(i)}(y) - F_k^{(i)}(y - \epsilon) \right], \tag{2}$$

where now $F_k^{(i)}(y)$ is the distribution function of the sum of k losses from the ith collection of k bonds.

In other words, $F_k^{(i)}(y)$ is the distribution function of the sum of k continuously distributed variables, here assumed to be uniformly distributed. We claim each such $F_k^{(i)}(y)$ is continuous for $y > 0$ and thus by (1), (2) and (1.71), so too is $F(y)$, and thus $F(y)$ is a mixed distribution as claimed.

By (2.13), if X and Y are independent random variables on a probability space $(\mathcal{S}, \mathcal{E}, \lambda)$ with continuously differentiable distribution functions F_X and F_Y with continuous density *functions* $F'_X = f_X$ *and* $F'_Y = f_Y$, *then the density function of* $Z = X + Y$ *is continuous and given by the Riemann integrals:*

$$f_Z(z) = \int f_X(z-y)f_Y(y)dy = \int f_Y(z-x)f_X(x)dx. \tag{3}$$

Thus by induction, each $F_k^{(i)}(y)$ has a continuous density $f_k^{(i)}(y)$ and is continuous as claimed.

Exercise 1.33 *Given two loans with loss given default* Y_j *uniformly distributed on* $[0, L_j]$ *for* $j = 1, 2$, *use (3) to determine* $f_Y(y)$ *for* $Y = Y_1 + Y_2$. *Assume* $L_1 < L_2$ *to avoid ambiguity.*

2

Transformed Random Variables

In this chapter, we investigate the distribution and density functions associated with various transformations of random variables. Many of these results will be generalized in Book VI with the aid of the general integration theory of Book V.

2.1 Monotonic Transformations

Given a random variable X on a probability space $(\mathcal{S}, \mathcal{E}, \lambda)$ with distribution function $F(x)$, we are sometimes interested in evaluating the distribution function of $Y \equiv g(X)$ for Borel measurable function $g : \mathbb{R} \to \mathbb{R}$. By Proposition I.3.33, the composite function $g(X)$ is then a λ-measurable function on $\mathcal{S}$ and is hence a random variable. By definition, $F_Y(y) \equiv \lambda\left[Y^{-1}(-\infty, y]\right]$ and since $Y^{-1} = X^{-1}(g^{-1})$:

$$F_Y(y) = \lambda\left[X^{-1}\left(g^{-1}(-\infty, y]\right)\right], \tag{2.1}$$

where $g^{-1}(-\infty, y] \equiv \{z | g(z) \leq y\}$.

When $g(x)$ is strictly monotonic, we obtain the following result. Monotonic implies Borel measurable, but we add this to the assumptions as a reminder for the more general cases.

Proposition 2.1 (Distribution of $g(X)$, g strictly monotonic) *Let X be a random variable on a probability space $(\mathcal{S}, \mathcal{E}, \lambda)$ with distribution function $F(x)$, and $g : \mathbb{R} \to \mathbb{R}$ a strictly monotonic Borel measurable function.*

The distribution function for $Y \equiv g(X)$ is given by:

1. ***Increasing*** $g(x)$:

$$F_Y(y) = F_X(g^{-1}(y)), \tag{2.2}$$

2. ***Decreasing*** $g(x)$:

$$F_Y(y) = 1 - F_X(g^{-1}(y)^-), \tag{2.3}$$

where $F_X(x^-) \equiv \lim_{y \to x-} F_X(y)$ denotes the left limit at x.

If $F_X(x)$ is continuously differentiable with density $f_X(x)$, and $g(x)$ is continuously differentiable with $g'(x) \neq 0$, then $F_Y(y)$ is continuously differentiable with density function $f_Y(y) = F_Y'(y)$ given by:

$$f_Y(y) = f_X(g^{-1}(y)) \left| \frac{dg^{-1}(y)}{dy} \right|. \tag{2.4}$$

Thus:

$$F_Y(y) = (\mathcal{R}) \int_{-\infty}^{y} f_X(g^{-1}(z)) \left| \frac{dg^{-1}(z)}{dz} \right| dz, \tag{2.5}$$

defined as a Riemann integral.

DOI: 10.1201/9781003264583-2

Proof. *If $g(x)$ is strictly increasing, $g^{-1}(-\infty, y] = (-\infty, g^{-1}(y)]$ and thus:*

$$F_Y(y) = \lambda\left[X^{-1}(-\infty, g^{-1}(y)]\right] = F_X(g^{-1}(y)).$$

If $g(x)$ is also continuously differentiable with $g'(x) \neq 0$ for all x, then $\frac{dg^{-1}(y)}{dy} = 1/g'(g^{-1}(y))$ is well defined. Further, $\frac{dg^{-1}(y)}{dy}$ is continuous since $g'(x) \neq 0$ and is continuous, and g^{-1} is continuous since g is continuous and strictly increasing. Thus if $F_X(x)$ is continuously differentiable with density function $f_X(x) = F'_X(x)$, then $F_Y(y)$ is differentiable and has an associated continuous density function $f_Y(y) \equiv F'_Y(y)$:

$$f_Y(y) = f_X(g^{-1}(y))\frac{dg^{-1}(y)}{dy}. \tag{1}$$

With $g(x)$ strictly decreasing, now $g^{-1}(y, \infty) = (-\infty, g^{-1}(y))$, so by complementarity:

$$F_Y(y) = 1 - \lambda\left[X^{-1}(g^{-1}(y, \infty))\right] = 1 - F_X(g^{-1}(y)^-).$$

When $F_X(x)$ is continuous, this becomes:

$$F_Y(y) = 1 - F_X(g^{-1}(y)).$$

If $F_X(x)$ and $g(x)$ are continuously differentiable, $f_Y(y) \equiv F'_Y(y)$ is continuous and given by:

$$f_Y(y) = -f_X(g^{-1}(y))\frac{dg^{-1}(y)}{dy}. \tag{2}$$

Since $\frac{dg^{-1}(y)}{dy} > 0$ for increasing $g(x)$, and $\frac{dg^{-1}(y)}{dy} < 0$ for decreasing $g(x)$, (2.4) follows from (1) and (2).

For the result in (2.5), it follows by the continuity of $f_Y(y)$ and Proposition III.1.32 that for every interval $[a, b]$:

$$F_Y(y) = F_Y(a) + (\mathcal{R})\int_a^y f_X(g^{-1}(z))\left|\frac{dg^{-1}(z)}{dz}\right| dz, \quad y \in [a, b],$$

and this can be iterated with $a \to -\infty$ since then $F_Y(a) \to 0$ by Proposition 1.3. ∎

Example 2.2 (Normal/lognormal distributions) *Recalling the density functions in (1.65) and (1.67):*

1. If X is normally distributed and $Y = e^X$, then by (2.4):

$$f_Y(y) = \frac{1}{\sigma y \sqrt{2\pi}}\exp\left(-(\ln y - \mu)^2/2\sigma^2\right),$$

the density function of the lognormal distribution.

2. Similarly, if X is lognormally distributed then $Y = \ln X$ is normally distributed.

Remark 2.3 (Absolutely continuous $F_X(x)$) *When $F_X(x)$ is absolutely continuous, (2.2) remains unchanged for increasing $g(x)$, while (2.3) simplifies for decreasing $g(x)$ to:*

$$F_Y(y) = 1 - F_X(g^{-1}(y)).$$

If $g(x)$ is continuously differentiable with $g'(x) \neq 0$, then $g^{-1}(y)$ is continuous as noted above, and thus uniformly continuous on any bounded interval $[a, b]$ by Exercise III.1.14. It *follows from this that $F_Y(y)$ as defined in (2.2), or (2.3) as modified above, is absolutely continuous on every interval $[a, b]$ by Definition 3.54. Thus $f_Y(y) \equiv F_Y'(y)$ exists a.e. by Proposition III.3.59, and of necessity is given by (2.4) a.e. Then by Proposition III.3.62, (2.5) is satisfied as a Lebesgue integral:*

$$F_Y(y) = (\mathcal{L}) \int_{-\infty}^{y} f_X(g^{-1}(z)) \left| \frac{dg^{-1}(z)}{dz} \right| dz,$$

with the same proof as seen above.

The above framework can sometimes be adapted by first principals in situations where $g(x)$ is not strictly monotonic. However, general measurable $g(x)$ will require better tools. See Book VI.

Example 2.4 (Normal squared is $\chi^2_{1\ d.f.}$) *If X is standard normal, consider $Y = X^2$. Here $g(x) = x^2$ is not monotonic on the range of X, but we can explicitly evaluate $F_Y(y)$ by (2.1). Because $\lambda\left[X^{-1}(x)\right] = 0$ for each x by continuity of $F_N(x)$ and (1.3), we can be casual about interval endpoints.*
Thus:

$$
\begin{aligned}
F_Y(y) &= \lambda\left[X^{-1}\left(\left[-\sqrt{y}, \sqrt{y}\right]\right)\right] \\
&= \lambda\left[X^{-1}\left((-\infty, \sqrt{y}]\right)\right] - \lambda\left[X^{-1}\left((-\infty, -\sqrt{y}]\right)\right] \\
&= F_X\left(\sqrt{y}\right) - F_X\left(-\sqrt{y}\right).
\end{aligned}
$$

The associated density function can then be calculated, and using the symmetry of f_X, we derive that for $y > 0$:

$$
\begin{aligned}
f_Y(y) &= f_X\left(\sqrt{y}\right)/\sqrt{y} \\
&= \frac{1}{\sqrt{2\pi}} y^{-1/2} \exp\left(-y/2\right).
\end{aligned}
$$

Comparing this to (1.55), we see that Y has a gamma density with $\lambda = \alpha = 1/2$ if:

$$\Gamma(1/2) = \sqrt{\pi}.$$

This is indeed the case, as noted in (1.59).
Recalling Remark 1.30, this gamma distribution is also known as $\chi^2_{1\ d.f.}$, or chi-squared with 1 degree of freedom.

Remark 2.5 (Sum of independent squared normals is $\chi^2_{n\ d.f.}$) *As noted in Remark 1.30, when a random variable Y has a gamma distribution with $\lambda = 1/2$ and $\alpha = n/2$, it is said to have a **chi-squared distribution with n degrees of freedom**, which is denoted $\chi^2_{n\ d.f.}$. The density function of the chi-squared distribution with n degrees of freedom is thus given by (1.55) with the above parameters and defined on $x \geq 0$ by:*

$$f_{\chi^2_{n\ d.f.}}(x) = \frac{1}{2^{n/2}\Gamma(n/2)} x^{n/2-1} e^{-x/2}. \tag{2.6}$$

As proved above, if X is standard normal then X^2 is $\chi^2_{1\ d.f.}$. We will seen in item 5 of Section 4.4.1 that if $\{X_i\}_{i=1}^{n}$ are independent standard normals, then $\sum_{i=1}^{n} X_i^2$ is $\chi^2_{n\ d.f.}$.

2.2   Sums of Independent Random Variables

In this section, we introduce a calculation that will be generalized in Book VI using the integration tools of Book V. This calculation is known as the **convolution of functions**. The motivating application is:

  Given independent random variables X and Y on a probability space $(\mathcal{S}, \mathcal{E}, \lambda)$ with respective distribution functions F_X and F_Y, what is the distribution function of the random variable $Z \equiv X + Y$? If X and Y have density functions, what is the density function of Z?

  In this section, we will assume that X and Y have distribution functions which are either both saltus functions, or both absolutely continuous with continuous densities:

$$\text{Saltus:} \quad F(x) = \sum_{y \leq x} f(y),$$

$$\text{Absolutely Continuous:} \quad F(x) = \int_{-\infty}^{x} f(y)dy,$$

and defer generalizations to Book VI.

2.2.1   Distribution Functions of Sums

Beginning with saltus or discrete distribution functions, assume we are given $\{x_n\}_{n=1}^N, \{y_m\}_{m=1}^M \subset \mathbb{R}$ with $N, M \leq \infty$, and real nonnegative sequences $\{u_n\}_{n=1}^N$ and $\{v_m\}_{m=1}^M$, each with a sum of 1. Let X, Y be independent (Definition 1.13) discrete random variables on a probability space $(\mathcal{S}, \mathcal{E}, \lambda)$ with respective distribution functions $F_X(x)$ and $F_Y(y)$, and associated density functions $f_X(x)$ and $f_Y(y)$:

$$F_X(x) = \sum_{x_n \leq x} u_n, \quad F_Y(y) = \sum_{y_m \leq y} v_m \tag{2.7}$$
$$f_X(x_n) = u_n, \quad f_Y(y_m) = v_m.$$

In other words, density functions are defined to be 0 outside the points $\{x_n\}_{n=1}^N$ and $\{y_m\}_{m=1}^M$.

Proposition 2.6 ($Z = X + Y$: $F_Z(z)$ as a sum, discrete F_X, F_Y) *Let X and Y be independent random variables on a probability space $(\mathcal{S}, \mathcal{E}, \lambda)$ with saltus distribution functions given in (2.7). Then the distribution function of $Z = X + Y$ is given by:*

$$F_Z(z) = \sum_{m=1}^{M} F_X(z - y_m) f_Y(y_m) = \sum_{n=1}^{N} F_Y(z - x_n) f_X(x_n). \tag{2.8}$$

Proof. *Defining $Z = X + Y$, then by definition:*

$$F_Z(z) = \lambda \left[\{ s | X + Y \leq z \} \right].$$

Letting $B_m = \{ s | Y = y_m \}$, then $\{B_m\}_{m=1}^\infty \subset \mathcal{E}$ and $\bigcup_{m=1}^\infty B_m = \mathcal{S}$. By the law of total probability of Proposition II.1.35:

$$F_Z(z) = \sum_{m=1}^{M} \lambda \left[\{ X + Y \leq z | Y = y_m \} \right] \lambda \left[B_m \right]. \tag{1}$$

Now $\lambda [B_m] = f_Y(y_m)$ by definition, and by independence and Proposition II.1.34:

$$\lambda \{ X + Y \leq z | Y = y_m \} = \lambda \{ X \leq z - y_m | Y = y_m \} = \lambda \{ X \leq z - y_m \}.$$

By (1) this obtains the first expression above.
  Reversing the roles of X and Y obtains the second. ∎

The next result transforms the above representation into a form that will be seen to generalize below. The reader is invited to derive the analogous result for the second expression in (2.8) as an exercise.

Corollary 2.7 ($Z = X + Y : F_Z(z)$ **as an R-S integral, discrete** F_X, F_Y) *Let X and Y be independent random variables on a probability space $(\mathcal{S}, \mathcal{E}, \lambda)$ with saltus distribution functions as given in (2.7). Given z, assume that $\{z - x_n\}_{n=1}^N \bigcap \{y_m\}_{m=1}^M = \emptyset$ and that $\{z - x_n\}_{n=1}^N \bigcup \{y_m\}_{m=1}^M$ has no accumulation points. Then the distribution function of $Z = X + Y$ is given at z by:*

$$F_Z(z) = \int_{-\infty}^{\infty} F_X(z - y) dF_Y(y), \tag{2.9}$$

defined as a Riemann-Stieltjes integral.

Proof. *Since these distributions functions are not continuous, the existence results of Proposition III.4.17 do not apply. We, therefore, must verify existence of this integral directly.*
Define:

$$a = \min \left\{ |u - w| \,\Big|\, u, w \in \{z - x_n\}_{n=1}^N \bigcup \{y_m\}_{m=1}^M \right\},$$

and note that $a > 0$ by the assumptions on disjointedness of domains and absence of accumulation points. Let $P = \{w_k\}_{k=-K'}^{K''}$ with $w_k < w_{k+1}$ for all k be a partition of mesh size $\mu < a$ of an interval I that contains all $z - x_n$, y_m, where $I = \mathbb{R}$ is possible. Since there are no accumulation points, we can assume that P does not contain any $z - x_n$ or y_m. For notational consistency, assume that $\{z - x_n\}_{n=1}^N$ has been reordered to increasing $\{z - x_n\}_{n=-N'}^{N''}$, and $\{y_m\}_{m=1}^M$ to $\{y_m\}_{m=-M'}^{M''}$, noting that any or all of these set index bounds may be infinite.
With z fixed as above, recall Definition III.4.6 for the lower and upper Darboux sums:

$$L(F_X(z - y), F_Y(y), P) = \sum_{k=-K'}^{K''} m_k(z) \left[F_Y(w_{k+1}) - F_Y(w_k) \right],$$

$$U(F_X(z - y), F_Y(y), P) = \sum_{k=-K'}^{K''} M_k(z) \left[F_Y(w_{k+1}) - F_Y(w_k) \right].$$

Here $m_k(z) = \inf\{F_X(z - w) | w \in [w_k, w_{k+1}]$, and $M_k(z)$ is similarly defined in terms of the supremum. These sums are convergent since $0 \leq m_k$, $M_k \leq 1$ and $\sum_{k=-K'}^{K''} [F_Y(w_{k+1}) - F_Y(w_k)] = 1$.
By construction, each interval (w_k, w_{k+1}) contains at most one y_m. If $y_m \in (w_k, w_{k+1})$, then $F_Y(w_{k+1}) - F_Y(w_k) = v_m \equiv f_Y(y_m)$. Conversely, $F_Y(w_{k+1}) - F_Y(w_k) = 0$ if (w_k, w_{k+1}) contains no y_m. Thus:

$$\begin{aligned} L(F_X(z - y), F_Y(y), P) &= \sum_{m=-M'}^{M''} m_{k(m)}(z) f_Y(y_m), \\ U(F_X(z - y), F_Y(y), P) &= \sum_{m=-M'}^{M''} M_{k(m)}(z) f_Y(y_m). \end{aligned} \tag{1}$$

Here $m_{k(m)}(z)$ and $M_{k(m)}(z)$ are defined on $(w_{k(m)}, w_{k(m)+1})$, where this notation implies that $y_m \in (w_{k(m)}, w_{k(m)+1})$.
Given such an interval $[w_k, w_{k+1}]$ which contains some y_m, as this interval cannot contain any $z - x_n$, it follows that $F_X(z - w)$ is constant and thus $m_k(z) = M_k(z) = F_X(z - y_m)$. It now follows from (1) that for any partition P with mesh size $\mu < a$, that:

$$L(F_X(z - y), F_Y(y), P) = U(F_X(z - y), F_Y(y), P),$$

and further, these Darboux sums agree with $F_Z(z)$ in (2.8) by Proposition III.4.12. ∎

Remark 2.8 (On $\{z - x_n\}_{n=1}^N \bigcap \{y_m\}_{m=1}^M = \emptyset$,) *Corollary 2.7 made the assumption that $\{z - x_n\}_{n=1}^N \bigcap \{y_m\}_{m=1}^M = \emptyset$, and this assumption cannot be weakened. By Proposition III.4.15, if a Riemann-Stieltjes integral over $[a, b]$ exists, then the integrand and integrator can have no common discontinuities. As $\{z - x_n\}_{n=1}^N$ and $\{y_m\}_{m=1}^M$ are the discontinuities of $F_X(z - y)$ and $F_Y(y)$, this restriction is necessary since existence as an integral over $\mathbb{R}$ implies existence over all such $[a, b]$. Thus the representation in (2.9) is valid for all z outside an at most countable set, defined by $\{y_m + x_n\}$ for all m and n.*

We next turn to continuous distribution functions. and prove the following.

Proposition 2.9 ($Z = X + Y : F_Z(z)$ as an R-S integral, continuous F_X, F_Y) *Let X and Y be independent random variables on a probability space $(\mathcal{S}, \mathcal{E}, \lambda)$ with continuous distribution functions F_X and F_Y. Then defined as Riemann-Stieltjes integrals, the distribution function of $Z = X + Y$ is given by:*

$$F_Z(z) = \int_{-\infty}^{\infty} F_X(z - y) dF_Y(y) = \int_{-\infty}^{\infty} F_Y(z - x) dF_X(x). \tag{2.10}$$

Proof. *By continuity of the distribution functions, these integrals are well defined over every bounded interval $[a, b]$ by Proposition III.4.17. By boundedness, these integrals have well defined limits as $a \to -\infty$ and $b \to \infty$ by Proposition III.4.21. It then follows from Proposition III.4.12 that integrals can be evaluated by any sequence of Riemann-Stieltjes summations with partition mesh size $\mu \to 0$.*

To this end, define $P_n \equiv \{j/2^n\}_{j \in \mathbb{Z}^+}$, a partition of $(-\infty, \infty)$ of mesh size $\mu = 2^{-n}$. Then the first integral in (2.10) equals the limit as $n \to \infty$ of the Riemann-Stieltjes sums:

$$\sum\nolimits_{j=-\infty}^{\infty} F_X(z - y_{j,n}) \left[F_Y((j+1)/2^n) - F_Y(j/2^n) \right], \tag{1}$$

for any interval tags $y_{j,n} \in [j/2^n, (j+1)/2^n]$.

Define $A_j^{(n)} \in \mathcal{E}$ by:

$$A_j^{(n)} = \{j/2^n < Y \le (j+1)/2^n\}.$$

Then for any n, $\{A_j^{(n)}\}_{j=-\infty}^{\infty}$ are disjoint and $\bigcup_{j=-\infty}^{\infty} A_j^{(n)} = \mathcal{S}$.

Given z, define $B_j^{(n)}(z) \in \mathcal{E}$ by:

$$B_j^{(n)}(z) = \{X \le z - j/2^n\} \bigcap A_j^{(n)},$$

and let $B^{(n)}(z) = \bigcup_{j=-\infty}^{\infty} B_j^{(n)}(z)$. We claim that $\{B^{(n)}(z)\}_{n=1}^{\infty}$ is a nested collection, $B^{(n+1)}(z) \subset B^{(n)}(z)$, and:

$$\bigcap_{n=1}^{\infty} B^{(n)}(z) = B(z) \equiv \{X + Y \le z\}. \tag{2}$$

For nesting, we prove that:

$$B_{2j}^{(n+1)}(z) \bigcup B_{2j+1}^{(n+1)}(z) \subset B_j^{(n)}(z).$$

If $s \in B_{2j}^{(n+1)}(z)$ then:

$$X(s) \le z - 2j/2^{n+1}, \quad 2j/2^{n+1} < Y(s) \le (2j+1)/2^{n+1},$$

while $s \in B_{2j+1}^{(n+1)}(z)$ *implies:*

$$X(s) \le z - (2j+1)/2^{n+1}, \quad (2j+1)/2^{n+1} < Y(s) \le (2j+2)/2^{n+1}.$$

In either case, such $s \in B_j^{(n)}(z)$ *by the definition above.*

For (2), assume that $s \in B(z)$ and so $X(s) + Y(s) \le z$. Given any n, there is a unique j such that $j/2^n < Y(s) \le (j+1)/2^n$ and hence $s \in A_j^{(n)}$. It then follows that $X(s) \le z - j/2^n$ and so $s \in B_j^{(n)}(z) \subset B^{(n)}(z)$. Thus $B(z) \subset \bigcap_{n=1}^{\infty} B^{(n)}(z)$.

Conversely, if $s \in \bigcap_{n=1}^{\infty} B^{(n)}(z)$, then for any n there exists j_n so that $s \in B_{j_n}^{(n)}(z)$ and thus $s \in \bigcap_{n=1}^{\infty} B_{j_n}^{(n)}(z)$. That is, for all n there exists j_n so that:

$$\{X(s) \le z - j_n/2^n\} \wedge \{j_n/2^n < Y(s) \le (j_n+1)/2^n\}.$$

Thus for all n:

$$X(s) + Y(s) \le z - j_n/2^n + (j_n+1)/2^n = z - 1/2^n.$$

This obtains that $s \in B(z)$, *and the proof of (2) is complete.*

By continuity from above of λ by Proposition I.2.45:

$$\lambda[B(z)] = \lim_{n \to \infty} \lambda\left[B^{(n)}(z)\right]. \tag{3}$$

Now $\lambda[B(z)] = F(z)$ by definition, and by countable additivity, $\lambda[B^{(n)}(z)] = \sum_{j=-\infty}^{\infty} \lambda\left[B_j^{(n)}(z)\right]$. So (3) obtains by independence:

$$
\begin{aligned}
F_Z(z) &= \lim_{n \to \infty} \left[\sum_{j=-\infty}^{\infty} \lambda\left[\{X \le z - j/2^n\} \bigcap A_j^{(n)}\right]\right] \\
&= \lim_{n \to \infty} \left[\sum_{j=-\infty}^{\infty} F_X(z - j/2^n)\left[F_Y((j+1)/n) - F_Y(j/n)\right]\right].
\end{aligned}
$$

Comparing to (1), this is the limit of Riemann-Stieltjes sums with tags $y_{j,n} = j/2^n$, and the proof is complete for the first integral in (2.10).

The second integral follows identically with a change of notation. ∎

Corollary 2.10 ($Z = X+Y$: **When** $F_Z(z)$ **is continuous; absolutely continuous**) *Let* X *and* Y *be independent random variables on a probability space* $(\mathcal{S}, \mathcal{E}, \lambda)$ *with continuous distribution functions* F_X *and* F_Y. *Then the distribution function* $F_Z(z)$ *of* $Z = X + Y$ *in (2.11) is continuous.*

Further, if one of F_X *and* F_Y *is absolutely continuous on* $\mathbb{R}$, *then so too is* $F_Z(z)$.
Proof. *Given* z, *assume that* $\{z_n\}_{n=1}^{\infty}$ *is given with* $z_n \to z$. *Then :*

$$F_Z(z) - F_Z(z_n) = \int_{-\infty}^{\infty} \left[F_Y(z - x) - F_Y(z_n - x)\right] dF_X(x),$$

and this integral is well defined as a limit of integrals over $[a, b]$ as $a \to -\infty$ and $b \to \infty$. This follows from Proposition III.4.21 since $F_Y(z - x) - F_Z(z_n - x)$ is continuous and bounded, and $F_X(x)$ is increasing and bounded.

Thus by item 3 of Proposition III.4.24 applied to such intervals and taking a limit, and then applying item 2 of that result:

$$
\begin{aligned}
|F_Z(z) - F_Z(z_n)| &\le \int_{-\infty}^{\infty} |F_Y(z - x) - F_Y(z_n - x)| \, dF_X(x) \\
&\le \sup |F_Y(z - x) - F_Y(z_n - x)|,
\end{aligned}
$$

since $\int dF_X(x) = 1$. To complete the proof, we claim that this supremum converges to 0 as $n \to \infty$.

To this end, let $\epsilon > 0$ be given and choose m such that $1/m < \epsilon/2$. For $0 \leq j \leq m - 1$ define:

$$A_j^{(m)} = F_Y^{-1} [j/m, (j+1)/m].$$

Then $\bigcup_{j=0}^{m-1} A_j^{(m)} = \mathbb{R}$ by definition, and each $A_j^{(m)}$ is a nonempty interval since F_Y is continuous.

Define $\delta = \min_j \left\{ \left| A_j^{(m)} \right| \right\}$, where $\left| A_j^{(m)} \right|$ denotes interval length. Note that $\delta > 0$ since this is a finite collection of intervals, and all $\left| A_j^{(m)} \right| > 0$ by continuity of $F_Y(y)$. Assume that $y < y'$ are given with $|y - y'| < \delta$. If $y, y' \in A_j^{(m)}$ for some j, then $|F_Y(y) - F_Y(y')| < 1/m < \epsilon/2$ by construction. Otherwise $y \in A_j^{(m)}$ and $y' \in A_{j+1}^{(m)}$ for some j, and:

$$|F_Y(y) - F_Y(y')| \leq |F_Y(y) - F_Y((j+1)/m)| + |F_Y((j+1)/m) - F_Y(y')| < \epsilon.$$

Thus if $|z_n - z| < \delta$, then $\sup |F_Y(z - x) - F_Y(z_n - x)| < \epsilon$ and the proof of continuity of $F_Z(z)$ is complete.

For absolute continuity, assume that $F_X(x)$ has this property and that $\epsilon > 0$ is given. Recalling Definition III.3.54, choose δ so $\sum_{i=1}^{n} |F_X(x_i) - F_X(x_i')| < \epsilon$ for any collection $\{(x_i', x_i)\}_{i=1}^{n}$ of disjoint intervals with $\sum_{i=1}^{n} |x_i - x_i'| < \delta$. Given disjoint $\{(z_i', z_i)\}_{i=1}^{n}$ with $\sum_{i=1}^{n} |z_i - z_i'| < \delta$, it then follows from properties 1 and 3 of Proposition III.4.24 that:

$$\sum_{i=1}^{n} |F_Z(z_i) - F_Z(z_i')| \leq \int_{-\infty}^{\infty} \sum_{i=1}^{n} |F_X(z_i - y) - F_X(z_i' - y)| \, dF_Y(y)$$

$$< \epsilon \int_{-\infty}^{\infty} dF_Y(y) = \epsilon.$$

∎

When one or both distribution functions in Proposition 2.9 has a continuous derivative, as is the common assumption in continuous probability theory, we can convert one or both of the above Riemann-Stieltjes integrals to Riemann integrals.

In Book VI, this result will be generalized using the general integration theory of Book V. There it will be proved that if one or both of these distribution functions is absolutely continuous, and thus only differentiable almost everywhere by Proposition III.3.59, a similar conclusion results, but we must then interpret the integrals below in the sense of Lebesgue.

Corollary 2.11 ($Z = X + Y$: $F_Z(z)$ as an R integral, continuous f_X, f_Y) *Let X and Y be independent random variables on a probability space $(S, \mathcal{E}, \lambda)$ with continuously differentiable distribution functions F_X and F_Y with continuous density functions $f_X \equiv F_X'$ and $f_Y \equiv F_Y'$. Then the distribution function of $Z = X + Y$ is continuously differentiable and given as Riemann integrals by:*

$$F_Z(z) = (\mathcal{R}) \int_{-\infty}^{\infty} F_X(z - y) f_Y(y) dy = (\mathcal{R}) \int_{-\infty}^{\infty} F_Y(z - x) f_X(x) dx. \tag{2.11}$$

If only one of $F_X(x)$ and $F_Y(y)$ are continuously differentiable with the other continuous, then this representation is valid for the associated density function, and then $F_Z(z)$ is continuous.

Proof. *This representation as a Riemann integral follows directly from Proposition III.4.28. Continuity follows from Corollary 2.10, while continuous differentiability will be addressed in Proposition 2.17.* ∎

Exercise 2.12 *When both distribution functions are continuously differentiable, show that these two integrals are equal by integration by parts of III.(1.30). Hint: Start with integrals over* $[-N, N]$ *and consider limits. Recall Remark III.1.35.*

Notation 2.13 (Convolution) *In the terminology and notation of Book V, the first expression in (2.8) and (2.11) for the distribution function* $F_Z(z)$ *equals the* **convolution of** F_X **and** f_Y. *Analogously, the second expression equals the* **convolution of** F_Y *and* f_X. *Notationally:*

$$F_Z(z) = F_X * f_Y(z) = F_Y * f_X(z).$$

Convolutions are commutative, meaning that:

$$F_X * f_Y(z) = f_Y * F_X(z).$$

For example in (2.11):

$$\int_{-\infty}^{\infty} F_X(z-y) f_Y(y) dy = \int_{-\infty}^{\infty} F_X(y) f_Y(z-y) dy,$$

as a change of variables in the integral verifies. The verification is similar for (2.8).

Convolution of three or more functions proves to be associative, and thus can be defined iteratively, by

$$f * g * h(z) \equiv (f * g) * h(z) = f * (g * h)(z).$$

Associativity will be proved with the aid of Fubini's theorem in Book V.

Example 2.14 *We apply the above formulas in (2.8) and (2.11) to previously introduced distribution functions.*

1. **Sums of Binomials are Binomial:** *Let* X *and* Y *be independent and have binomial distributions as in (1.40) with common parameter* p, *but with respective parameters* n *and* m. *Then by (2.8):*

$$
\begin{aligned}
F_Z(z) &= \sum_{j=0}^{z} F_X(z-j) \binom{m}{j} p^j (1-p)^{m-j} \\
&= \sum_{j=0}^{z} \left[\sum_{k=0}^{z-j} \binom{n}{k} p^k (1-p)^{n-k} \right] \binom{m}{j} p^j (1-p)^{m-j} \\
&= \sum_{j=0}^{z} \sum_{k=0}^{z-j} \binom{n}{k} \binom{m}{j} p^{j+k} (1-p)^{n+m-(j+k)}.
\end{aligned}
$$

This double summation can now be rearranged.

Defining $i = j+k$, *then* $k = i-j$ *and the* k-*sum becomes an* i-*sum from* j *to* z. *Reversing the resulting double summations:*

$$\sum_{j=0}^{z} \sum_{k=0}^{z-j} = \sum_{j=0}^{z} \sum_{i=j}^{z} = \sum_{i=0}^{z} \sum_{j=0}^{i},$$

obtains:

$$F_Z(z) = \sum_{i=0}^{z} \sum_{j=0}^{i} \binom{n}{i-j} \binom{m}{j} p^i (1-p)^{n+m-i}.$$

Now:

$$\sum_{j=0}^{i} \binom{n}{i-j} \binom{m}{j} = \binom{m+n}{i}$$

as can be verified as an exercise by evaluating and comparing the coefficients of x^i in the identity:

$$(1+x)^{n+m} = (1+x)^n (1+x)^m.$$

Thus:

$$F_Z(z) = \sum_{i=0}^{z} \binom{m+n}{i} p^i (1-p)^{n+m-i}.$$

In other words, $Z = X + Y$ has a binomial distribution with parameters p and $n+m$.

2. **Sums of Exponentials are Gamma:** Let X and Y be independent and have exponential distributions as in (1.53) with the same parameter λ, so $f_E(x) = \lambda e^{-\lambda x}$ and $F_E(x) = 1 - e^{-\lambda x}$ for $x \geq 0$. Then by (2.11), noting the range of integration in y :

$$
\begin{aligned}
F_Z(z) &= \int_0^z \left(1 - e^{-\lambda(z-y)}\right) \lambda e^{-\lambda y} dy \\
&= 1 - e^{-\lambda z} (1 + \lambda z).
\end{aligned}
$$

This formula is satisfied for $z \geq 0$, and hence by differentiation:

$$f_Z(z) = \lambda^2 z e^{-\lambda z}, \quad z \geq 0.$$

It then follows that $f_Z(z)$ is the gamma density function in (1.55) with parameters λ and $\alpha = 2$.

3. **Sums of Exponentials and the Poisson:** By Exercise 2.15, the sum of k independent exponentials with common parameter λ produces a gamma with parameters λ and $\alpha = k$, and this motivates an interesting connection between sums of such exponentials and the Poisson distribution.

Given such independent exponentials $\{X_j\}_{j=1}^{\infty}$, let $S_n = \sum_{j=1}^{n} X_j$ and define a new random variable N by:

$$N \equiv \max\{n | S_n \leq 1\}. \tag{1}$$

Then $N = n$ if and only if $S_n \leq 1 < S_{n+1}$, and hence $N \geq n$ if and only if $S_n \leq 1$. By Exercise 2.15, the distribution function of S_n is given in (1.60), and since $\Pr[S_n \leq 1] = F_{S_n}(1)$ we obtain:

$$\Pr[N \geq n] = e^{-\lambda} \sum_{j=n}^{\infty} \frac{\lambda^j}{j!}.$$

Now:

$$\Pr[N = n] = \Pr[N \geq n] - \Pr[N \geq n+1],$$

and thus:

$$\Pr[N = n] = e^{-\lambda} \frac{\lambda^n}{n!}.$$

In other words, N in (1) has a Poisson distribution with parameter λ.

Exercise 2.15 *Two exercises that generalize the above:*

1. **Sums of Binomials are Binomial:** *Generalize item 1 above and prove by induction that the sum of N such independent binomials with parameters p and n_i has a binomial distribution with parameters p and $n = \sum_{i=1}^{N} n_i$.*

2. **Sums of Exponentials are Gamma:** *Generalize item 2 above and prove by induction that the sum of k independent exponentials with the same parameter λ has a gamma distribution with parameters λ and $\alpha = k$.*

Note: *This result will be further generalized in Section 4.4.1 to the statement that sums of independent gamma random variables are gamma as long as they have a common λ parameter. Then the resultant α parameter of the sum satisfies $\alpha = \sum_i \alpha_i$, where $\{\alpha_i\}_i$ are the parameters of the individual gamma variates.*

2.2.2 Density Functions of Sums

In this section, we derive density functions associated with the discrete distribution functions of Proposition 2.6, and the continuously differentiable distribution functions of Corollary 2.11. These results will be expanded upon in Book VI.

Proposition 2.16 ($Z = X + Y$: $f_Z(z)$ as a sum, discrete F_X, F_Y) *Let X and Y be independent random variables on a probability space $(\mathcal{S}, \mathcal{E}, \lambda)$ with saltus distribution and density functions given in (2.7). Then the density function of $Z = X + Y$ is given by:*

$$f_Z(z) = \sum_{m=1}^{M} f_X(z - y_m) f_Y(y_m) = \sum_{n=1}^{N} f_Y(z - x_n) f_X(x_n). \tag{2.12}$$

Proof. *Rewriting the first expression in (2.8):*

$$
\begin{aligned}
F_Z(z) &= \sum_{m=1}^{M} F_X(z - y_m) f_Y(y_m) \\
&\equiv \sum_{m=1}^{M} \sum_{x_n \leq z - y_m} f_X(x_n) f_Y(y_m).
\end{aligned} \tag{1}
$$

Now $Z(\mathcal{S}) = \{x_n + y_m\}_{n,m}$, and so if $z \notin Z(\mathcal{S})$ then $f_Z(z) = 0$ and (2.12) is satisfied. Otherwise given $z_{n,m} \equiv x'_n + y'_m$, (1) and (1.3) obtain the first expression in (2.12):

$$
\begin{aligned}
f_Z(z_{n,m}) &= F_Z(z_{n,m}) - \lim_{z \to z_{n,m}^-} F_Z(z) \\
&= \sum_{m=1}^{M} \left[\sum_{x_n \leq z_{n,m} - y_m} f_X(x_n) - \lim_{z \to z_{n,m}^-} \sum_{x_n \leq z - y_m} f_X(x_n) \right] f_Y(y_m) \\
&= \sum_{m=1}^{M} \left[\lim_{z \to z_{n,m}^-} \sum_{z - y_m < x_n \leq z_{n,m} - y_m} f_X(x_n) \right] f_Y(y_m) \\
&= \sum_{m=1}^{M} f_X(z_{n,m} - y_m) f_Y(y_m).
\end{aligned}
$$

The other sum is similarly derived. ∎

Turning to continuously differentiable distribution functions, we have the following perhaps unsurprising result, recalling the Leibniz integral rule of Proposition III.1.40. However, the direct proof of this result is surprisingly subtle.

The first challenge that arises for this result is the existence of the integrals in (2.13). If f_X and f_Y are continuous density functions that thus integrate to 1, are these integrals well-defined as Riemann integrals? Moreover, is $f(z)$ continuous or at least Riemann integrable, and does the integral of $f_Z(z)$ then equal 1 as it must for $f_Z(z)$ to be a density function? Finally, is this function the density function of the distribution function in (2.10)?

We provide a proof with some reference to Book V results, where a generalized Leibniz integral will also be found that will make this result transparent.

Proposition 2.17 ($Z = X + Y : f_Z(z)$ **as an R integral, continuous** f_X, f_Y) *Let* X *and* Y *be independent random variables on a probability space* $(\mathcal{S}, \mathcal{E}, \lambda)$ *with continuously differentiable distribution functions* F_X *and* F_Y, *and associated continuous density functions* $f_X \equiv F_X'$ *and* $f_Y \equiv F_Y'$. *Then the density function of* $Z = X + Y$ *is continuous and given by the Riemann integrals:*

$$f_Z(z) = (\mathcal{R}) \int_{-\infty}^{\infty} f_X(z - y) f_Y(y) dy = (\mathcal{R}) \int_{-\infty}^{\infty} f_Y(z - x) f_X(x) dx. \qquad (2.13)$$

Proof. *We derive the first representation, with the second derived by a change of variable. To first investigate* $f_Z(z)$ *as defined above, let:*

$$f_Z^{(n)}(z) = (\mathcal{R}) \int_{-n}^{n} f_X(z - y) f_Y(y) dy. \qquad (1)$$

Then $f_Z^{(n)}(z)$ *is well-defined as a Riemann integral by continuity of the integrand and Proposition III.1.15.*

To see that $f_Z^{(n)}(z)$ *is continuous for all* n, *let* $z_m \to z$ *and assume without loss of generality that* $|z_m - z| < 1$ *for all* m. *Then by Proposition III.1.23:*

$$\left| f_Z^{(n)}(z) - f_Z^{(n)}(z_m) \right| \leq \int_{-n}^{n} |f_X(z - y) - f_X(z_m - y)| \, f_Y(y) dy.$$

Since continuous, $f_X(x)$ *is uniformly continuous on* $[-n - 1, n + 1]$ *by Exercise III.1.14. Thus given* $\epsilon > 0$, *there exists* δ *so that* $|x - x'| < \delta$ *assures* $|f_X(x) - f_X(x')| < \epsilon$. *This obtains that if* $|z_m - z| < \delta$:

$$\left| f_Z^{(n)}(z) - f_Z^{(n)}(z_m) \right| \leq \epsilon \int_{-n}^{n} f_Y(y) dy \leq \epsilon.$$

Hence $f_Z^{(n)}(z)$ *is continuous and Riemann integrable on* $[-N, N]$ *for all* N *by Proposition III.1.15, and the improper integral is defined when it exists by:*

$$\int_{-\infty}^{\infty} f_Z^{(n)}(z) dz \equiv \lim_{N \to \infty} \int_{-N}^{N} \int_{-n}^{n} f_X(z - y) f_Y(y) dy dz.$$

By continuity of the integrand, the Riemann integral of $f_X(z - y) f_Y(y)$ *is well defined as an integral over the rectangle* $[-N, N] \times [-n, n]$ *by Proposition III.1.63. Corollary III.1.77 then justifies reversing the iterated integrals:*

$$\int_{-\infty}^{\infty} f_Z^{(n)}(z) dz \equiv \lim_{N \to \infty} \int_{-n}^{n} \int_{-N}^{N} f_X(z - y) dz f_Y(y) dy$$

$$= \lim_{N \to \infty} \int_{-n}^{n} [F_X(N - y) - F_X(-N - y)] f_Y(y) dy.$$

Defining $G_N(y) \equiv [F_X(N - y) - F_X(-N - y)] f_Y(y)$, *note that* $G_N(y) \to f_Y(y)$ *pointwise and this sequence is monotonically increasing. Thus interpreted as a Lebesgue integral (Proposition III.2.18), Lebesgue's monotone convergence theorem obtains:*

$$\lim_{N \to \infty} (\mathcal{L}) \int_{-n}^{n} G_N(y) dy = (\mathcal{L}) \int_{-n}^{n} f_Y(y) dy.$$

Each of these integrals equals the Riemann counterparts by Proposition III.2.18 and thus:

$$(\mathcal{R}) \int_{-\infty}^{\infty} f_Z^{(n)}(z) dz = \int_{-n}^{n} f_Y(y) dy = F_Y(n) - F_Y(-n). \qquad (2)$$

For the integral of $f_Z(z)$, it follows by definition that $f_Z^{(n)}(z) \to f_Z(z)$ pointwise, and again the sequence is monotonically increasing by nonnegativity of the integrand in (1). Another application of monotone convergence and (2) obtains:

$$(\mathcal{L}) \int_{-\infty}^{\infty} f_Z(z)dz = \lim_{n \to \infty} \int_{-\infty}^{\infty} f_Z^{(n)}(z)dz = 1.$$

Although we will not prove this until Book V, $f_Z(z)$ is in fact continuous, and thus the Riemann and Lebesgue integrals agree over every bounded interval $[-N, N]$ by Proposition III.2.18, so taking the limit as $N \to \infty$:

$$(\mathcal{R}) \int_{-\infty}^{\infty} f_Z(z)dz = 1. \tag{3}$$

In summary, $f_Z(z)$ is continuous, nonnegative, and integrates to 1, and thus has all the properties of a density function.

Now define the distribution function $\tilde{F}_Z(z)$ associated with $f_Z(z)$:

$$\tilde{F}_Z(z) = (\mathcal{R}) \int_{-\infty}^{z} f_Z(w)dw \equiv (\mathcal{R}) \int_{-\infty}^{z} \int_{-\infty}^{\infty} f_X(w-y)f_Y(y)dydw. \tag{4}$$

Note that $\tilde{F}_Z(z)$ is continuously differentiable by Proposition III.1.33 and $\tilde{F}_Z'(z) = f_Z(z)$. If reversing the order of integration in (4) is justified, it will then follow that $\tilde{F}_Z(z) = F_Z(z)$.

When interpreted as Lebesgue integrals using Proposition III.2.56, this justification is found in Book V in either Fubini's theorem or Tonelli's theorem. This reversal of iterated integrals can also be seen as a generalization of the Riemann result in Corollary III.1.77.

To prove it here with the same approach as above, note that by definition of the improper Riemann integral:

$$\tilde{F}_Z(z) = \int_{-\infty}^{z} \left[\lim_{N \to \infty} H_N(w) \right] dw,$$

where:

$$H_N(w) \equiv \int_{-N}^{N} f_X(w-y)f_Y(y)dy.$$

Switching back and forth between Riemann and Lebesgue integrals, we have by Lebesgue's monotone convergence theorem that:

$$\tilde{F}_Z(z) = \lim_{N \to \infty} \int_{-\infty}^{z} \int_{-N}^{N} f_X(w-y)f_Y(y)dydw. \tag{5}$$

But by Corollary III.1.77 and monotone convergence:

$$\int_{-\infty}^{z} \int_{-N}^{N} f_X(w-y)f_Y(y)dydw \equiv \lim_{M \to \infty} \int_{-M}^{z} \int_{-N}^{N} f_X(w-y)f_Y(y)dydw$$

$$= \lim_{M \to \infty} \int_{-N}^{N} \int_{-M}^{z} f_X(w-y)dw f_Y(y)dy$$

$$= \int_{-N}^{N} F_X(w-y)f_Y(y)dy.$$

Then by (5) and Corollary 2.11, it follows that $\tilde{F}_Z(z) = F_Z(z)$, and the proof is complete. ∎

Example 2.18 *In this example, we illustrate two applications of Proposition 2.17.*

   *1. **Average of two independent standard normals is normal, but not standard:***
*Let X and Y be independent and have standard normal distributions as in (1.66). We derive
that $Z \equiv X/2 + Y/2$ also has a normal distribution, but not the standard normal.*
   By (2.13):

$$f_Z(z) = \frac{2}{\pi} \int_{-\infty}^{\infty} \exp\left(-2\left[x^2 + (z-x)^2\right]\right) dx.$$

*This follows from the observation that if $U = V/2$, then $f_U(u) = 2f_v(2u)$, which can be
derived by first principles or by (2.4).*
   A little algebra and the substitution $w = 2^{3/2}(x - z/2)$ obtains:

$$
\begin{aligned}
f_Z(z) &= \frac{2}{\pi} \exp\left(-z^2\right) \int_{-\infty}^{\infty} \exp\left(-4\left(x - z/2\right)^2\right) dx \\
&= \frac{1}{\pi\sqrt{2}} \exp\left(-z^2\right) \int_{-\infty}^{\infty} \exp\left(-w^2/2\right) dw \\
&= \frac{1}{\sqrt{\pi}} \exp\left(-z^2\right).
\end{aligned}
$$

Comparing with (1.65), Z is seen to have a normal distribution with $\mu = 0$ and $\sigma^2 = 1/2$.
   *2. **Average of two independent standard Cauchy is standard Cauchy:** Let X
and Y be independent and have standard Cauchy distributions as in (1.69). In this example,
we derive that $Z \equiv X/2 + Y/2$ also has a standard Cauchy distribution.*
   As in item 1, by (2.13):

$$f_Z(z) = \frac{4}{\pi^2} \int_{-\infty}^{\infty} \frac{1}{\left[1 + 4x^2\right]\left[1 + 4(z-x)^2\right]} dx,$$

To prove that Z is Cauchy and:

$$f_Z(z) = \frac{1}{\pi}\frac{1}{1 + z^2}, \tag{1}$$

*a direct integral is quite challenging. Instead, we finesse the result and prove that $g(z) = 1$
for all z where:*

$$g(z) \equiv \frac{4}{\pi} \int_{-\infty}^{\infty} \frac{1 + z^2}{\left[1 + 4x^2\right]\left[1 + 4(z-x)^2\right]} dx.$$

   *To better exploit the symmetries of this integrand, substitute $x = y + \frac{z}{2}$ and evaluate
$g(2z)$ to obtain:*

$$g(2z) = \frac{4}{\pi} \int_{-\infty}^{\infty} \frac{1 + 4z^2}{\left[1 + 4(y+z)^2\right]\left[1 + 4(y-z)^2\right]} dy. \tag{2}$$

To prove that $g(2z) = 1$ for all z, we prove that $g(0) = 1$ and $g'(2z) = 0$ for all z.

   a. $g(0) = 1$: From (2) and the substitution $y = x/2$:

$$g(0) = \frac{2}{\pi} \int_{-\infty}^{\infty} \frac{1}{\left[1 + x^2\right]^2} dx.$$

Substituting $y = \tan\theta$ obtains:

$$g(0) = \frac{2}{\pi} \int_{-\pi/2}^{\pi/2} \cos^2\theta d\theta,$$

and this is seen to equal 1 after two applications of integration by parts, obtaining the antiderivative:

$$\int \cos^2\theta d\theta = \frac{1}{2}\left(\cos\theta\sin\theta + \theta\right) + C.$$

b. $g'(2z) = 0$ *for all* z : *Rewriting* (2) *with* $A(y,z) \equiv 1 + 4(z+y)^2$:

$$g(2z) = \frac{4}{\pi}\int_{-\infty}^{\infty} \frac{1+4z^2}{A(y,z)A(-y,z)}dy.$$

This integral is absolutely convergent for all z since the integrand is $O(1/y^2)$ at $\pm\infty$. In addition, the derivative of this integrand:

$$\frac{\partial}{\partial z}\left[\frac{1+4z^2}{A(y,z)A(-y,z)}\right]$$

$$= \frac{8yA(y,z)A(-y,z) - 8\left(1+4z^2\right)\left[(z+y)A(-y,z) - (z-y)A(y,z)\right]}{\left[A(y,z)A(-y,z)\right]^2}, \quad (3)$$

is absolutely integrable for all z, since it is $O(1/y^3)$ at $\pm\infty$.

Thus we can apply the general Leibniz integral rule of Book V and differentiate under the integral. Denoting the derivative of the integrand in (3) by $h(y,z)$:

$$\frac{dg(2z)}{dz} = \frac{4}{\pi}\int_{-\infty}^{\infty} h(y,z)dy.$$

*For any z, $h(y,z)$ is an **odd function of** y, meaning that $h(-y,z) = -h(y,z)$. The integral of any odd function is 0 by substitution, and thus the proof of (1) is complete.*

Remark 2.19 (Generalizations for n independent Cauchy/normal) *It will be seen in item 4 of Section 4.4.1 that a sum of independent normal variates $\{X_i\}_{i=1}^n$ with parameters $\{\mu_i\}_{i=1}^n$ and $\{\sigma_i^2\}_{i=1}^n$ is normal with parameters $\mu = \sum_{i=1}^n \mu_i$ and $\sigma^2 = \sum_{i=1}^n \sigma_i^2$. In the above example, $\mu_i = 0$ and $\sigma_i^2 = 1/4$ for X and Y, and this reproduces the above result. This general result will be readily obtained using moment-generating functions and the method of moments. It can also be obtained by induction and (2.13), and is left as an exercise.*

*Since the Cauchy distribution has no moments by item 4 of Section 4.2.7, and thus certainly no moment-generating function by Proposition 4.25, no generalization of the above result will be seen in this book. But in Book VI, the above result on the Cauchy distribution will be far more easily obtained using **characteristic functions**, which generalize the notion of a moment generating function in important ways. This approach will also readily generalize to obtain the same result for the average of any finite number of Cauchy variates.*

2.3 Ratios of Random Variables

In this section, we alter the question of the prior section somewhat. Given random variables X and Y on a probability space $(\mathcal{S}, \mathcal{E}, \lambda)$ with respective distribution functions F_X and F_Y, what is the distribution function of the random variable $Z \equiv X/Y$? What is the density function of Z? To avoid definitional problems, it will generally be assumed that Y has range $(0, \infty)$ and thus $F_Y(0) = 0$. This assumption can be weakened somewhat when Y is discrete.

2.3.1 Independent Random Variables

Beginning with the case of discrete random variables.

Proposition 2.20 ($Z = X/Y$, $F_Z(z)$ as a sum, discrete F_X, F_Y, $F_Y(0) = 0$) *Let X and Y be independent random variables on a probability space $(\mathcal{S}, \mathcal{E}, \lambda)$ with saltus distribution functions given in (2.7) and $F_Y(0) = 0$. Then the distribution function of $Z = X/Y$ is given by:*

$$F_Z(z) = \sum_{m=1}^{M} F_X(zy_m) f_Y(y_m). \qquad (2.14)$$

Proof. *By definition:*

$$F_Z(z) = \lambda\left[\{s | X/Y \le z\}\right].$$

Letting $B_m = \{s | Y = y_m\}$, then $\{B_m\}_{m=1}^{M} \subset \mathcal{E}$ and $\bigcup_{m=1}^{M} B_m = \mathcal{S}$. Since these sets are disjoint, the law of total probability of Proposition II.1.35 obtains:

$$F_Z(z) = \sum_{m=1}^{M} \lambda\left[\{X/Y \le z | Y = y_m\}\right] \lambda\left[B_m\right]. \qquad (1)$$

Now $\lambda[B_m] = f_Y(y_m)$ by definition. By independence and Proposition II.1.34, and since $y_m > 0$ for all m :

$$\lambda\{X/Y \le z | Y = y_m\} = \lambda\{X \le zy_m | Y = y_m\} = \lambda\{X \le zy_m\}.$$

The proof is complete by (1) since $\lambda\{X \le zy_m\} = F_X(zy_m)$. ∎

The random variable Z in this case is well-defined under a more general assumption than $F_Y(0) = 0$ as assumed above, and it is apparent that what is necessary is that $f_Y(0) = 0$ for the associated density function. To simplify the development, we generalize the notation underlying (2.7) to $\{y_m\}_{m=-M'}^{M} \subset \mathbb{R}$ with $M, M' \le \infty$, and associated real nonnegative sequence $\{\nu_m\}_{m=-M'}^{M}$ with sum of 1. Here we assume that $y_m < 0$ for $m \le -1, y_0 = v_0 = 0$, and $y_m > 0$ for $m \ge 1$.

Proposition 2.21 ($Z = X/Y$, $F_Z(z)$ as a sum, discrete F_X, F_Y, $f_Y(0) = 0$) *Let X and Y be independent random variables on a probability space $(\mathcal{S}, \mathcal{E}, \lambda)$ with saltus distribution functions given in (2.7) and $f_Y(0) = 0$. Then the distribution function of $Z = X/Y$ is given by:*

$$F_Z(z) = \sum_{m=1}^{M} F_X(zy_m) f_Y(y_m) \qquad (2.15)$$
$$+ \sum_{m=-1}^{-M'} \left[1 - F_X(zy_m) + f_Y(zy_m)\right] f_Y(y_m).$$

The symmetry of this expression is better revealed by:

$$F_Z(z) = \sum_{m=1}^{M} \sum_{x_n \le zy_m} f_X(x_n) f_Y(y_m) \qquad (2.16)$$
$$+ \sum_{m=-1}^{-M'} \sum_{x_n \ge zy_m} f_X(x_n) f_Y(y_m).$$

Proof. *Letting* $B_m = \{s | Y = y_m\}$, *then* $\{B_m\}_{m=-M'}^{M} \subset \mathcal{E}$ *and* $\bigcup_{m=-M'}^{M} B_m = \mathcal{S}$. *As noted above,* $y_m < 0$ *for* $m \leq -1$ *and* $y_m > 0$ *for* $m \geq 1$. *Then by disjointness of* $\{B_m\}_{m=-M'}^{M}$ *and the law of total probability of Proposition II.1.35:*

$$F_Z(z) = \sum_{m=1}^{M} \lambda\left[\{X/Y \leq z | Y = y_m\}\right] \lambda\left[B_m\right] + \sum_{m=-1}^{-M'} \lambda\left[\{X/Y \leq z | Y = y_m\}\right] \lambda\left[B_m\right]. \tag{1}$$

The first summation in (1) *becomes the first summation in* (2.15) *and* (2.16) *as in the prior proof, so we focus on the second summation.*

For $m \leq -1$, *again* $\lambda\left[B_m\right] = f_Y(y_m)$ *by definition. By independence and Proposition II.1.34:*

$$\lambda\{X/Y \leq z | Y = y_m\} = \lambda\{X \geq z y_m | Y = y_m\} = \lambda\{X \geq z y_m\}.$$

The proof is complete since $\lambda\{X \geq z y_m\} = 1 - F_X(z y_m) + f_Y(z y_m)$. ∎

The next result provides the density function $f_Z(z)$ in the more general setting of Proposition 2.21, but reduces to that for Proposition 2.20 by changing the summation to $m = 1$ to M.

Corollary 2.22 ($Z = X/Y$, $f_Z(z)$ **as a sum, discrete** F_X, F_Y, $f_Y(0) = 0$) *Let* X *and* Y *be independent random variables on a probability space* $(\mathcal{S}, \mathcal{E}, \lambda)$ *with saltus distribution functions given in* (2.7) *and* $f_Y(0) = 0$. *Then the density function of* $Z = X/Y$ *is given by:*

$$f_Z(z) = \sum_{m=-M'}^{M} f_X(z y_m) f_Y(y_m). \tag{2.17}$$

Proof. *By definition,* $Z(\mathcal{S}) = \{x_n/y_m\}_{n,m}$, *and so if* $z \notin Z(\mathcal{S})$ *then* $f_Z(z) = 0$ *and* (2.17) *is satisfied.*

Otherwise, given $z_{n,m} \equiv x'_n/y'_m$, *it follows from* (1.3) *and* (2.16) *that:*

$$
\begin{aligned}
f_Z(z_{n,m}) &= F_Z(z_{n,m}) - \lim_{z \to z_{n,m}^-} F_Z(z) \tag{1} \\
&= \sum_{m=1}^{M} \left[\sum_{x_n \leq z_{n,m} y_m} f_X(x_n) - \lim_{z \to z_{n,m}^-} \sum_{x_n \leq z y_m} f_X(x_n) \right] f_Y(y_m) \\
&\quad + \sum_{m=-1}^{-M'} \left[\sum_{x_n \geq z_{n,m} y_m} f_X(x_n) - \lim_{z \to z_{n,m}^-} \sum_{x_n \geq z y_m} f_X(x_n) \right] f_Y(y_m).
\end{aligned}
$$

If $y_m > 0$:

$$
\begin{aligned}
\sum_{x_n \leq z_{n,m} y_m} f_X(x_n) - \lim_{z \to z_{n,m}^-} \sum_{x_n \leq z y_m} f_X(x_n) &= \lim_{z \to z_{n,m}^-} \sum_{z y_m < x_n \leq z_{n,m} y_m} f_X(x_n) \\
&= f_X(z_{n,m} y_m).
\end{aligned}
$$

For $y_m < 0$:

$$
\begin{aligned}
\sum_{x_n \geq z_{n,m} y_m} f_X(x_n) - \lim_{z \to z_{n,m}^-} \sum_{x_n \geq z y_m} f_X(x_n) &= \lim_{z \to z_{n,m}^-} \sum_{z_{n,m} y_m \leq x_n < z y_m} f_X(x_n) \\
&= f_X(z_{n,m} y_m).
\end{aligned}
$$

This and (1) *complete the proof.* ∎

Turning next to continuous distribution functions, we begin with a Riemann-Stieltjes representation. For simplicity, we again initially assume that $F_Y(0) = 0$.

Proposition 2.23 ($Z = X/Y$, $F_Z(z)$ **for continuous** F_X, F_Y, $F_Y(0) = 0$) *Let X and Y be independent random variables on a probability space $(\mathcal{S}, \mathcal{E}, \lambda)$ with continuous distribution functions F_X and F_Y and $F_Y(0) = 0$. Then defined as a Riemann-Stieltjes integral, the distribution function of $Z = X/Y$ is given by:*

$$F_Z(z) = \int_0^\infty F_X(zy) dF_Y(y). \tag{2.18}$$

Proof. *If $z = 0$ then $F_Z(0) = F_X(0)$ by definition, and this is obtained from (2.18). We now prove this result for $z > 0$ and assign $z < 0$ as an exercise.*

As F_X and F_Y are continuous and increasing functions, this integral is well defined over every bounded interval $[0, b]$ by Proposition III.4.17. By boundedness of these functions, this integral has a well defined limit as $b \to \infty$ by Proposition III.4.21. It then follows by Proposition III.4.12 that integrals can be evaluated by any sequence of Riemann-Stieltjes summations with partition mesh sizes $\mu \to 0$.

To this end, define $P_n \equiv \{j/2^n\}_{j \in \mathbb{Z}^+}$, a partition of $[0, \infty)$ of mesh size $\mu = 2^{-n}$, where $\mathbb{Z}^+$ includes 0. Then the integral in (2.18) equals the limit as $n \to \infty$ of the Riemann-Stieltjes sums:

$$\sum_{j=0}^\infty F_X(zy_{j,n}) \left[F_Y((j+1)/2^n) - F_Y(j/2^n) \right], \tag{1}$$

for any interval tags $y_{j,n} \in [j/2^n, (j+1)/2^n]$.

Define $A_j^{(n)} \in \mathcal{E}$ by:

$$A_j^{(n)} = \{j/2^n < Y \le (j+1)/2^n\}.$$

Then for any n, $\{A_j^{(n)}\}_{j=0}^\infty$ are disjoint and $\bigcup_{j=0}^\infty A_j^{(n)} = \mathcal{S}$, recalling that $F_Y(0) = 0$.

Given $z > 0$, define $B_j^{(n)}(z) \in \mathcal{E}$ by:

$$B_j^{(n)}(z) = \{X \le z(j+1)/2^n\} \bigcap A_j^{(n)},$$

and let $B^{(n)}(z) = \bigcup_{j=0}^\infty B_j^{(n)}(z)$.

We claim that $\{B^{(n)}(z)\}_{n=1}^\infty$ is a nested collection, $B^{(n+1)}(z) \subset B^{(n)}(z)$, and:

$$\bigcap_{n=1}^\infty B^{(n)}(z) = B(z) \equiv \{X/Y \le z\}. \tag{2}$$

For nesting, this follows from:

$$B_{2j}^{(n+1)}(z) \bigcup B_{2j+1}^{(n+1)}(z) \subset B_j^{(n)}(z),$$

which in turn follows from the definitions. If $s \in B_{2j}^{(n+1)}(z)$ then:

$$X(s) \le z(2j+1)/2^{n+1}, \quad 2j/2^{n+1} < Y(s) \le (2j+1)/2^{n+1},$$

while $s \in B_{2j+1}^{(n+1)}(z)$ implies:

$$X(s) \le z(2j+2)/2^{n+1}, \quad (2j+1)/2^{n+1} < Y(s) \le (2j+2)/2^{n+1},$$

and these sets are contained in $B_j^{(n)}(z)$.

For (2), let $s \in B(z)$. Then for any n there exists j so that $j/2^n < Y(s) \le (j+1)/2^n$ and hence $s \in A_j^{(n)}$. Since $X(s) = zY(s)$ obtains $X(s) \le z(j+1)/2^n$, it follows that $s \in B_j^{(n)}(z) \subset B^{(n)}(z)$. Thus $B(z) \subset \bigcap_{n=1}^\infty B^{(n)}(z)$. Conversely, if $s \in \bigcap_{n=1}^\infty B^{(n)}(z)$ then *for any n there exists j_n so that $s \in B_{j_n}^{(n)}(z)$ and thus $s \in \bigcap_{n=1}^{\infty} B_{j_n}^{(n)}(z)$. That is, for all n there exists j_n so that:*

$$\{X(s) \le z(j_n + 1)/2^n\} \wedge \{j_n/2^n < Y(s) \le (j_n + 1)/2^n\}.$$

This obtains that for all n :

$$X(s)/Y(s) < [z(j_n + 1)/2^n] / \backslash [j_n/2^n] = z(j_n + 1)/j_n.$$

As $j_n \to \infty$ by construction, $s \in B(z)$ and the proof of (2) is complete.
By continuity from above of λ by Proposition I.2.45:

$$\lambda[B(z)] = \lim_{n\to\infty} \lambda\left[B^{(n)}(z)\right].$$

Rewriting, this obtains by countable additivity, and then independence:

$$
\begin{aligned}
F_Z(z) &= \lim_{n\to\infty}\left[\sum_{j=0}^{\infty} \lambda\left[\{X \le z(j+1)/2^n\}\bigcap A_j^{(n)}\right]\right] \\
&= \lim_{n\to\infty}\left[\sum_{j=0}^{\infty} F_X(z(j+1)/2^n)\left[F_Y((j+1)/2^n) - F_Y(j/2^n)\right]\right].
\end{aligned}
$$

Comparing to (1), this is the limit of Riemann-Stieltjes sums with tags $y_{j,n} = (j+1)/2^n$, and the proof is complete for $z > 0$. ∎

Exercise 2.24 *($z < 0$) Complete the above proof for the case $z < 0$. Hint: Define $B_j^{(n)}(z) = \{X \le zj/2^n\}\bigcap A_j^{(n)}$.*

As was noted above Proposition 2.21, the assumption that $F_Y(0) = 0$ is more than is needed. Since $\lambda\left[Y^{-1}(0)\right] = 0$ by (1.3) for continuous distribution functions, we can generalize the above result with no additional assumption on $F_Y(y)$.

Proposition 2.25 *($Z = X/Y$, $F_Z(z)$ for continuous F_X, F_Y) Let X and Y be independent random variables on a probability space $(\mathcal{S}, \mathcal{E}, \lambda)$ with continuous distribution functions F_X and F_Y. Then defined as Riemann-Stieltjes integrals, the distribution function of $Z = X/Y$ is given by:*

$$F_Z(z) = \int_0^{\infty} F_X(zy)dF_Y(y) + \int_{-\infty}^0 [1 - F_X(zy)]dF_Y(y). \tag{2.19}$$

This can equivalently be expressed:

$$F_Z(z) = F_Y(0) + \int_{-\infty}^{\infty} sgn(y)F_X(zy)dF_Y(y), \tag{2.20}$$

where $sgn(y)$ equals 1 for $y > 0$ and -1 for $y < 0$.
Proof. *If $z = 0$ then by definition :*

$$F_Z(0) = F_X(0)(1 - F_Y(0)) + (1 - F_X(0))F_Y(0),$$

and this is obtained from (2.19). The proof of this result for $z > 0$ and $z < 0$ are similar to those above, so we omit some of the details.
By continuity and monotonicity of the distribution functions, this integral is well defined over every bounded interval $[a, b]$ by Proposition III.4.17. By boundedness of these functions, this integral has a well defined limit as $a \to -\infty$ and $b \to \infty$ by Proposition III.4.21. It then

follows by Proposition III.4.12 that integrals can be evaluated by any sequence of Riemann-Stieltjes summations with partition mesh size $\mu \to 0$.

To this end, define $P_n \equiv \{j/2^n\}_{j \in \mathbb{Z}}$, a partition of $(-\infty, \infty)$ of mesh size $\mu = 2^{-n}$. Then the integrals in (2.19) equal the limit as $n \to \infty$ of the Riemann-Stieltjes sums:

$$\sum_{j=0}^{\infty} F_X(z y_{j,n}) \left[F_Y((j+1)/2^n) - F_Y(j/2^n)\right] \tag{1}$$
$$+ \sum_{j=-1}^{-\infty} \left[1 - F_X(z y_{j,n})\right] \left[F_Y((j+1)/2^n) - F_Y(j/2^n)\right],$$

for any interval tags $y_{j,n} \in [j/2^n, (j+1)/2^n]$.

As in the prior proof, define $A_j^{(n)} \in \mathcal{E}$ by:

$$A_j^{(n)} = \{j/2^n < Y \le (j+1)/2^n\}.$$

Then for any n, $\{A_j^{(n)}\}_{j=-\infty}^{\infty}$ are disjoint and $\bigcup_{j=-\infty}^{\infty} A_j^{(n)} = \mathcal{S}$.

For $z > 0$, define $B_j^{(n)}(z) \in \mathcal{E}$ by:

$$B_j^{(n)}(z) = \begin{cases} \{X \le z(j+1)/2^n\} \cap A_j^{(n)}, & j \ge 0, \\ \{X \ge zj/2^n\} \cap A_j^{(n)}, & j < 0, \end{cases}$$

and let $B_+^{(n)}(z) = \bigcup_{j=0}^{\infty} B_j^{(n)}(z)$, $B_-^{(n)}(z) = \bigcup_{j=-1}^{-\infty} B_j^{(n)}(z)$. As above, $\{B_\pm^{(n)}(z)\}_{n=1}^{\infty}$ are nested collections, $B_\pm^{(n+1)}(z) \subset B_\pm^{(n)}(z)$, and:

$$\bigcap_{n=1}^{\infty} B_+^{(n)}(z) = B_+(z) \equiv \{X/Y \le z\} \bigcap \{Y > 0\}, \tag{2}$$
$$\bigcap_{n=1}^{\infty} B_-^{(n)}(z) = B_-(z) \equiv \{X/Y \le z\} \bigcap \{Y < 0\}.$$

Further:

$$B(z) \equiv B_+(z) \bigcup B_-(z) = \{X/Y \le z\}. \tag{3}$$

Similarly for $z < 0$, define $B_j^{(n)}(z) \in \mathcal{E}$ by:

$$B_j^{(n)}(z) = \begin{cases} \{X \le zj/2^n\} \cap A_j^{(n)}, & j \ge 0, \\ \{X \ge z(j+1)/2^n\} \cap A_j^{(n)}, & j < 0, \end{cases}$$

and $B_\pm^{(n)}(z)$, $B_\pm(z)$ and $B(z)$ as above. Again the $B_\pm^{(n)}(z)$-collections are nested and (2) and (3) are verified.

For either case of $z < 0$ or $z > 0$, we have by finite additivity and continuity from above:

$$\begin{aligned} F(z) &= \lambda[B(z)] \\ &= \lambda[B_+(z)] + \lambda[B_-(z)] \\ &= \lim_{n \to \infty} \lambda\left[B_+^{(n)}(z)\right] + \lim_{n \to \infty} \lambda\left[B_-^{(n)}(z)\right]. \end{aligned}$$

As in Proposition 2.23 and Exercise 2.24, $\lambda\left[B_+^{(n)}(z)\right]$ equals the first Riemann-Stieltjes sum in (1) for $y_{j,n}$ that reflects the X-boundaries in the definitions of $B_j^{(n)}(z)$. Thus this limit agrees with the first Riemann-Stieltjes integral in (2.19).

For the second limit, assume that $z > 0$. Then by countable additivity and then independence, recalling that $\lambda\left[X^{-1}(zj/2^n)\right] = 0$ for all such points by continuity of F_X:

$$\begin{aligned} \lambda\left[B_-^{(n)}(z)\right] &= \sum_{j=-1}^{-\infty} \lambda\left[\{X \ge zj/2^n\} \bigcap A_j^{(n)}\right] \\ &= \sum_{j=-1}^{-\infty} \left[1 - F_X(zj/2^n)\right] \left[F_Y((j+1)/2^n) - F_Y(j/2^n)\right]. \end{aligned}$$

The same derivation holds for $z < 0$. Thus this limit agrees with the second Riemann-Stieltjes integral in (2.19).

The equivalence of (2.19) and (2.20) follows from $\int_{-\infty}^{0} dF_Y(y) = F_Y(0)$, which in turn follows from Riemann-Stieltjes sums. ∎

When the distribution function of Y in Proposition 2.25 has a continuous derivative, as is the common assumption in continuous probability theory, we can convert the above Riemann-Stieltjes integral to a Riemann integral.

Corollary 2.26 ($Z = X/Y$, $F_Z(z)$ **as an R integral, continuous** F_X, f_Y) *Let X and Y be independent random variables on a probability space $(\mathcal{S}, \mathcal{E}, \lambda)$ with continuous distribution function F_X, and continuously differentiable F_Y with continuous density function $f_Y \equiv F_Y'$. Then the distribution function of $Z = X/Y$ is given by Riemann integral:*

$$F_Z(z) = (\mathcal{R}) \int_0^\infty F_X(zy) f_Y(y) dy + (\mathcal{R}) \int_{-\infty}^0 [1 - F_X(zy)] f_Y(y) dy. \qquad (2.21)$$

Proof. *This follows directly from Proposition III.4.28.* ∎

We next address the density function of Z in the case of continuously differentiable distribution functions F_X and F_Y, for which we require another use of Tonelli's theorem of Book V. While in applications it is usually relatively simple to prove that $f_Z(z)$ in (2.22) is a continuous function, a general proof remains elusive.

Corollary 2.27 ($Z = X/Y$, $f_Z(z)$ **as an R/L integral, continuous** f_X, f_Y) *Let X and Y be independent random variables on a probability space $(\mathcal{S}, \mathcal{E}, \lambda)$ with continuously differentiable distribution functions F_X and F_Y, and associated continuous density functions $f_X \equiv F_X'$ and $f_Y \equiv F_Y'$. Then the density function $f_Z(z)$ of $Z = X/Y$ in the sense of Lebesgue:*

$$F_Z(z) = (\mathcal{L}) \int_{-\infty}^z f_Z(w) dw, \quad F_Z'(z) = f_Z(z), \ a.e.,$$

is given by Riemann integral:

$$f_Z(z) = (\mathcal{R}) \int_{-\infty}^\infty |y| f_X(zy) f_Y(y) dy. \qquad (2.22)$$

When $f_Z(z)$ so defined is continuous, then $F_Z(z)$ is definable as a Riemann integral and $F_Z'(z) = f_Z(z)$.

Proof. *As was the case for Proposition 2.17, the subtlety here relates to the existence of this integral, For this proof, we use Tonelli's theorem of Book V to justify reversing iterated integrals.*

From (2.21) and a substitution into this Riemann integral, recalling Remark III.1.35:

$$
\begin{aligned}
F_Z(z) &= \int_0^\infty \int_{-\infty}^{zy} f_X(w) dw f_Y(y) dy + \int_{-\infty}^0 \int_{zy}^\infty f_X(w) dw f_Y(y) dy \\
&= \int_0^\infty \int_{-\infty}^z y f_X(vy) f_Y(y) dv dy + \int_{-\infty}^0 \int_z^{-\infty} y f_X(vy) f_Y(y) dv dy \\
&= \int_0^\infty \int_{-\infty}^z y f_X(vy) f_Y(y) dv dy - \int_{-\infty}^0 \int_{-\infty}^z y f_X(vy) f_Y(y) dv dy \\
&= \int_{-\infty}^\infty \int_{-\infty}^z |y| f_X(zy) f_Y(y) dv dy.
\end{aligned}
$$

As this iterated integral equals $F_Z(z)$, it is well-defined and finite as a Riemann integral, though Corollary III.1.77 does not apply due to the unbounded domain of integration. However, by continuity of the integrand, this integral is well-defined and finite when interpreted as a Lebesgue integral by Proposition III.2.56. Tonelli's theorem then obtains that this iterated Lebesgue integral can be reversed:

$$F_Z(z) = (\mathcal{L}) \int_{-\infty}^{z} \left[\int_{-\infty}^{\infty} |y| \, f_X(vy) f_Y(y) dy \right] dv. \tag{1}$$

Tonelli's theorem also proves that:

$$g(v) \equiv (\mathcal{L}) \int_{-\infty}^{\infty} |y| \, f_X(vy) f_Y(y) dy,$$

is Lebesgue integrable. By Proposition III.3.39, (1) obtains $F_Z'(z) = g(z)$ for almost all z. Thus from the Lebesgue theory we obtain that:

$$F_Z'(z) = g(z), \quad a.e., \tag{2}$$

and $g(z)$ so defined is indeed a density function associated with $F_Z(z)$ in the Lebesgue sense:

$$F_Z(z) = (\mathcal{L}) \int_{-\infty}^{z} g(w) dw. \tag{3}$$

Finally, the integrand in (2.22) is continuous for any z by assumption. It is thus Riemann integrable over any bounded interval $[a, b]$ by Proposition III.1.15, and further by Proposition III.2.18:

$$(\mathcal{R}) \int_{a}^{b} y f_X(zy) f_Y(y) dy = (\mathcal{L}) \int_{a}^{b} y f_X(zy) f_Y(y) dy.$$

As this is true for every interval $[a, b]$, and the integral on the right is well defined as $a \to -\infty$ and $b \to \infty$, it follows that $f_Z(z)$ in (2.22) satisfies $f_Z(z) = g(z)$ for all z and so from (1) and (2):

$$F_Z(z) = (\mathcal{L}) \int_{-\infty}^{z} f_Z(w) dw, \quad F_Z'(z) = f_Z(z), \ a.e.$$

If $f_Z(z)$ is continuous, then $F_Z(z)$ is definable as a Riemann integral by the same argument as above, and $F_Z'(z) = f_Z(z)$ by Proposition III.1.33. ∎

We now turn to some examples, all of which have continuous $f_Z(z)$.

Example 2.28 **1. Ratio of gammas and the beta distribution:** *Let X and Y be independent gamma random variables with common λ parameter and respective parameters of α_1 and α_2. If $Z = X/Y$, we have from substituting (1.55) into (2.22) that for $z > 0$:*

$$
\begin{aligned}
f_Z(z) &= \frac{\lambda^{\alpha_1 + \alpha_2} z^{\alpha_1 - 1}}{\Gamma(\alpha_1) \Gamma(\alpha_2)} \int_0^{\infty} y^{\alpha_1 + \alpha_2 - 1} e^{-\lambda(1+z)y} dy \\
&= \frac{\Gamma(\alpha_1 + \alpha_2)}{\Gamma(\alpha_1) \Gamma(\alpha_2)} \frac{z^{\alpha_1 - 1}}{(1 + z)^{\alpha_1 + \alpha_2}}.
\end{aligned}
$$

As a continuous function, this is the density of Z in the Riemann sense.

Comparing this result to (1.61) and recalling the identity in (1.63), the density function of Z is very closely related to the beta distribution with parameters $v = \alpha_1$ and $w = \alpha_2$.

Specifically, let $F_\beta(w)$ denote the beta distribution function with parameters $v = \alpha_1$ and $w = \alpha_2$. Integrating $f_Z(z)$ and substituting $y = x/(1+x)$ obtains:

$$
\begin{aligned}
F_Z(z) &= \frac{\Gamma(\alpha_1 + \alpha_2)}{\Gamma(\alpha_1)\Gamma(\alpha_2)} \int_0^z \frac{x^{\alpha_1 - 1}}{(1+x)^{\alpha_1 + \alpha_2}} dx \\
&= \frac{\Gamma(\alpha_1 + \alpha_2)}{\Gamma(\alpha_1)\Gamma(\alpha_2)} \int_0^{z/(1+z)} y^{\alpha_1 - 1}(1-y)^{\alpha_2 - 1} dy \\
&= F_\beta\left(\frac{z}{1+z}\right).
\end{aligned}
$$

2. **Ratio of chi-squared and the F-distribution:** *In the special case of item 1 where $\lambda = 1/2$, $\alpha_1 = n/2$, and $\alpha_2 = m/2$, these X and Y gamma variates are called **chi-squared variables** with respective **degrees of freedom of n and m** as noted in Remark 1.30. In this special case, the random variable:*

$$
F \equiv \frac{m}{n} Z = \frac{X/n}{Y/m},
$$

*is said to have an **F-distribution with n and m degrees of freedom**. It is sometimes called **Snedecor's F-distribution** or a **Fisher-Snedecor distribution**, named for **R. A. Fisher (1890–1962)** and **George W. Snedecor (1881–1974)**.*

The distribution function for the F-variate satisfies:

$$
f_F(x) = f_Z(\frac{n}{m}x),
$$

and so from the above calculation, the density function of an F-variate with n and m degrees of freedom is:

$$
f_{F_{n,m}}(x) = \frac{\Gamma((n+m)/2)}{\Gamma(n/2)\Gamma(m/2)} \frac{(nx/m)^{n/2-1}}{(1+nx/m)^{(n+m)/2}}. \tag{2.23}
$$

As will be seen in Section 4.2.7, n is the mean of chi-squared X with n degrees of freedom, and similarly m is the mean of Y. So the scalings in the definition of $F \equiv \frac{X/n}{Y/m}$ produce a ratio of chi-squared variates which have been normalized to each have a mean of 1.

As an application of the F-distribution, the sample variance of a normal variate with known mean has a chi-squared distribution as seen in Example 2.29. Thus the F-distribution can be used to establish confidence intervals for the ratio of sample variances, and then test the hypothesis that the sample variances are equal.

3. **Ratio of normal and "chi" and Student's T distribution:** *Another important example of a variate defined as the ratio of independent variates is developed as follows. By Example 2.29, if X is standard normal, and Y is chi-squared with n degrees of freedom, then the distribution function of:*

$$
T \equiv \frac{X}{\sqrt{Y/n}},
$$

is useful as discussed there. From item 2, Y/n is chi-squared, but normalized to have a mean of 1, and the square root of this variate is here fittingly labeled a "chi" variate.

To find the density function of T we apply (2.22), writing $T = X\sqrt{n}/\sqrt{Y}$, as the ratio of a normal variate with $\mu = 0$ and $\sigma^2 = n$, and the square root of a chi-squared variate. The density of the normal is found in (1.65), while for the latter we apply (2.4). This states that if f_Y denotes the distribution function of the chi-squared with n degrees of freedom, then the density function of $\sqrt{Y}$ is given by:

$$f_{\sqrt{Y}}(y) = 2y f_Y(y^2).$$

With a little algebra, the density function for T is derived:

$$
\begin{aligned}
f_T(t) &= \int_0^\infty y f_X(ty) f_{\sqrt{Y}}(y) dy \\
&= \frac{1}{\sqrt{\pi n} 2^{(n-1)/2} \Gamma(n/2)} \int_0^\infty y^n \exp\left(-\frac{1}{2}\left(1 + \frac{t^2}{n}\right) y^2\right) dy.
\end{aligned}
$$

The substitution $s = \frac{1}{2}\left(1 + \frac{t^2}{n}\right) y^2$ produces:

$$f_T(t) = \frac{\Gamma((n+1)/2)}{\sqrt{\pi n} \Gamma(n/2)} \left(1 + \frac{t^2}{n}\right)^{-(n+1)/2}, \tag{2.24}$$

where $\Gamma(x)$ is the gamma function of (1.56).

The distribution function of T defined in (2.25) is known as the **Student T distribution** or **Student's T distribution,** with n **degrees of freedom**. This distribution is named for **William Sealy Gosset (1876–1937)** who published under the pen name of Student.

While the above derivation and many applications call for n to be an integer, this distribution is defined more generally with $\nu > 0$ degrees of freedom, by:

$$f_T(t) = \frac{\Gamma((\nu+1)/2)}{\sqrt{\pi \nu} \Gamma(\nu/2)} \left(1 + \frac{t^2}{\nu}\right)^{-(\nu+1)/2}. \tag{2.25}$$

Using the identities in (1.63) and (1.59), this density function is also written:

$$f_T(t) = \frac{1}{\sqrt{\nu} B(\nu/2, 1/2)} \left(1 + \frac{t^2}{\nu}\right)^{-(\nu+1)/2},$$

with $B(u, w)$ the beta function of (1.62).

See Example 2.29 for an application of this distribution.

4. **Ratio of standard normals and the Cauchy distribution:** Assume that X and Y are independent standard normal variables with density given in (1.66) by $\phi(x) = \exp\left(-x^2/2\right)/\sqrt{2\pi}$, and define $Z = X/Y$. By (2.22) and then symmetry of the integrand:

$$
\begin{aligned}
f_Z(z) &= \frac{1}{2\pi} \int_{-\infty}^\infty |y| \exp\left(-\left(1 + z^2\right) y^2/2\right) dy \\
&= \frac{1}{\pi} \int_0^\infty y \exp\left(-\left(1 + z^2\right) y^2/2\right) dy.
\end{aligned}
$$

This integral can be evaluated using a substitution of $u = \left(1 + z^2\right) y^2/2$, and this obtains the standard **Cauchy density function** in (1.69):

$$
\begin{aligned}
f_Z(z) &= \frac{1}{\pi\left(1 + z^2\right)} \int_0^\infty \exp\left(-u\right) du \\
&= \frac{1}{\pi\left(1 + z^2\right)}.
\end{aligned}
$$

Thus the ratio of independent standard normal variates is standard Cauchy.

Example 2.29 (Application of Student's T) *An important application of Student's T distribution is for the determination of confidence intervals for the mean μ of a normal distribution based on a given sample, where the variance parameter σ^2 is unknown. In this application, given a sample $\{Z_i\}_{i=1}^n$ from a normal distribution with unknown parameters (μ, σ^2), define the "sample mean" $\bar{Z}$ and the "sample variance" s^2 :*

$$\bar{Z} = \sum\nolimits_{i=1}^{n} Z_i/n, \quad s^2 = \sum\nolimits_{i=1}^{n} \left(Z_i - \bar{Z}\right)^2 /n.$$

Now let:

$$X = \frac{\bar{Z} - \mu}{\sigma/\sqrt{n}}, \qquad Y^2 = \frac{ns^2}{\sigma^2}.$$

In a given application, $\bar{Z}$ and s^2 are simply numbers. But as discussed in Chapter II.4 and continued in Chapter 5, we can interpret $\bar{Z}$ and s^2 as random variables by interpreting $\{Z_i\}_{i=1}^n$ as independent, identically distributed random variables on some probability space. By Section 4.2.4, the mean of $\bar{Z}$ is μ, and its variance is σ^2/n, and thus X has mean 0 and variance 1. Indeed X is standard normal by item 4 of Section 4.4.1.

We now motivate the fact that Y^2 is chi-squared with $n-1$ degrees of freedom. To this end:

$$
\begin{aligned}
ns^2 &= \sum\nolimits_{i=1}^{n} \left[(Z_i - \mu) - (\bar{Z} - \mu)\right]^2 \\
&= \sum\nolimits_{i=1}^{n} (Z_i - \mu)^2 - n\left(\bar{Z} - \mu\right)^2,
\end{aligned}
$$

and this obtains:

$$Y^2 + \left(\frac{\bar{Z} - \mu}{\sigma/\sqrt{n}}\right)^2 = \sum\nolimits_{i=1}^{n} \left(\frac{Z_i - \mu}{\sigma}\right)^2.$$

As noted in Remark 2.5 on the chi-squared distribution, the summation on the right is chi-squared with n degrees of freedom, and similarly, $\left(\frac{\bar{Z}-\mu}{\sigma/\sqrt{n}}\right)^2$ is chi-squared with 1 degree of freedom since $X = \frac{\bar{Z}-\mu}{\sigma/\sqrt{n}}$ is standard normal. Independence of s^2 and $\bar{Z}$ for normal distributions will be proved in Book VI in the study of the multivariate normal distribution, and this obtains independence of $Y^2 = \frac{ns^2}{\sigma^2}$ and $\left(\frac{\bar{Z}-\mu}{\sigma/\sqrt{n}}\right)^2$ by Proposition II.3.56. It will then follow from item 6 of Section 4.4.1 that Y^2 is chi-squared with $n-1$ degrees of freedom.

Hence, $T = \frac{X\sqrt{n-1}}{\sqrt{Y}}$ is Student T with $n-1$ degrees of freedom. Rewriting to reflect sample variates:

$$T = \frac{(\bar{Z} - \mu)}{s\sqrt{n-1}}.$$

Consequently, confidence intervals for such T can then be translated to confidence intervals for μ. See Example 6.56 for an application.

2.3.2 Example without Independence

For independent, nonnegative X and Y, it is intuitively apparent that the random variable $Z = \frac{X}{X+Y}$ is not a ratio of independent variates, and the reader is invited to formalize this with the definitions and an example. Though not a ratio of independent variates, the development for $F_Z(z)$ here requires only a modest adaptation of the prior section's results.

With $W \equiv X/Y$, it follows from nonnegativity of variates that $Z \leq z$ if and only if $W \leq \frac{z}{1-z}$. Thus:

$$F_Z(z) = F_W\left(\frac{z}{1-z}\right), \tag{2.26}$$

and if F_W has a density function, so too does F_Z by differentiation:

$$f_Z(z) = \frac{1}{(1-z)^2} f_W\left(\frac{z}{1-z}\right). \tag{2.27}$$

The first result will be applied in Chapter 5 on simulating samples of random variables.

Proposition 2.30 ($Z = \frac{X}{X+Y}$ **is beta for** F_X, F_Y **gamma**) *Let X and Y be independent gamma random variables defined on a probability space $(\mathcal{S}, \mathcal{E}, \lambda)$, with parameters α_1, λ and α_2, λ respectively. Define the random variable*

$$Z = \frac{X}{X+Y}.$$

Then Z is defined on the interval $(0,1)$, and has density function $f_Z(z)$ that is independent of the parameter λ:

$$f_Z(z) = \frac{\Gamma(\alpha_1 + \alpha_2)}{\Gamma(\alpha_1)\Gamma(\alpha_2)} z^{\alpha_1 - 1} (1-z)^{\alpha_2 - 1}. \tag{2.28}$$

By (1.61) and (1.63), Z is a beta random variable with parameters $v = \alpha_1$ and $w = \alpha_2$.
Proof. *Since X and Y have range of $[0, \infty)$, Z is potentially undefined when $X = Y = 0$. But this event has probability 0 by independence and (1.54). In detail, as $\epsilon_j \to 0$:*

$$\Pr[X \leq \epsilon_1, Y \leq \epsilon_2] = [1 - \exp(-\lambda\epsilon_1)][1 - \exp(-\lambda\epsilon_2)] \to 0.$$

Thus we can redefine X and Y to have range $(0, \infty)$ without changing their distributions or the calculations below.
From (1.55),

$$\begin{aligned} f_X(x) &= \lambda^{\alpha_1} x^{\alpha_1 - 1} e^{-\lambda x} / \Gamma(\alpha_1), \\ f_Y(y) &= \lambda^{\alpha_2} y^{\alpha_2 - 1} e^{-\lambda y} / \Gamma(\alpha_2), \end{aligned}$$

which are continuous on $[0, \infty)$. By (2.27) and (2.22):

$$f_Z(z) = \frac{\lambda^{\alpha_1 + \alpha_2}}{\Gamma(\alpha_1)\Gamma(\alpha_2)} \frac{1}{(1-z)^2} \left(\frac{z}{1-z}\right)^{\alpha_1 - 1} \int_0^\infty y^{\alpha_1 + \alpha_2 - 1} e^{-\lambda y/(1-z)} dy.$$

The substitution $y = w(1-z)/\lambda$ then obtains:

$$\begin{aligned} f_Z(z) &= \frac{\lambda^{\alpha_1 + \alpha_2}}{\Gamma(\alpha_1)\Gamma(\alpha_2)} \frac{1}{(1-z)^2} \left(\frac{z}{1-z}\right)^{\alpha_1 - 1} \left(\frac{1-z}{\lambda}\right)^{\alpha_1 + \alpha_2} \int_0^\infty w^{\alpha_1 + \alpha_2 - 1} e^{-w} dw \\ &= \frac{1}{\Gamma(\alpha_1)\Gamma(\alpha_2)} z^{\alpha_1 - 1} (1-z)^{\alpha_2 - 1} \int_0^\infty w^{\alpha_1 + \alpha_2 - 1} e^{-w} dw. \end{aligned}$$

Comparing with (1.56), this integral equals $\Gamma(\alpha_1 + \alpha_2)$ and the derivation of (2.28) is complete. ■

If $\alpha_1 = \alpha_2 = 1$, so X and Y have the same standard exponential distribution, then $f_Z(z)$ reduces to the uniform distribution on $(0, 1)$. For the general case of exponentials, we have the following.

Proposition 2.31 ($Z = \frac{X}{X+Y}$, F_X, F_Y **exponential**) *Let X and Y be independent exponential random variables defined on a probability space $(\mathcal{S}, \mathcal{E}, \lambda)$, with parameters λ_1 and λ_2, respectively. Then the random variable $Z = \frac{X}{X+Y}$ is defined on the interval $(0, 1)$ with density function:*

$$f_Z(z) = \frac{\lambda_1 \lambda_2}{[\lambda_1 z + \lambda_2(1 - z)]^2}. \tag{2.29}$$

In particular, if $\lambda_1 = \lambda_2$, then $f_Z(z) = 1$, the uniform distribution.
Proof. *As in the above proof, we begin with:*

$$f_Z(z) = \frac{1}{(1 - z)^2} \int_0^\infty y f_X\left(\frac{zy}{1 - z}\right) f_Y(y) dy.$$

From (1.53), $f_X(x) = \lambda_1 e^{-\lambda_1 x}$ and $f_Y(y) = \lambda_2 e^{-\lambda_2 y}$, which are continuous on $[0, \infty)$. This obtains:

$$f_Z(z) = \frac{\lambda_1 \lambda_2}{(1 - z)^2} \int_0^\infty y e^{-\alpha y} dy,$$

where $\alpha = \lambda_1 \left(\frac{z}{1-z}\right) + \lambda_2$.
 Using integration by parts,

$$\int_0^\infty y e^{-\alpha y} dy = \frac{1}{\alpha^2},$$

and combining obtains (2.29). ∎

For an example using the distributional approach of (2.18), we have the following.

Example 2.32 ($Z = \frac{X}{X+Y}$, F_X, F_Y **uniform**) *Let X and Y be independent random variables defined on a probability space $(\mathcal{S}, \mathcal{E}, \lambda)$ with continuous uniform distributions. Then by (2.26) and (2.18), the distribution function for $Z = \frac{X}{X+Y}$ is given by:*

$$F_Z(z) = \int_0^\infty F_X\left(\frac{zy}{1 - z}\right) dF_Y(y).$$

Now $dF_Y = 0$ for $y > 1$ since $F_Y(y) = 1$, while on $[0, 1]$ both distributions functions are the identity functions, $F_X(x) = x$, $F_Y(y) = y$. Hence $F_X\left(\frac{zy}{1-z}\right) = \left(\frac{zy}{1-z}\right)$ if $0 \leq \frac{zy}{1-z} \leq 1$, or $0 \leq y \leq \frac{1-z}{z}$. Thus:

1. *If $0 \leq z \leq \frac{1}{2}$:*

$$\begin{aligned} F_Z(z) &= \left(\frac{z}{1 - z}\right) \int_0^1 y dy \\ &= \frac{1}{2}\left(\frac{z}{1 - z}\right). \end{aligned}$$

2. *If $\frac{1}{2} \leq z \leq 1$:*

$$\begin{aligned} F_Z(z) &= \left(\frac{z}{1 - z}\right) \int_0^{(1-z)/z} y dy + \int_{(1-z)/z}^1 dy \\ &= 1 - \frac{1}{2}\left(\frac{1 - z}{z}\right) \end{aligned}$$

3

Order Statistics

Given a collection of independent, identically distributed variables $\{X_j\}_{j=1}^M$ defined on $(\mathcal{S}, \mathcal{E}, \lambda)$, these can be pointwise though not uniquely reordered into $\{X_{(k)}\}_{k=1}^M$ with $X_{(k)} \leq X_{(k+1)}$ for all k. Each $X_{(k)}$ is then called a kth **order statistic** when M is apparent from the context, or the kth **order statistic from a sample of** M, otherwise. To further emphasize M, some authors use notation such as:

$$X_{(k,M)} \equiv X_{(k)}.$$

See below for a discussion of samples of random variables.

Notation 3.1 (Order of order statistics) *Perhaps ironically, there is no universal notational convention for the **order** of **order statistics**. In some references, order statistics are ordered in the natural numerical order, so $X_{(k)} \leq X_{(k+1)}$ as above. However, it is not uncommon to see order statistics denoted so that $X_{(1)}$ is the largest, and hence $X_{(k+1)} \leq X_{(k)}$.*

In this section, we derive the distribution and density functions of kth **order statistics**, including various multivariate results, and introduce the **Rényi representation theorem** for exponential order statistics which will be applied in Section 7.2 on extreme value theory.

3.1 M-Samples and Order Statistics

Chapter II.4 introduced the notion of an N-**sample** of a given random variable, which we here denote an M-sample. Recalling Definitions II.4.1 and II.4.3:

Definition 3.2 (M-Sample) *Let a probability space $(\mathcal{S}, \mathcal{E}, \lambda)$ and random variable $X : \mathcal{S} \longrightarrow \mathbb{R}$ be given. With M finite or infinite, a collection of random variables $\{X_j\}_{j=1}^M$ defined on a probability space $(\mathcal{S}', \mathcal{E}', \lambda')$ is said to be an M-**sample of** X, or a **sample of** X when M is implied, if this collection is **independent, and identically distributed with** X (i.i.d.-X):*

*1. $\{X_j\}_{j=1}^M$ are **independent** if given $(i_1, ..., i_m) \subset (1, 2, ..., M)$ and $\{A_j\}_{j=1}^m \subset \mathcal{B}(\mathbb{R})$:*

$$\lambda' \left[\bigcap_{j=1}^m X_{i_j}^{-1}(A_j) \right] = \prod_{j=1}^m \lambda'[X_{i_j}^{-1}(A_j)]. \tag{3.1}$$

*2. $\{X_j\}_{j=1}^M$ are **identically distributed with** X if for all j, and all $A \in \mathcal{B}(\mathbb{R})$:*

$$\lambda'[X_j^{-1}(A)] = \lambda[X^{-1}(A)]. \tag{3.2}$$

DOI: 10.1201/9781003264583-3

Equivalently, denoting the distribution function of X by $F(x)$, and those of $\{X_j\}_{j=1}^M$ by $\{F_j(x)\}_{j=1}^M$:

1. $\{X_j\}_{j=1}^M$ *are* **independent** *if given* $(i_1, ..., i_m) \subset (1, 2, ..., M)$, *then for all* $x = (x_{i_1}, ..., x_{i_m}) \in \mathbb{R}^m$:

$$F(x_{i_1}, ..., x_{i_m}) = \prod_{j=1}^m F_j(x_{i_j}), \tag{3.3}$$

   where $F(x_{i_1}, ..., x_{i_m})$ *is the joint distribution function of* $\{X_{i_j}\}_{j=1}^m$.

2. X_j *are* **identically distributed with** X *if for any j and all x:*

$$F_j(x) = F(x). \tag{3.4}$$

The constructions in Chapter II.4 defined $(\mathcal{S}', \mathcal{E}', \lambda')$ as M-dimensional product spaces. In two cases, the product probability space was defined by $\mathcal{S}' \equiv \mathcal{S}^M$ using $(\mathcal{S}, \mathcal{E}, \lambda)$ and the product measure space constructions of Chapter I.7 ($M < \infty$) or I.9 ($M = \infty$), while in the third construction, $\mathcal{S}' \equiv (0, 1)^M$ where $(0, 1)$ is endowed with the Lebesgue measure m and Borel sigma algebra $\mathcal{B}(0, 1)$.

The M-samples on these spaces were then defined in terms of the components of $s = (s_1, ..., s_M)$ of this space $\mathcal{S}'$:

1. $\mathcal{S}' \equiv \mathcal{S}^M$: $\{X_j\}_{j=1}^M$ is defined by the projection mapping:

$$X_j(s) = s_i.$$

Independence and distributional properties of $\{X_j\}_{j=1}^M$ were achieved by the construction of the product measure on $\mathcal{S}^M$.

2. $\mathcal{S}' \equiv (0, 1)^M$: $\{X_j\}_{j=1}^M$ is defined in terms of the left-continuous inverse $F^*(y)$ of $F(x)$ (Definition 1.29):

$$X_j(s) = F^*(U_j(s)),$$

where $U_j(s)$ is the projection mapping of item 1. Independence and uniform distributional properties of $\{U_j\}_{j=1}^M$ were achieved by the construction of the product measure on $(0, 1)^M$, while these variates were transformed to the required properties of $\{X_j\}_{j=1}^M$ by $F^*(y)$ (Proposition II.4.9; see Chapter 5 introduction).

The terminology **random sample** from a random variable X implies the result of some experimental or other empirical process by which **numerical** values $\{X_j(s_j)\}_{j=1}^M$ are observed, generated, or otherwise obtained. For this collection to be called a "random sample" requires that these variates be deemed independent, and each governed by the distribution function underlying X. In practice, such determinations are made with a mix of science and judgment.

Thus $\{X_j\}_{j=1}^M$ as an M-sample, and $\{X_j(s)\}_{j=1}^M$ as a random sample, are intimated related. In theory, the collection $\{X_j(s)\}_{j=1}^M$ as defined above has the **potential** to be a random sample for any $s \in \mathcal{S}'$, but not all random samples are equally useful.

For example, if $\{X_j\}_{j=1}^M$ represents the outcomes of M coin flips, then certainly there is an $s \in \mathcal{S}'$ for which $X_j(s) = H$ for all j. In isolation, this would seem to be an odd example of a "random sample," especially for M large, but it could be an example of such a random sample from the experimental or other empirical process by which it was obtained.

In order to be deemed random samples, it must be the case that the experimental or other empirical process employed to produce these, if repeated N times, would result in a collection of random samples $\{\{X_j(s_k)\}_{j=1}^M\}_{k=1}^N$ which at least approximately satisfied the requirements of an M-sample. These requirements are defined in terms of independence and distribution as noted in Definition 3.2. If such a collection is generated for the M-flip model, then one would anticipate that the all-H random sample would appear for N large. For N small, the above distributional requirement should assure that such a sample had at best a remote chance of an appearance for M large.

Returning to **order statistics**, if $\{X_j(s)\}_{j=1}^M$ is a given random sample, these can be reordered, though not necessarily uniquely, into $\{X_{(k)}(s)\}_{k=1}^M$ with $X_{(k)}(s) \leq X_{(k+1)}(s)$ for all k. But we also want to be able to make distributional statements about $\{X_{(k)}\}_{k=1}^M$ as **random variables** defined on the same space as the i.i.d. random variables $\{X_j\}_{j=1}^M$. These ordered random variables are defined as follows.

Definition 3.3 (Order statistics) *Let $\{X_j\}_{j=1}^M$ be independent, identically distributed random variables on a probability space $(\mathcal{S}', \mathcal{E}', \lambda')$. Define the collection of kth **order statistics** $\{X_{(k)}\}_{k=1}^M$ by:*

- $X_{(1)} \equiv \min\{X_j\}_{j=1}^M$;

- $X_{(k)} \equiv \min\{X_j | X_j \geq X_{(k-1)}\}$, *for* $2 \leq k \leq M-1$;

- $X_{(M)} \equiv \min\{X_j | X_j \geq X_{(M-1)}\} = \max\{X_j\}_{j=1}^M$.

Proposition 3.4 (Order statistics are random variables) *The collection of kth order statistics $\{X_{(k)}\}_{k=1}^M$ of Definition 3.3 are random variables on $(\mathcal{S}', \mathcal{E}', \lambda')$.*
Proof. *That $X_{(1)}$ and $X_{(M)}$ are random variables on $(\mathcal{S}', \mathcal{E}', \lambda')$ follows from Proposition I.3.47. For other $X_{(k)}$, define $Y_k = \min_j \left[\max(X_j - X_{(k-1)}, 0)\right]$, which a random variable by this proposition. Then $X_{(k)} = X_{(k-1)} + Y_k$ is a random variable by Proposition I.3.30.* ∎

Notation 3.5 $((\mathcal{S}', \mathcal{E}', \lambda') \to (\mathcal{S}, \mathcal{E}, \lambda))$ *We now drop the added notation of $(\mathcal{S}', \mathcal{E}', \lambda')$, which was largely used to distinguish this space from the space $(\mathcal{S}, \mathcal{E}, \lambda)$ on which the original X was defined. But for the rest of this chapter we will only be investigating an i.i.d collection $\{X_j\}_{j=1}^M$ and its associated kth order statistics $\{X_{(k)}\}_{k=1}^M$, and for this we resort back to the simpler notation $(\mathcal{S}, \mathcal{E}, \lambda)$.*

3.2 Distribution Functions of Order Statistics

Let $\{X_j\}_{j=1}^M$ be a collection of independent, identically distributed random variables defined on $(\mathcal{S}, \mathcal{E}, \lambda)$ with distribution function $F(x)$. Then the distribution function of the kth order statistic $X_{(k)}$ is relatively straightforward to derive. Indeed if $X_{(k)} \leq x$, then **at least** k of the variates satisfy this constraint and at most $n - k$ variates exceed x. Denoting this distribution function by $F_{(k)}(x)$, we have the following:

Proposition 3.6 ($F_{(k)}(x)$) *Given independent random variables $\{X_j\}_{j=1}^M$ with common distribution function $F(x)$, the distribution function of the kth order statistic $F_{(k)}(x)$ is defined on the same domain as is $F(x)$ and given by:*

$$F_{(k)}(x) = \sum_{j=k}^M \binom{M}{j} F^j(x) \left(1 - F(x)\right)^{M-j}. \tag{3.5}$$

Proof. *For a given ordering of independent variates $(X_1, ..., X_M)$ and j subscripts specified, the probability that exactly these j variates are less than or equal to x and the remaining*

$M - j$ *variates greater than* x *is* $F^j(x)(1 - F(x))^{M-j}$. *There are* $\binom{M}{j}$ *such specifications possible, so by independence, the probability that exactly* j *variates are less than or equal to* x *is* $\binom{M}{j}F^j(x)(1 - F(x))^{M-j}$. *If* $A_j \subset \mathcal{S}$ *denotes the event on which exactly* j *variates are less than or equal to* x, *then* $\{A_j\}_{j=1}^M$ *are disjoint and union to* $\mathcal{S}$.

As noted above, the event $X_{(k)} \leq x$ *is the union of such* A_j-*events for* $j \geq k$. *Addition of probabilities is justified by the disjointness of this collection and finite additivity of* λ. $\blacksquare$

Example 3.7 *The result in (3.5) is easy to apply.*

1. **Extreme Value Distributions**

   When $k = M$, $F_{(M)}(x) = F^M(x)$ *is the distribution function introduced and characterized in Section II.9.2, as the distribution function of* $\max\{X_j\}$. *This study is continued in Section 7.2.*

2. **Uniform Continuous Distribution on** $[0, 1]$

   If $F_U(x) = x$ *on* $[0, 1]$, *the distribution function of the* kth *order statistic is given by:*

$$F_{(k)}(x) = \sum_{j=k}^{M} \binom{M}{j} x^j (1 - x)^{M-j}.$$

   In particular, the distribution functions for the smallest and largest uniform variate are respectively given on $[0, 1]$ *by:*

$$F_{(1)}(x) = 1 - (1 - x)^M; \qquad F_{(M)}(x) = x^M.$$

3. **Exponential Distribution, Parameter** λ

   If $F_\Gamma(x) = 1 - e^{-\lambda x}$ *on* $[0, \infty)$, *the distribution function of the* kth *order statistic is given by:*

$$F_{(k)}(x) = \sum_{j=k}^{M} \binom{M}{j} \left(1 - e^{-\lambda x}\right)^j \left(e^{-\lambda x}\right)^{M-j}.$$

   In particular, the distribution functions for the smallest and largest exponential variates are given on $[0, \infty)$ *by:*

$$F_{(1)}(x) = 1 - e^{-\lambda M x}; \qquad F_{(M)}(x) = \left(1 - e^{-\lambda x}\right)^M.$$

Consequently, the first order statistic $X_{(1)}$ *from a sample of* M *exponentials with parameter* λ *is exponentially distributed with parameter* λM. *This observation will be expanded upon in the section below on the* **Rényi representation theorem on order statistics.**

3.3   Density Functions of Order Statistics

If $F(x)$ is absolutely continuous, then by the Lebesgue version of the fundamental theorem of calculus of Proposition III.3.62, $F(x)$ has an associated measurable density function $f(x)$ with $f(x) = F'(x)$ almost everywhere, and for all x:

$$F(x) = (\mathcal{L}) \int_{-\infty}^{x} f(x) dx.$$

If $F(x)$ is continuously differentiable, then $f(x) = F'(x)$ for all x by the Riemann version of this result in Proposition III.1.32, and:

$$F(x) = (\mathcal{R}) \int_{-\infty}^{x} f(x)dx.$$

In either case, the given assumption on $F(x)$ yields the analogous property on the distribution function of $F_{(k)}(x)$ in (3.5).

To simplify the next statement, we assume $F(x)$ is continuously differentiable. The absolutely continuous result is derived by qualifying that the density function is defined almost everywhere, and obtains the associated distribution function as a Lebesgue integral as above.

Proposition 3.8 ($f_{(k)}(x)$, **continuous** $f(x)$) *If $F(x)$ is continuously differentiable with associated continuous density $f(x)$, then the density function $f_{(k)}(x)$ of the kth order statistic is continuous and given by:*

$$f_{(k)}(x) = c_{(k)} F^{k-1}(x) \left(1 - F(x)\right)^{M-k} f(x). \tag{3.6}$$

The constant $c_{(k)}$ can be alternatively expressed:

$$c_{(k)} = \frac{M!}{(k-1)!\,(M-k)!} = M\binom{M-1}{k-1} = k\binom{M}{k} = \frac{\Gamma(M+1)}{\Gamma(k)\Gamma(M-k+1)}. \tag{3.7}$$

Proof. *As $F_{(k)}(x)$ in (3.5) is continuously differentiable, the associated density satisfies $f_{(k)}(x) = F'_{(k)}(x)$ by Proposition III.1.32, and is given by:*

$$F'_{(k)}(x) = \sum_{j=k}^{M} \binom{M}{j} \left[jF^{j-1}(x)\left(1 - F(x)\right)^{M-j} - (M-j)F^{j}(x)\left(1 - F(x)\right)^{M-j-1} \right] f(x).$$

This summation "telescopes" by noting that:

$$j\binom{M}{j} = M\binom{M-1}{j-1}, \qquad (M-j)\binom{M}{j} = M\binom{M-1}{j},$$

and this then produces (3.6).

The equivalent formulations for the constant $c_{(k)}$ can be verified as an exercise, recalling (1.58). ∎

One application of (3.6) is the following.

Example 3.9 ($f_{(k)}(x)$ **beta for** $f(x)$ **uniform**) *Let U be continuous and uniformly distributed on $[0, 1]$. Then the density function of the kth order statistic of a sample of M is a Beta density with parameters $v = k$ and $w = M - k + 1$:*

$$f_{(k)}(x) = \frac{\Gamma(M+1)}{\Gamma(k)\Gamma(M-k+1)} x^{k-1} \left(1 - x\right)^{M-k}. \tag{3.8}$$

This follows from (3.6) since $F(x) = x$ here, by (1.61) and the last expression in (3.7).

3.4   Joint Distribution of All Order Statistics

We next derive the joint distribution function $F_{(1,...,M)}(x_1, ..., x_M)$ of all order statistics. By definition, for $x_1 \leq x_2 \leq \cdots \leq x_M$:

$$F_{(1,...,M)}(x_1, ..., x_M) = \Pr\{X_{(j)} \leq x_j \text{ for all } j = 1, ..., M\}.$$

In terms of the probability space $(\mathcal{S}, \mathcal{E}, \lambda)$, this can be expressed:

$$F_{(1,...,M)}(x_1, ..., x_M) = \lambda\left[\bigcap_{j=1}^{M} \left\{X_{(j)}^{-1}(-\infty, x_j]\right\}\right].$$

For this investigation, we will transform this probability statement into probability statements on the various X_j, and then obtain the connection between $F_{(1,...,M)}(x_1, ..., x_M)$ and the common distribution $F(x)$. To do this, we introduce a partitioning of $\mathbb{R}^M$ into nearly disjoint sets with various implied orderings of components.

Recall the following notion of a permutation. This definition also applies to $M = \infty$, but we will not use this generalization except in Exercise 3.11.

Definition 3.10 (Permutation) *Given an ordered set $A \equiv (x_1, x_2, ..., x_M)$, a **permutation** $\pi : A \to A$ is a one-to-one and onto mapping:*

$$\pi : (x_1, x_2, ..., x_M) \to (\pi(x_1), \pi(x_2), ..., \pi(x_M)). \tag{3.9}$$

Exercise 3.11 (Number of permutations) *Show that for M finite, there are $M!$ possible permutations including the identity permutation, $\pi(x_j) = x_j$.*

If $M = \infty$, then there are uncountably many. Hint for $M = \infty$: Given a binary expansion for $b \in [0, 1)$, $b = b_1 b_2, ...$ with each $b_j \in \{0, 1\}$, define a permutation π_b on $\{x_j\}_{j=1}^{\infty}$ so that for each j :

- *If $b_j = 0$: $\pi_b(x_{2j}) = x_{2j}$, $\pi_b(x_{2j-1}) = x_{2j-1}$.*

- *If $b_j = 1$: $\pi_b(x_{2j}) = x_{2j-1}$, $\pi_b(x_{2j-1}) = x_{2j}$.*

Given a permutation $\pi : (1, 2, ..., M) \to (\pi(1), ..., \pi(M))$ where $M < \infty$, define the set $D_\pi \subset \mathbb{R}^M$ by:

$$D_\pi = \{(x_1, ..., x_M) | x_{\pi(1)} \leq x_{\pi(2)} \leq \cdots \leq x_{\pi(M)}\}.$$

The set D_π is thus the collection of points in $\mathbb{R}^M$ which are **weakly ordered** by π, meaning that equality is allowed. Further:

$$\bigcup_\pi D_\pi = \mathbb{R}^M,$$

where this union is over all $M!$ permutations.

Define the set $D_\pi^o \subset D_\pi$:

$$D_\pi^o = \{((x_1, ..., x_M) | x_{\pi(1)} < x_{\pi(2)} < \cdots < x_{\pi(M)}\}.$$

The set D_π^o is thus the collection of points in $\mathbb{R}^M$ which are **strongly ordered** by π.

Example 3.12 (D_π and order statistics) *For points in $\mathbb{R}^3$, there are $3! = 6$ permutations as noted in Exercise 3.11. These can be identified in terms of the results of π on the coordinate indexes,*

$$\pi : (1, 2, 3) \to (\pi(1), \pi(2), \pi(3)).$$

The six permutations are:

$$(1, 2, 3), (1, 3, 2), (2, 1, 3), (2, 3, 1), (3, 2, 1), (3, 1, 2),$$

and for example, if $\pi = (2, 3, 1)$ then:

$$D_\pi = \{(x_1, x_2, x_3) | x_2 \leq x_3 \leq x_1\}$$

To connect with order statistics, assume that $\{X_j\}_{j=1}^3$ is a collection of independent, identically distributed random variables defined on $(\mathcal{S}, \mathcal{E}, \lambda)$, and let $X \equiv (X_1 X_2, X_3)$ be the associated random vector. Given π, define A_π by:

$$A_\pi \equiv X^{-1}(D_\pi) = \{s | X_{\pi(1)} \leq X_{\pi(2)} \leq X_{\pi(3)}\}.$$

Thus on A_π:

$$X_{(1)} = X_{\pi(1)}, \quad X_{(2)} = X_{\pi(2)}, \quad X_{(3)} = X_{\pi(3)}.$$

For example, if $\pi = (2, 3, 1)$ then:

$$A_\pi = \{s | X_2 \leq X_3 \leq X_1\},$$

and on A_π:

$$X_{(1)} = X_2, \quad X_{(2)} = X_3, \quad X_{(3)} = X_1.$$

We will see below that $D_\pi \in \mathcal{B}\left(\mathbb{R}^3\right)$. As X is a random vector, this obtains $A_\pi \in \mathcal{E}$ by (1.9). Then since $\mathcal{S} = \bigcup_\pi A_\pi$, the D_π-sets provide a way of decomposing $\mathcal{S}$ into measurable sets in which the ordering of $\{X_j\}_{j=1}^3$ is known.

But this ordering is based on $\leq$ rather than $<$, and thus it is not uniquely defined. In general, $D_{\pi_1} \cap D_{\pi_2} \neq \emptyset$.

In the next result we investigate the intersection sets $D_\pi \cap D_{\pi'}$, and will show that these have Lebesgue measure 0 in $\mathbb{R}^M$. For certain distribution functions $F(x)$, this will then imply that $X^{-1}(D_{\pi_1} \cap D_{\pi_2}) = A_{\pi_1} \cap A_{\pi_2}$ has λ-measure 0 in $\mathcal{S}$. This will then yield detailed information on a decomposition of $\mathcal{S}$ into measurable sets with well-defined orderings of $\{X_j\}_{j=1}^M$.

Proposition 3.13 $(D_\pi, D_\pi^o \in \mathcal{B}(\mathbb{R}^M), \{D_\pi^o\}_\pi$ **are disjoint,** $m(D_{\pi_1} \cap D_{\pi_2}) = 0)$ *With the notation above, D_π is closed and D_π^o is open. Thus $D_\pi, D_\pi^o \in \mathcal{B}(\mathbb{R}^M)$ and are Lebesgue measurable for all π.*

If $\pi_1 \neq \pi_2$, then:

$$D_{\pi_1}^o \cap D_{\pi_2}^o = 0, \quad m(D_{\pi_1} \cap D_{\pi_2}) = 0.$$

Proof. The set D_π is closed by Exercises III.4.46 and 3.14 as the intersection of $M - 1$ closed sets:

$$\{x_{\pi(j)} \leq x_{\pi(j+1)}\}_{j=1}^{M-1}.$$

Closed sets are Borel measurable by Definition I.2.13, and hence Lebesgue measurable by Proposition I.2.38.

Each D_π^o is open by the same exercises as the intersection of $M - 1$ open sets:

$$\{x_{\pi(j)} < x_{\pi(j+1)}\}_{j=1}^{M-1},$$

and again these are Borel and Lebesgue measurable.

Given any two permutations $\pi_1 \neq \pi_2$:

$$D_{\pi_1}^o \cap D_{\pi_2}^o = \emptyset,$$

because if $x \in D_{\pi_1}^o \cap D_{\pi_2}^o$, then both:

$$x_{\pi_1(1)} < \cdots < x_{\pi_1(M)} \text{ and } x_{\pi_2(1)} < \cdots < x_{\pi_2(M)}.$$

This is only possible if $\pi_1 = \pi_2$, and thus $\{D_\pi^o\}_\pi$ are disjoint.

Further, the set $D_{\pi_1} \cap D_{\pi_2}$ is contained in an $(M-1)$-dimensional subset of $\mathbb{R}^M$. If $x \in D_{\pi_1} \cap D_{\pi_2}$, then for some $j \neq k$ both $x_j \leq x_k$ and $x_k \leq x_j$ are true and thus $x_j = x_k$. Hence:

$$D_{\pi_1} \cap D_{\pi_2} \subset \{((x_1, ..., x_M)|x_j = x_k\}, \tag{1}$$

a set of dimension $M - 1$. The final step is in Exercise 3.14. ∎

Exercise 3.14 *Prove that $\{x_{\pi(j)} \leq x_{\pi(j+1)}\}$ is closed and $\{x_{\pi(j)} < x_{\pi(j+1)}\}$ is open in $\mathbb{R}^M$. Hint: Recall Definition I.2.10.*

Verify that any intersection set $D_{\pi_1} \cap D_{\pi_2}$ for $\pi_1 \neq \pi_2$ has Lebesgue measure 0. Hint: Recall Definition III.1.64 and use (1).

Before proving the general result for $F_{(1,...,M)}(x_1, ..., x_M)$, we develop the idea in an example.

Example 3.15 *(Pr $[(X_1, X_2, X_3) \in D_\pi]$ and Pr $[(X_1, X_2, X_3) \in D_\pi^o]$, A.C. F) Assume that the random variable X defined on $(\mathcal{S}, \mathcal{E}, \lambda)$ has an absolutely continuous distribution function $F(x)$. By Proposition III.3.59, $F'(x)$ exists almost everywhere and is Lebesgue measurable, and then by Proposition III.3.62, if $f(x) \equiv F'(x)$ then, suppressing $(\mathcal{L})$:*

$$F(x) = \int_{-\infty}^x f(y)dy.$$

This appears a little different from III.(3.35), where $F(x) = F(a) + \int_a^x f(y)dy$ for any a. But as $F(a) \to 0$ as $a \to -\infty$ by Proposition 1.3, the above characterization is derived.

Let $\pi : (1, 2, 3) \to (1, 2, 3)$ be the identity permutation to simplify notation. The assumed independence of the random variables (X_1, X_2, X_3) and Proposition 1.14 obtains that the joint density is given by $f(x_1, x_2, x_3) = f(x_1)f(x_2)f(x_3)$. Here each $f(x_j)$ is the density associated with $F(x)$ noted above, since $\{X_i\}_{i=1}^3$ are identically distributed. This follows for continuous densities by Corollary III.1.77, and in the general case by Tonelli's theorem in Book V. See also Section 3.5.

Recall the notational convention that for any $A \in \mathcal{B}(\mathbb{R}^3)$:

$$\Pr[(X_1, X_2, X_3) \in A] \equiv \lambda\left[(X_1, X_2, X_3)^{-1}A\right].$$

As noted in the remarks before (1.30), it will be proved in Book V that for any Borel measurable set $A \in \mathcal{B}(\mathbb{R}^3)$ and random vector $X = (X_1, X_2, X_3)$ defined on $(\mathcal{S}, \mathcal{E}, \lambda)$:

$$\Pr[(X_1, X_2, X_3) \in A] = \int_A f(x_1, x_2, x_3)dx. \tag{3.10}$$

where dx denotes Lebesgue measure on $\mathbb{R}^3$.

This identity is true for general M, and extends the result that for $x = (x_1, ..., x_M)$ and $A = \prod_{i=1}^M (-\infty, x_i]$, this integral obtains $F(x) = \lambda\left[(X_1, ..., X_M)^{-1}A\right]$. By (1.14), with λ_F is the probability measure on $(\mathbb{R}^M, \mathcal{B}(\mathbb{R}^M))$ induced by the distribution function $F(x)$, the result in (3.10) can also be expressed:

$$\lambda_F(A) = \int_A f(x)dx.$$

The Lebesgue integral of a Lebesgue measurable function $f(x_1, x_2, x_3)$ over A in (3.10) can be expressed in terms of iterated integrals. A special case of this was seen for the Riemann integral in Corollary III.1.77, and the general case comes from either Fubini's theorem or Tonelli's theorem of Book V. See item 3 of Remark III.1.79 for a summary of these results.

With dx denoting Lebesgue measure on $\mathbb{R}^3$, the bounds for the iterated integrals over $A = D_\pi = \{-\infty < x_1 \leq x_2 \leq x_3 < \infty\}$ become:

$$\iiint_{D_\pi} f(x_1, x_2, x_3)dx = \int_\mathbb{R} \int_{x_2 \leq x_3} \int_{x_1 \leq x_2} f(x_1)f(x_2)f(x_3)dx_1 dx_2 dx_3. \tag{1}$$

In the special case where $f(x_1, x_2, x_3)$ is continuous almost everywhere, these integrals can also be interpreted as Riemann integrals by Propositions III.1.68 (extended with III.(1.53)) and III.2.56.

By (3.10) and (1):

$$
\begin{aligned}
\Pr\left[(X_1, X_2, X_3) \in D_\pi\right] &= \int_\mathbb{R} \int_{x_2 \leq x_3} \int_{x_1 \leq x_2} f(x_1)f(x_2)f(x_3)dx_1 dx_2 dx_3 \\
&= \int_\mathbb{R} \int_{x_2 \leq x_3} F(x_2)f(x_2)f(x_3)dx_2 dx_3 \\
&= \frac{1}{2} \int_\mathbb{R} F^2(x_3)f(x_3)dx_3 \\
&= 1/3!
\end{aligned}
$$

This same result is produced for any permutation π, with only a change in the definition of the iterated integrals and the order of the integrations.

For any π, since $F(x)$ is absolutely continuous and thus continuous, the same calculation produces:

$$\Pr\left[(X_1, X_2, X_3) \in D_\pi^o\right] = \Pr\left[(X_1, X_2, X_3) \in D_\pi\right].$$

For example:

$$\int_{x_1 < x_2} f(x_1)dx_1 = F(x_2^-) = F(x_2).$$

Recalling Exercise III.1.71, this can also be understood in terms of the boundaries of D_π which have measure zero by Exercise 3.14.

In M dimensions, the same calculations obtain that for any π,

$$\Pr\left[(X_1, ..., X_M) \in D_\pi\right] = \Pr\left[(X_1, ..., X_M) \in D_\pi^o\right] = 1/M!,$$

and thus:

$$\sum_\pi \Pr\left[(X_1, ..., X_M) \in D_\pi\right] = 1. \tag{3.11}$$

Since $\{D_\pi\}_\pi$ is not a disjoint collection, it is important to recognize that (3.11) is not a consequence of finite additivity of λ. Instead, it is absolute continuity of the distribution function $F(x)$ that produced this result because this assumption assured that for any permutation,

$$\Pr\left[(X_1, ..., X_M) \in D_\pi\right] = \Pr\left[(X_1, ..., X_M) \in D_\pi^o\right],$$

or equivalently,

$$\Pr\left[(X_1, ..., X_M) \in (D_\pi - D_\pi^o)\right] = 0.$$

That (3.11) can fail in the absence of absolute continuity is exemplified next.

Example 3.16 (Discrete $F(x)$) *If $X : \mathcal{S} \to \{0, 1\}$ is binomial, with $\lambda[X^{-1}(1)] = p$ with $0 < p < 1$, then with $M = 2$ there are only 2 permutations:*

- *If $\pi_1 : (1, 2) \to (1, 2)$, then $(X_1, X_2)^{-1} D_{\pi_1} = \{X_1 \leq X_2\}$ and $D_{\pi_1} = \{(0, 0), (0, 1), (1, 1)\}$.*

- *If $\pi_2 : (1, 2) \to (2, 1)$, then $(X_1, X_2)^{-1} D_{\pi_2} = \{X_2 \leq X_1\}$ and $D_{\pi_2} = \{(0, 0), (1, 0), (1, 1)\}$.*

In either case:

$$\lambda \left[(X_1, X_2)^{-1} D_\pi \right] = 1 - p(1 - p),$$

and so:

$$\sum_\pi \lambda \left[(X_1, X_2)^{-1} D_\pi \right] = 2 \left[1 - p(1 - p) \right] > 1.$$

This sum exceeds 1 by exactly $p^2 + (1 - p)^2$, and this is because $D_{\pi_1} \cap D_{\pi_2} = \{(0, 0), (1, 1)\}$ and:

$$\lambda \left[(X_1, X_2)^{-1} \left(D_{\pi_1} \cap D_{\pi_2} \right) \right] = p^2 + (1 - p)^2 \neq 0.$$

Put another way, for either permutation:

$$\lambda \left[(X_1, X_2)^{-1} D_\pi \right] \neq \lambda \left[(X_1, X_2)^{-1} D_\pi^o \right].$$

With the above warm-up, we now turn to the main result. The reader may want to supplement this proof with a graphical depiction of the various sets in $\mathbb{R}^2$.

Notation 3.17 *Note that given random variables $(X_1, ..., X_M)$, that $(X_{\pi(1)}, ..., X_{\pi(M)})$ denotes the reordering given a permutation π, while $(X_{(1)}, ..., X_{(M)})$ denotes the associated vector of order statistics.*

Proposition 3.18 ($F_{(1, ..., M)}(x_1, ..., x_M)$, A.C. $F(x)$) *Given independent, identically distributed random variables $(X_1, ..., X_M)$ defined on $(\mathcal{S}, \mathcal{E}, \lambda)$ with absolutely continuous distribution function $F(x)$, the joint distribution function $F_{(1, ..., M)}$ for all order statistics $(X_{(1)}, ..., X_{(M)})$ is defined on $x_1 \leq x_2 \leq \cdots \leq x_M$ by:*

$$F_{(1, ..., M)}(x_1, ..., x_M) = M! F(x_1) \prod_{j=2}^M \left[F(x_j) - F(x_{j-1}) \right]. \tag{3.12}$$

Proof. *Given $x \equiv (x_1, ..., x_M)$ with $x_1 \leq x_2 \leq \cdots \leq x_M$, define the right semi-closed rectangle:*

$$R_x = \prod_{j=1}^M (-\infty, x_j].$$

Then as sets in $\mathcal{S}$:

$$
\begin{aligned}
\{(X_{(1)}, ..., X_{(M)}) \in R_x\} &= \bigcup_\pi \{(X_1, ..., X_M) \leq (x_{\pi(1)}, ..., x_{\pi(M)})\} \\
&= \bigcup_\pi \{(X_1, ..., X_M) \in D_\pi \cap R_x\}.
\end{aligned}
$$

Thus:

$$
\begin{aligned}
F_{(1, ..., M)}(x_1, ..., x_M) &\equiv \Pr \{(X_{(1)}, ..., X_{(M)}) \in R_x\} \\
&= \lambda \left[(X_1, ..., X_M)^{-1} \left(\bigcup_\pi [D_\pi \cap R_x] \right) \right] \\
&= \lambda \left[\bigcup_\pi (X_1, ..., X_M)^{-1} (D_\pi \cap R_x) \right].
\end{aligned}
$$

By (3.10) and absolute continuity of $F(x)$:

$$\lambda\left[\bigcup_{\pi}(X_1,...,X_M)^{-1}(D_\pi - D_\pi^o)\right] = 0, \tag{1}$$

and hence by finite additivity:

$$F_{(1,...,M)}(x_1,...,x_M) = \lambda\left[\bigcup_{\pi}(X_1,...,X_M)^{-1}[D_\pi^o \cap R_x]\right].$$

As $\{(X_1,...,X_M)^{-1}[D_\pi^o \cap R_x]\}_\pi$ *are disjoint sets by Proposition 3.13, finite additivity of* λ, (1) *and independence obtain:*

$$
\begin{aligned}
F_{(1,...,M)}(x_1,...,x_M) &= \sum_{\pi}\lambda\left[(X_1,...,X_M)^{-1}[D_\pi^o \cap R_x]\right]\\
&= \sum_{\pi}\lambda\left[(X_1,...,X_M)^{-1}[D_\pi \cap R_x]\right]\\
&= \sum_{\pi}\lambda\{X_1(s) \le x_{\pi(1)}\}\prod_{j=2}^{M}\lambda\{x_{\pi(j-1)} \le X_j(s) \le x_{\pi(j)}\}\\
&= \sum_{\pi}\lambda\{X_{\pi^{-1}(1)}(s) \le x_1\}\prod_{j=2}^{M}\lambda\{x_{j-1} \le X_{\pi^{-1}(j)}(s) \le x_j\}.
\end{aligned}
$$

In this last expression, π^{-1} *denotes the inverse permutation for* π, *so that* $\pi\pi^{-1} = \pi^{-1}\pi =$ *identity.*

As $\{X_{\pi^{-1}(j)}\}_{j=1}^{M}$ *is just a reordering of* $\{X_j\}_{j=1}^{M}$, *this obtains:*

$$
\begin{aligned}
F_{(1,...,M)}(x_1,...,x_M) &= \sum_{\pi}F(x_1)\left[F(x_2) - F(x_1)\right]...\left[F(x_M) - F(x_{M-1})\right]\\
&= M!F(x_1)\left[F(x_2) - F(x_1)\right]...\left[F(x_M) - F(x_{M-1})\right].
\end{aligned}
$$

∎

Example 3.19 (Discrete $F(y)$**)** *When the distribution function is not absolutely continuous, the above result is not valid. For* X *binomial as in Example 3.16, a calculation shows that* $F_{(1,2)}(0,1) = 1 - p^2$, *since this is the probability that the smaller variate is less than or equal to 0, and the larger is less than or equal to 1. Only* $(1,1)$ *fails this criterion. On the other hand,* $2F(0)\left[F(1) - F(0)\right] = 2p(1-p)$.

3.5   Density Functions on $\mathbb{R}^n$

The notion of a multivariate density function is a definitional generalization of the one variable density function of Section 1.3. Such a function was initially introduced in Example II.3.42. In this section, we outline the major ideas needed for the current chapter, and defer some details to Books V and VI.

If $F(x)$ is the joint distribution function of a random vector $X \equiv (X_1, X_2, ..., X_n)$ defined on $(\mathcal{S}, \mathcal{E}, \lambda)$ with range in $\mathbb{R}^n$, we say that $f(x)$ is **an associated joint density function** if for all $x \equiv (x_1, x_2, ..., x_n)$ and $R_x \equiv \prod_{j=1}^{n}(-\infty, x_j]$.

$$F(x) = \int_{R_x} f(y)dy. \tag{3.13}$$

The domain of integration is also denoted $R_x \equiv \{y \le x\}$ and is shorthand for $\{y_j \le x_j$ for all $j\}$. This domain of integration also appears implicitly in Definition 1.6 for $F(x)$:

$$F(x_1, x_2, ..., x_n) = \lambda\left[X^{-1}R_x\right].$$

As in Section 1.3, the integral in (3.13) may be defined in the sense of Riemann or Lebesgue, depending on the properties of the function $f(x)$. In either case:

$$\int_{\mathbb{R}^n} f(y)dy = 1,$$

and such densities are not unique since in both integration theories, integrands may be changed pointwise in various ways without changing the value of the integral. Recalling Exercise 1.24, the primary exception to this observation is for the Riemann theory when $f(x)$ is continuous. Then one can say that this density function is unique among continuous functions.

In one variable, singular and saltus distribution functions are examples that illustrate that not all distribution functions have associated density functions in the above sense. The same is true for joint distribution functions. For example if $F(x)$ is the joint distribution function of independent random variables with a common singular distribution function, then by Proposition 1.14:

$$F(x) = \prod_{j=1}^{n} F(x_i).$$

If $F(x)$ had an associated density function $f(x)$, then it would follow by integration to obtain marginal distributions, that such distributions had density functions, which contradicts that $f(x_i)$ does not exist.

In one dimension, the existence of a density function $f(x)$ in the sense of a Lebesgue integral required absolute continuity of the distribution function $F(x)$ by Proposition III.3.62. Then $F'(x)$ exists almost everywhere, and we can take as a density any function $f(x)$ with $f(x) = F'(x)$ a.e. When such a distribution function is also continuously differentiable, then a density function exists in the sense of a Riemann integral by Proposition III.1.32. Now $F'(x)$ is continuous, and we can again take $f(x) = F'(x)$ as a density.

For joint distribution functions, the Riemann theory is essentially the same for a continuously differentiable joint distribution functions $F(x)$, by which is meant that $f(x)$ as defined in (3.14) below is a continuous function. For densities in the Lebesgue sense, it will be seen in Book V that a density exists only then the distribution function is absolutely continuous. Generalizing the discussion in Summary 1.20 and Remark 1.21, absolute continuity will there be defined in terms of the associated induced probability measure λ_F defined on $\mathbb{R}^n$, recalling Proposition 1.9.

Turning to some details, assume that such $f(x)$ exists in the Riemann sense. Then by Corollary III.1.77, the integral for $F(x)$ in (3.13) can be expressed as an iterated integral:

$$F(x) = (\mathcal{R}) \int_{-\infty}^{x_n} \cdots \int_{-\infty}^{x_1} f(y_1, y_2, ..., y_n)dy_1 dy_2...dy_n,$$

where by "iterated" is meant that these integrals can be evaluated one at a time, in this or in any given order. While this corollary was stated in terms of integrals over bounded rectangles $R = \prod_{j=1}^{n}(a_i, x_j]$, since $F(x) \to 0$ as all $a_i \to -\infty$, this representation is valid over R_x.

If $f(y_1, y_2, ..., y_n)$ is continuous, Proposition III.1.76 states with the same generalization that for any $(x_1, ..., x_{n-1})$, the function:

$$g(x_1, ..., x_{n-1}, y_n) \equiv (\mathcal{R}) \int_{-\infty}^{x_{n-1}} \cdots \int_{-\infty}^{x_1} f(y_1, y_2, ..., y_n)dy_1...dy_{n-1},$$

is a continuous function of y_n. Thus by Proposition III.1.33:

$$G(x_n) \equiv (\mathcal{R}) \int_{-\infty}^{x_n} g(x_1, ..., x_{n-1}, y_n)dy_n$$

is differentiable and:

$$\frac{\partial G}{\partial x_n} = g(x_1, x_2, ..., x_n).$$

In other words:

$$\frac{\partial F}{\partial x_n} = (\mathcal{R}) \int_{-\infty}^{x_{n-1}} \cdots \int_{-\infty}^{x_1} f(y_1, y_2, ..., x_n) dy_1 dy_2 ... dy_{n-1},$$

and this process can then be implemented iteratively.

In this Riemann case, we then obtain that given a distribution function $F(x)$ with continuous density $f(y)$ in (3.13), this density is given by:

$$f(x_1, x_2, ..., x_n) = \frac{\partial^n F}{\partial x_1 \partial x_2 \cdots \partial x_n}. \tag{3.14}$$

When $f(x)$ exists in the Lebesgue sense, the Lebesgue integral in (3.13) can again be "iterated," but now, as was noted in Example 3.15, with the aid of Fubini's theorem or Tonelli's theorem of Book V. Omitting some details, the function $g(x_1, ..., x_{n-1}, y_n)$ is now Lebesgue integrable. Thus $G(x_n)$ is differentiable almost everywhere by Proposition III.3.38, and:

$$\frac{\partial G}{\partial x_n} = g(x_1, x_2, ..., x_n), \text{ a.e.}$$

Iterating this process obtains that if $f(x)$ is a density for $F(x)$ in the sense of Lebesgue, then (3.14) is true almost everywhere.

See Books V and VI for general results on the existence of density functions.

3.6 Multivariate Order Functions

In this section, we derive density functions associated with various multivariate distribution functions of order statistics. Since all formulas will ultimately flow from $F_{(1,...,M)}(x_1, ..., x_M)$ in (3.12), we continue to assume that the distribution function $F(x)$ for the random variable X is absolutely continuous. Moreover, to keep these derivations in the realm of Riemann integration, we assume a little more, that such $F(x)$ is in fact continuously differentiable, with a continuous density $f(x) = F'(x)$.

This last assumption can be eliminated. In the absolutely continuous case, $f(x) = F'(x)$ a.e. and all integrals below become Lebesgue integrals. This will then require the results from Book V on the connection between multivariate Lebesgue integrals and iterated Lebesgue integrals as was illustrated in Example 3.15. The reader is encouraged to fill in these details after reviewing Book V.

3.6.1 Joint Density of All Order Statistics

In this section, we derive the **joint density function** of all order statistics associated with the joint distribution function $F_{(1,...,M)}(x_1, ..., x_M)$ of (3.12). As noted above, we assume that the distribution function $F(x)$ for the random variable X is continuously differentiable, which is more than the absolute continuity assumed for (3.12).

Before beginning, note that for order statistics, the relationship in (3.13) between the integral of the density function and the distribution function must be modified because

$x_j \leq x_{j+1}$ by definition. In detail, assume that continuous $f_{(1,...,M)}(x_1, ..., x_M)$ exists. Then for $x_1 \leq x_2 \leq \cdots \leq x_M$:

$$F_{(1,...,M)}(x_1, ..., x_M) = (\mathcal{R}) \int_{x_{M-1}}^{x_M} \cdots \int_{x_1}^{x_2} \int_{-\infty}^{x_1} f_{(1,...,M)}(y_1, y_2, ..., y_M) dy_1 dy_2 ... dy_M. \quad (3.15)$$

This formula significantly complicates the relationship between $f_{(1,...,M)}(x_1, ..., x_M)$ and derivatives of $F_{(1,...,M)}(x_1, ..., x_M)$ compared with the result in (3.14). Fortunately for the current application, this approach need not be followed.

An application of (3.16) will be given below for the Rényi representation theorem.

Proposition 3.20 ($f_{(1,...,M)}(x_1, ..., x_M)$, **continuous** $f(x)$) *Given independent, identically distributed random variables* $(X_1, ..., X_M)$ *defined on* $(\mathcal{S}, \mathcal{E}, \lambda)$ *with continuously differentiable distribution function* $F(x)$ *and associated continuous density function* $f(x)$, *the joint density function* $f_{(1,...,M)}$ *of all order statistics is continuous and given for* $x_1 \leq x_2 \leq \cdots \leq x_M$ *by:*

$$f_{(1,...,M)}(x_1, ..., x_M) = M! f(x_1) f(x_2) ... f(x_M). \quad (3.16)$$

Proof. *It is apparent from (3.16) that* $f_{(1,...,M)}(x_1, ..., x_M)$ *is continuous, so we need only show that (3.15) is satisfied.*

From (3.12) it must be proved that for $x_1 \leq x_2 \leq \cdots \leq x_M$:

$$M! F(x_1) \prod_{j=2}^{M} [F(x_j) - F(x_{j-1})] = (\mathcal{R}) \int_{x_{M-1}}^{x_M} \cdots \int_{x_1}^{x_2} \int_{-\infty}^{x_1} M! f(y_1) f(y_2) ... f(y_M) dy_1 dy_2 ... dy_M. \quad (1)$$

Now for $j \geq 2$:

$$\int_{x_{j-1}}^{x_j} f(y_j) dy_j = F(x_j) - F(x_{j-1}),$$

while for $j = 1$:

$$\int_{-\infty}^{x_1} f(y_1) dy_1 = F(x_1),$$

and the result follows. ∎

Exercise 3.21 ($\int f_{(1,...,M)}(x_1, ..., x_M) dx = 1$) *Prove that* $f_{(1,...,M)}(x_1, ..., x_M)$ *in (3.16) is indeed a density function and integrates to 1. Hint: Decompose the integral using the* D_π *sets and generalize Example 3.15.*

3.6.2 Marginal Densities and Distributions

Recall Definition 1.11, which introduced the **marginal distribution functions** given a joint distribution function $F(x_1, ..., x_n)$. Given $F(x_1, x_2, ..., x_n)$, there are $2^n - 2$ **proper** marginal distribution functions defined by the $2^n - 2$ proper subsets of $\{1, 2, ..., n\}$, meaning excluding $I = \{1, 2, ..., n\}$ and $I = \emptyset$.

For a general distribution function defined on $\mathbb{R}^n$ with a density $f(x_1, x_2, ..., x_n) \equiv f(x_I, x_J)$, (1.16) implies by Proposition III.1.80 with apparent notation:

$$F_I(x_{i_1}, x_{i_2}, ..., x_{i_m}) \equiv (\mathcal{R}) \int_{-\infty}^{x_I} \int_{-\infty}^{\infty} f(y_I, y_J) dy_J dy_I.$$

Thus:

$$F_I(x_{i_1}, x_{i_2}, ..., x_{i_m}) = (\mathcal{R}) \int_{-\infty}^{x_I} f_I(y_I) dy_I,$$

where the marginal density function $f_I(x_I)$ is defined:

$$f_I(x_I) = (\mathcal{R}) \int_{-\infty}^{\infty} f(x_I, y_J) dy_J. \tag{3.17}$$

For the current application to $F_{(1,...,M)}(x_1, ..., x_M)$, we must again take more care with this integration since the variables are ordered. For example with $x_J = x_1$, it would make no sense to let $x_1 \to \infty$ in the distribution function since $x_1 \leq x_2 \leq \cdots \leq x_M$ and so $x_1 \to \infty$ here really means $x_1 \to x_2$. Thus integrating the x_1 variate of the density function over $(-\infty, \infty)$ must be interpreted as the integral over $(-\infty, x_2]$.

In general, for the definition of marginal distribution function we must interpret "$infty$" as the upper boundary point of the domain of the variable, and this differs depending on which indexes are in the x_J vector. Similar comments apply to the lower limit of integration, that $-\infty$ must also be interpreted in terms of the lower boundary point of the domain of the given variable.

The resulting calculations can become tedious, remembering there are $2^n - 2$ possibilities, so we provide the results for marginal densities when:

1. $I = (1, ..., j)$ with $1 \leq j < M$, deriving the marginal density functions $f_{(1,...,j)}(x_1, ..., x_j)$;

2. $I = (i, ..., j)$ $1 \leq i < j \leq M$, deriving the marginal density functions $f_{(i,...,j)}(x_i, ..., x_j)$;

3. $I = (i, j)$ with $1 \leq i < j \leq M$, deriving the marginal density functions $f_{(i,j)}(x_i, x_j)$.

4. $I = (i)$ with $1 \leq i \leq M$, deriving the marginal density functions $f_{(i)}(x_i)$.

Proposition 3.22 ($f_{(1,...,j)}(x_1, ..., x_j)$, $1 \leq j \leq M$, **continuous** $f(x)$) *Given independent, identically distributed random variables* $(X_1, ..., X_M)$ *defined on* $(\mathcal{S}, \mathcal{E}, \lambda)$ *with continuously differentiable distribution function* $F(x)$ *and associated continuous density function* $f(x)$, *and* $I = (1, ..., j)$ *with* $1 \leq j < M$, *the marginal density function* $f_{(1,...,j)}(x_1, ..., x_j)$ *is continuous and given on* $x_1 \leq ... \leq x_j$ *by:*

$$f_{(1,...,j)}(x_1, ..., x_j) = \frac{M!}{(M-j)!} f(x_1) f(x_2) ... f(x_j) [1 - F(x_j)]^{M-j}. \tag{3.18}$$

Proof. *We calculate the marginal density* $f_{(1,...,j)}(x_1, ..., x_i)$ *from* $f_{(1,...,M)}(x_1, ..., x_M)$, *first dividing by* $M! f(x_1) f(x_2) ... f(x_j)$ *and suppressing the* $(\mathcal{R})$ *to simplify notation:*

$$\frac{f_{(1,...,j)}(x_1, ..., x_j)}{M! f(x_1) f(x_2) ... f(x_j)} = \int_{x_j}^{\infty} \left[\int_{y_{j+1}}^{\infty} \cdots \int_{y_{M-2}}^{\infty} \left[\int_{y_{M-1}}^{\infty} f(y_M) dx_M \right] f(y_{M-1}) dy_{M-1} \cdots \right] f(y_{j+1}) dy_{j+1}$$

$$= \int_{x_j}^{\infty} \left[\int_{y_{j+1}}^{\infty} \cdots \int_{y_{M-2}}^{\infty} [1 - F(y_{M-1})] f(y_{M-1}) dy_{M-1} \cdots \right] f(y_{j+1}) dy_{j+1}$$

$$\vdots$$

$$= \frac{1}{(M-j)!} [1 - F(x_j)]^{M-j},$$

which obtains (3.18). ∎

Proposition 3.23 ($f_{(i,...,j)}(x_i, ..., x_j)$, $1 \leq i < j \leq M$, **continuous** $f(x)$) *Given independent, identically distributed random variables* $(X_1, ..., X_M)$ *defined on* $(\mathcal{S}, \mathcal{E}, \lambda)$ *with continuously differentiable distribution function* $F(x)$ *and associated continuous density function*

$f(x)$, and $I = (i, ..., j)$ with $1 \le i < j \le M$, the marginal density function $f_{(i,...,j)}(x_i, ..., x_j)$ is continuous and given on $x_i \le ... \le x_j$ by:

$$f_{(i,...,j)}(x_i, ..., x_j) = \frac{M!}{(M-j)!(i-1)!} f(x_i)f(x_{i+1})...f(x_j) \left[1 - F(x_j)\right]^{M-j} F^{i-1}(x_i). \quad (3.19)$$

Proof. The density $f_{(i,...,j)}(x_i, ..., x_j)$ is derived from $f_{(1,...,j)}(x_1, ..., x_j)$ in (3.18), first dividing by $\frac{M!}{(M-j)!} f(x_i)f(x_{i+1})...f(x_j) \left[1 - F(x_j)\right]^{M-j}$ to simplify notation:

$$\frac{(M-j)! f_{(i,...,j)}(x_i, ..., x_j)}{M! f(x_i)f(x_{i+1})...f(x_j) \left[1 - F(x_j)\right]^{M-j}}$$

$$= \int_{-\infty}^{x_i} \left[\int_{-\infty}^{y_{i-1}} \cdots \int_{-\infty}^{y_3} \left[\int_{-\infty}^{y_2} f(y_1)dy_1 \right] f(y_2)dy_2 \cdots \right] f(y_{i-1})dy_{i-1}$$

$$= \int_{-\infty}^{x_i} \left[\int_{-\infty}^{y_{i-1}} \cdots \int_{-\infty}^{y_3} F(y_2)f(y_2)dy_2 \cdots \right] f(y_{i-1})dy_{i-1}$$

$$\vdots$$

$$= \frac{1}{(i-1)!} F^{i-1}(x_i),$$

which obtains (3.19). ■

Proposition 3.24 ($f_{(i,j)}(x_i, x_j)$, $1 \le i < j \le M$, **continuous** $f(x)$) *Given independent, identically distributed random variables* $(X_1, ..., X_M)$ *defined on* $(\mathcal{S}, \mathcal{E}, \lambda)$ *with continuously differentiable distribution function* $F(x)$ *and associated continuous density function* $f(x)$, *and* $I = (i, j)$ *with* $1 \le i < j \le M$, *the marginal density function* $f_{(i,j)}(x_i, x_j)$ *is continuous and given on* $x_i \le x_j$ *by:*

$$f_{(i,j)}(x_i, x_j) = \frac{M!}{(M-j)!(j-i-1)!(i-1)!} f(x_i)f(x_j) \quad (3.20)$$
$$\times \left[1 - F(x_j)\right]^{M-j} \left[F(x_j) - F(x_i)\right]^{j-i-1} F^{i-1}(x_i).$$

Proof. The density $f_{(i,j)}(x_i, x_j)$ is derived from $f_{(i,...,j)}(x_i, ..., x_j)$ in (3.19) by integrating the variates x_k with $i < k < j$ from x_{k-1} to x_j. Dividing by $\frac{M!}{(M-j)!(i-1)!} f(x_i)f(x_j) \left[1 - F(x_j)\right]^{M-j} F^{i-1}(x_i)$ obtains:

$$\frac{(M-j)!(i-1)! f_{(i,j)}(x_i, x_j)}{M! f(x_i)f(x_j) \left[1 - F(x_j)\right]^{M-j} F^{i-1}(x_i)}$$

$$= \int_{x_i}^{x_j} \left[\cdots \int_{y_{j-3}}^{x_j} \left[\int_{y_{j-2}}^{x_j} f(y_{j-1})dy_{j-1} \right] f(y_{j-2})dy_{j-2} \cdots \right] f(y_{i+1})dy_{i+1}$$

$$= \int_{x_i}^{x_j} \left[\cdots \int_{y_{j-3}}^{x_j} \left[F(x_j) - F(y_{j-2})\right] f(y_{j-2})dy_{j-2} \cdots \right] f(y_{i+1})dy_{i+1}$$

$$\vdots$$

$$= \frac{\left[F(x_j) - F(x_i)\right]^{j-i-1}}{(j-i-1)!},$$

which obtains (3.20). ■

Recall Remark II.3.37, that marginal distributions have no "memory" of the variates $x_J \to \infty$, and thus the marginal $F_I(x_{i_1}, x_{i_2}, ..., x_{i_m})$ is the joint distribution function for the random vector $(X_{i_1}, X_{i_2}, ..., X_{i_m})$. The same is true for densities of course, and the following result is no surprise. That is, the marginal density $f_{(i)}(x_i)$ agrees with the density function for the *i*th order statistic in (3.16).

Proposition 3.25 ($f_{(i)}(x_i)$, $1 \le i \le M$, **continuous** $f(x)$ **)** *Given independent, identically distributed random variables* $(X_1, ..., X_M)$ *defined on* $(\mathcal{S}, \mathcal{E}, \lambda)$ *with continuously differentiable distribution function* $F(x)$ *and associated continuous density function* $f(x)$, *and* $I = (i)$ *with* $1 \le i \le M$, *the marginal density function* $f_{(i)}(x_i)$ *is given by:*

$$f_{(i)}(x_i) = \frac{M!}{(M-i)!(i-1)!} f(x_i) F^{i-1}(x_i) \left[1 - F(x_i)\right]^{M-i}. \tag{3.21}$$

Proof. *From (3.20) with* $1 \le i < M$ *and* $j = i+1$, $f_{(i,i+1)}(x_i, x_{i+1})$ *is defined on* $x_i \le x_{i+1}$ *by:*

$$f_{(i,i+1)}(x_i, x_{i+1}) = \frac{M!}{(M-i-1)!(i-1)!} f(x_i) f(x_{i+1}) \left[1 - F(x_{i+1})\right]^{M-i-1} F^{i-1}(x_i).$$

Integrating x_{i+1} *over* $[x_i, \infty)$ *obtains:*

$$
\begin{aligned}
f_{(i)}(x_i) &= \frac{M!}{(M-i-1)!(i-1)!} f(x_i) F^{i-1}(x_i) \int_{x_i}^{\infty} f(x_{i+1}) \left[1 - F(x_{i+1})\right]^{M-i-1} dx_{i+1} \\
&= \frac{M!}{(M-i)!(i-1)!} f(x_i) F^{i-1}(x_i) \left[1 - F(x_i)\right]^{M-i}.
\end{aligned}
$$

The result for $i = M$ *is left as an exercise.* ∎

Remark 3.26 (Marginal distribution functions) *The corresponding* **marginal distribution functions** *can be defined from the marginal density functions with the usual variate restrictions. For example,* $F_{(i,j)}(x_i, x_j)$ *for* $x_i \le x_j$ *is defined:*

$$F_{(i,j)}(x_i, x_j) = \int_{-\infty}^{x_i} \int_{y_i}^{x_j} f_{(i,j)}(y_i, y_j) dy_j dy_i. \tag{3.22}$$

3.6.3 Conditional Densities and Distributions

Recall Definition 1.12 which introduced a **conditional distribution function** given the joint distribution function $F(x_1, ..., x_n)$. There, if $X \equiv (X_1, X_2, ..., X_n)$ is a random vector defined on $(\mathcal{S}, \mathcal{E}, \lambda)$, a conditioning random vector $X_J \equiv (X_{j_1}, X_{j_2}, ..., X_{j_m})$ is identified for $J \equiv \{j_1, ..., j_m\} \subset \{1, 2, ..., n\}$. Then given a Borel set $B \in \mathcal{B}(\mathbb{R}^m)$ with $\lambda \left[X_J^{-1}(B)\right] \ne 0$, the conditional distribution function of X given $X_J \in B$, denoted $F(x_1, x_2, ..., x_n | X_J \in B)$, is defined in terms of a conditional probability, which reduced to:

$$F(x_1, x_2, ..., x_n | X_J \in B) = \lambda \left[X^{-1}\left(\prod_{i=1}^{n}(-\infty, x_i]\right) \bigcap X_J^{-1}(B)\right] \Big/ \lambda \left[X_J^{-1}(B)\right].$$

For the next result we recall a commonly encountered and related notion from elementary probability theory. Given a bivariate distribution function $F(x, y)$, we seek to define a conditional distribution function where the y-conditional set B is replaced by a single point $B \equiv y_0$. In many applications of interest it will be the case that $\lambda \left[Y^{-1}(y_0)\right] = 0$, for example when the the marginal distribution function $F_Y(y)$ is continuous, and thus the above definition is not applicable. However, the intuition is compelling, that given the distribution function $F(x, y)$ and $Y = y_0$, there must be a distribution function of x:

$$F(x|y_0) \equiv F(x, y|y = y_0),$$

such that this distribution is parametrized by y_0.

We will return to a very general model for this notion and related ideas in Book VI in the study of **conditional probability measures** and **conditional expectations,** but for the current application it is enough to recall Example II.3.42. There was developed an approach to defining the conditional distribution function $F(x|y_0)$ from the bivariate continuous joint distribution function $F(x, y)$.

Dropping the subscript on y, this derivation defined $F(x|y)$ as the limit:

$$F(x|y) \equiv \lim_{\Delta y \to 0} F(x, y|Y \in [y, y + \Delta y]).$$

Assuming that $\frac{\partial F(y)}{\partial y} \equiv f(y) \neq 0$, where $F(y)$, $f(y)$ are the associated marginal distribution and density functions, the result derived there was:

$$F(x|y) = \frac{\partial F(x, y)}{\partial y} \bigg/ \frac{\partial F(y)}{\partial y}, \qquad f(x|y) = f(x, y) / f(y). \tag{3.23}$$

The goal of this section is to apply the result from this example to derive the conditional density function $f_{(i+1|i)}(x_{i+1}|x_i)$ from the conditional distribution function $F_{(i+1|i)}(x_{i+1}|x_i)$ of $X_{(i+1)}$ given $X_{(i)}$. The same analysis can be applied to $F_{(j|i)}(x_j|x_i)$ for $j > i$ and is left as an exercise.

Proposition 3.27 ($F_{(i+1|i)}(x_{i+1}|x_i)$, $f_{(i+1|i)}(x_{i+1}|x_i)$, $1 \leq i < M$, **continuous** $f(x)$) *Given independent, identically distributed random variables $(X_1, ..., X_M)$ defined on $(\mathcal{S}, \mathcal{E}, \lambda)$ with continuously differentiable distribution function $F(x)$ and associated continuous density function $f(x)$, and i with $1 \leq i < M$, the conditional distribution function $F_{(i+1|i)}(x_{i+1}|x_i)$ is given on $x_{i+1} \geq x_i$ by:*

$$F_{(i+1|i)}(x_{i+1}|x_i) = 1 - \left(\frac{1 - F(x_{i+1})}{1 - F(x_i)} \right)^{M-i}, \tag{3.24}$$

and the associated conditional density function $f_{(i+1|i)}(x_{i+1}|x_i)$ is:

$$f_{(i+1|i)}(x_{i+1}|x_i) = (M - i)f(x_{i+1}) \frac{(1 - F(x_{i+1}))^{M-i-1}}{(1 - F(x_i))^{M-i}}. \tag{3.25}$$

Proof. *The marginal distribution function $F_{(i,i+1)}(x_i, x_{i+1})$ can be expressed as in (3.22) using the marginal density function $f_{(i,i+1)}(x_i, x_{i+1})$ given in (3.20) for $x_i \leq x_{i+1}$:*

$$F_{(i,i+1)}(x_i, x_{i+1}) = \int_{-\infty}^{x_i} \int_x^{x_{i+1}} f_{(i,i+1)}(x, y) dy dx.$$

Using (3.23) and then the fundamental theorem of calculus of Proposition III.1.33:

$$
\begin{aligned}
F_{(i+1|i)}(x_{i+1}|x_i) &= \frac{\partial F_{(i,i+1)}(x_i, x_{i+1})}{\partial x_i} \bigg/ \frac{\partial F_{(i,i+1)}(x_i, \infty)}{\partial x_i} \\
&= \int_{x_i}^{x_{i+1}} f_{(i,i+1)}(x_i, y) dy \bigg/ \int_{x_i}^{\infty} f_{(i,i+1)}(x_i, y) dy \\
&= \int_{x_i}^{x_{i+1}} f(y) \left[1 - F(y) \right]^{M-i-1} dy \bigg/ \int_{x_i}^{\infty} f(y) \left[1 - F(y) \right]^{M-i-1} dy.
\end{aligned}
$$

In the last step, the factorial constants and the common factor of $f(x_i)F^{i-1}(x_i)$ cancel from numerator and denominator. These integrals can be evaluated by substitution to produce (3.24).

Given $x_i \leq x_{i+1}$:

$$F_{(i+1|i)}(x_{i+1}|x_i) = \int_{x_i}^{x_{i+1}} f_{(i+1|i)}(y|x_i)dy.$$

By Proposition III.1.33 again, the density function in (3.25) is obtained by differentiation:

$$f_{(i+1|i)}(x_{i+1}|x_i) = \frac{\partial}{\partial x_{i+1}} F_{(i+1|i)}(x_{i+1}|x_i),$$

using (3.24). ∎

That the distribution function of $X_{(i+1)}$ depends on the value of $X_{(i)}$ here is logically expected because it must be the case that $X_{(i+1)} \geq X_{(i)}$. We will investigate this further in the study of the **Rényi representation theorem** on order statistics, but here look at an example.

Example 3.28 (*$F(x)$ exponential*) *If $F(x)$ is the exponential distribution of (1.54) with parameter λ, then (3.24) becomes:*

$$F_{(i+1|i)}^{E}(x_{i+1}|x_i) = 1 - e^{-\lambda(M-i)(x_{i+1}-x_i)}.$$

While the distribution function of $X_{(i+1)}$ depends on the value of $X_{(i)}$, the distribution function of the difference, $X_{(i+1)} - X_{(i)}$ does not.

This formula states that whatever is the value of $X_{(i)}$, the value of $X_{(i+1)}$ is given by:

$$X_{(i+1)} = X_{(i)} + Y_i,$$

where $Y_i \equiv X_{(i+1)} - X_{(i)}$ is exponentially distributed with parameter $\lambda(M - i)$. And this is true for all i.

*The remarkable insight in the development of the Rényi representation theorem is that $\{Y_i\}_{i=0}^{M-1}$ so defined are **independent exponentials**.*

3.7 The Rényi Representation Theorem

As will be seen in Section 5.2, the standard approach to generating random samples of a random variable X using uniformly distributed U-samples provides a framework for generating ordered X-samples. However this approach is potentially costly in computer time to generate the necessary U-samples, and even more costly to evaluate the necessary inversions of the given distribution function $F(x)$ using the left-continuous inverse F^*. An alternative approach using **ordered exponential variables** is developed in this section. This approach is based on the **Rényi representation theorem on order statistics**, named for **Alfréd Rényi** (1921–1970).

The Rényi representation theorem derives the surprising conclusion that if F is an exponential distribution and $\{X_{(k)}\}_{k=1}^{M}$ its *kth* **order statistics**, then $\{X_{(k+1)} - X_{(k)}\}_{k=0}^{M-1}$ are **independent**, exponentially distributed random variables. As it turns out, the independence of $\{X_{(k+1)} - X_{(k)}\}_{k=0}^{M-1}$ is unique to the exponential distribution, and reflects the following insight. Here and below, we define $X_{(0)} = 0$ to simplify notation.

Recall the definition of conditional probability of Definition 1.12. When applied to the event $\Pr\{X \le x + y | X > x\}$ for exponential X with $F_E(x) = 1 - e^{-\lambda x}$:

$$
\begin{aligned}
\Pr\{X \le x + y | X > x\} &\equiv \frac{\Pr\{x < X \le x + y\}}{\Pr\{X > x\}} \\
&= \frac{F_E(x + y) - F_E(x)}{1 - F_E(x)} \\
&= F_E(y).
\end{aligned}
\tag{1}
$$

In other words, letting $x = X_{(k)}$, this calculation states that the distribution function of the excess variate $Y \equiv X_{(k+1)} - X_{(k)}$ is independent of $X_{(k)}$.

On an intuitive level it is clear that such a statement could not possibly be true for many distribution functions. Indeed, if $F_U(x) = x$ is the uniform distribution function, then a calculation produces:

$$
\frac{F_U(x + y) - F_U(x)}{1 - F_U(x)} = \frac{\min(x + y, 1) - x}{1 - x},
$$

and the distribution of allowable y values is compressed into $[0, 1 - x]$.

A famous result proved by **Augustin-Louis Cauchy** (1789–1857) states that if $f(x)$ is a continuous function on $\mathbb{R}$ that satisfies **Cauchy's functional equation:**

$$
f(x + y) = f(x) + f(y),
$$

then there is a constant c so that $f(x) = cx$. Defining $f(x) = \ln[1 - F(x)]$, Cauchy's functional equation is equivalent to:

$$
F(x + y) - F(x) = F(y)(1 - F(x)),
$$

and his conclusion is then that $F(x) = 1 - e^{cx}$. This proves that only the exponential distribution $F(x) = F_E(x)$ satisfies (1).

However, perhaps it is possible that for a given distribution function $F(x)$ that x and y are again independent, but with:

$$
\frac{F(x + y) - F(x)}{1 - F(x)} = F_1(y),
\tag{2}
$$

with $F_1(y)$ is a different distribution function. This would not contradict Cauchy's result, but would provide another example for which the excess variable y was independent of x.

Exercise 3.29 ($(2) \Rightarrow F_1 = F = F_E$) *Show that if (2) is satisfied with $F(x)$ a differentiable distribution function, then $F(x)$ is the exponential distribution and thus $F_1 = F$. Hint: By (2), since $F_1(y)$ is independent of x, the x-derivative of $[F(x + y) - F(x)] / [1 - F(x)]$ is zero, and this obtains that $F'(x)/[1 - F(x)]$ is constant.*

We now prove Rényi's result, that $\{X_{(k)} - X_{(k-1)}\}_{k=1}^{M}$ are independent, exponential random variables with respective parameters $\{\lambda(M - k + 1)\}_{k=1}^{M}$.

Although perhaps apparent, to say that two random variables U and V are **equal in distribution**, denoted $U =_d V$, means that for the respective distribution functions:

$$
F_U(x) = F_V(x),
$$

for all x. Sometimes for expediency, as in the next statement, one states simply that two random variables are equal. But this always means "equal in distribution."

Proposition 3.30 (Rényi Representation Theorem) *Let* $\{X_k\}_{k=1}^M$ *be independent random variables from an exponential distribution with parameter* λ, *and* $\{X_{(k)}\}_{k=1}^M$ *the associated ordered random variables. Then with* $X_{(0)} \equiv 0$, $\{X_{(j)} - X_{(j-1)}\}_{j=1}^M$ *are independent, exponentially distributed random variables with respective parameters* $\{\lambda(M - j + 1)\}_{j=1}^M$.

Thus in distribution:

$$X_{(k)} =_d \sum_{j=1}^k Y_j, \tag{3.26}$$

where $\{Y_j\}_{j=1}^M$ *are independent exponential variates with respective parameters* $\{\lambda(M - j + 1)\}_{k=1}^M$.

Proof. *This proof requires another integral of the joint density function* $f_{(1,...,M)}(x_1, ..., x_M)$ *of (3.16). The goal is to show that the joint distribution function* $G_Y(a_1, a_2, ..., a_M)$ *of*

$$Y = (Y_1, ..., Y_M) \equiv (X_{(1)}, X_{(2)} - X_{(1)}, ..., X_{(M)} - X_{(M-1)}),$$

satisfies:

$$G_Y(a_1, a_2, ..., a_M) = \prod_{k=1}^M G_k(a_k),$$

where $G_k(x)$ *is the distribution function of an exponential variate with parameter* $\lambda(M - k + 1)$. *This then proves that* $\{X_{(k)} - X_{(k-1)}\}_{k=1}^M$ *are independent random variables with these distributions by Proposition 1.14.*

Since $x_k \geq x_{k-1}$ *for all* k:

$$G_Y(a_1, a_2, ..., a_M)$$
$$= \Pr\left[X_{(M)} - X_{(M-1)} \leq a_M, ..., X_{(2)} - X_{(1)} \leq a_2, \ X_{(1)} \leq a_1\right]$$
$$= M! \int_{-\infty}^{a_1} \cdots \int_{x_{M-2}}^{x_{M-2}+a_{M-1}} \left(\int_{x_{M-1}}^{x_{M-1}+a_M} f(x_M)dx_M \right) f(x_{M-1})dx_{M-1} \cdots f(x_1)dx_1$$
$$= M! \int_{-\infty}^{a_1} .. \int_{x_{M-2}}^{x_{M-2}+a_{M-1}} \left[F(x_{M-1} + a_M) - F(x_{M-1})\right] f(x_{M-1})dx_{M-1}..f(x_1)dx_1.$$

Since F *is exponential with parameter* λ:

$$\left[F(x_{M-1} + a_M) - F(x_{M-1})\right] f(x_{M-1}) = \left[e^{-\lambda x_{M-1}} - e^{-\lambda(x_{M-1}+a_M)}\right] \lambda e^{-\lambda x_{M-1}}$$
$$= \left[1 - e^{-\lambda a_M}\right] \lambda e^{-2\lambda x_{M-1}}$$
$$= \frac{1}{2} \left[1 - e^{-\lambda a_M}\right] f_2(x_{M-1}),$$

where $f_2(x_{M-1})$ *is the exponential density with parameter* 2λ.

To complete this calculation by induction, let $f_k(x_{M-(k-1)})$ *denote the exponential density with parameter* $k\lambda$, *and note that in the next step:*

$$\int_{x_{M-k}}^{x_{M-k}+a_{M-k+1}} f_k(x_{M-k+1})dx_{M-k+1}f(x_{M-k})$$
$$= \left[F_k(x_{M-k} + a_{M-k+1}) - F_k(x_{M-k})\right] f(x_{M-k})$$
$$= \left[e^{-\lambda k x_{M-k}} - e^{-\lambda k(x_{M-k}+a_{M-k+1})}\right] \lambda e^{-\lambda x_{M-k}}$$
$$= \left[1 - e^{-\lambda k a_{M-k+1}}\right] \lambda e^{-\lambda(k+1)x_{M-k}}$$
$$= \frac{1}{k+1} \left[1 - e^{-\lambda k a_{M-k+1}}\right] f_{k+1}(x_{M-k}).$$

Thus:

$$
G_Y(a_1, a_2, ..., a_M) = \frac{M!}{2}\left[1 - e^{-\lambda a_M}\right] \int_{-\infty}^{a_1} \left[\cdots \left[\int_{x_{M-2}}^{x_{M-2}+a_{M-1}} f_2(x_{M-1})dx_{M-1}\right]\cdots\right] f(x_1)dx_1
$$

$$
= \frac{M!}{3!}\left[1 - e^{-\lambda a_M}\right]\left[1 - e^{-2\lambda a_{M-1}}\right] \times
$$

$$
\int_{-\infty}^{a_1}\left[\cdots \int_{-\infty}^{x_{M-3}+a_{M-2}} f_3(x_{M-2})dx_{M-2}\cdots\right] f(x_1)dx_1
$$

$$
\vdots
$$

$$
= \frac{M!}{(M-1)!}\prod_{k=2}^{M}\left[1 - e^{-(M-k+1)\lambda a_k}\right]\int_{-\infty}^{a_1} f_{M-1}(x_1)dx_1
$$

$$
= \prod_{k=1}^{M}\left[1 - e^{-(M-k+1)\lambda a_k}\right].
$$

∎

An important application of this representation theorem is that any of the *kth* order statistics, or any sequential grouping of *kth* order statistics, can be generated directly as a sum of independent exponential random variables. This is in contrast to the definitional procedure whereby the entire collection $\{X_j\}_{j=1}^{M}$ would need to be generated, then reordered to $\{X_{(j)}\}_{j=1}^{M}$ to identify each variate or grouping.

To generate all $\{X_{(j)}\}_{j=1}^{M}$ requires the independent $\{Y_k\}_{k=1}^{M}$ defined above, with Y_k exponential with parameter $\lambda(M - k + 1)$. To then generate a larger ordered collection of $M' > M$ variates requires $\{Y_i'\}_{i=1}^{M'}$ with Y_i' exponential with parameter $\lambda(M' - i + 1)$. However, given $\{Y_k\}_{k=1}^{M}$, only independent $\{Y_i'\}_{i=1}^{M'-M}$ need be so generated. For $k > 0$, $Y_{M'-M+k}'$ is exponential with parameter $\lambda(M' - [M' - M + k] + 1) = \lambda(M - k + 1)$, so $Y_{M'-M+k}' = Y_k$.

The following corollary provides a simpler version of this representation, in that now all independent exponentials are standard exponentials.

Corollary 3.31 (Rényi Representation Theorem) *Let* $\{X_k\}_{k=1}^{M}$ *denote independent random variables from an exponential distribution with parameter* λ, *and* $\{X_{(k)}\}_{k=1}^{M}$ *the associated ordered random variables. Then in distribution:*

$$
X_{(k)} =_d \sum_{j=1}^{k} \frac{E_j}{\lambda(M - j + 1)}, \tag{3.27}
$$

where $\{E_j\}_{j=1}^{M}$ *are independent standard exponential random variables.*
Proof. *The result follows from Proposition 3.30, and the observation that is left as an exercise, that if* E *is an exponential variable with parameter* $\lambda = 1$, *then* E/α *is an exponential variable with parameter* $\lambda = \alpha$. ∎

The final result reflects definitions from the next chapter on expectations of random variables, but is included here for completeness. If unfamiliar with these notions, the reader should read ahead and come back to this result.

Corollary 3.32 (Rényi Representation Theorem) *Let* $\{X_k\}_{k=1}^{M}$ *denote independent random variables from an exponential distribution with parameter* λ, *and* $\{X_{(k)}\}_{k=1}^{M}$ *the associated ordered random variables. Then denoting by* $\mu_{(k)}$ *and* $\sigma_{(k)}^2$ *the mean and variance of* $X_{(k)}$:

$$
\mu_{(k)} = \frac{1}{\lambda}\sum_{j=1}^{k}\frac{1}{M - j + 1} = \frac{1}{\lambda}\sum_{j=M-k+1}^{M}\frac{1}{j}, \tag{3.28}
$$

$$\sigma^2_{(k)} = \frac{1}{\lambda^2} \sum_{j=1}^{k} \frac{1}{(M-j+1)^2} = \frac{1}{\lambda^2} \sum_{j=M-k+1}^{M} \frac{1}{j^2}. \tag{3.29}$$

Further, with $M_{(k)}(t)$ *denoting the moment generating function of* $X_{(k)}$:

$$M_{(k)}(t) = \prod_{j=1}^{k} \left(1 - \frac{t}{\lambda(M-j+1)}\right)^{-1}, \qquad |t| < \lambda(M-k+1). \tag{3.30}$$

Proof. *By the prior proposition* $X_{(k)}$ *is the sum of* k *independent exponentials with parameters* $\lambda(M-j+1)$ *for* $j = 1$ *to* k. *These results then follow from Section 4.2.4 on moments of sums of random variables, using (4.72) and (4.73) with* $\alpha = 1$. ∎

Remark 3.33 *A few observations on Rényi's result:*

1. *The insight afforded by the* **Rényi representation theorem** *can be appreciated by attempting to derive the formulas in (3.28), (3.29) and (3.30) directly from the density function of* $X_{(k)}$ *in (3.6) and the formulas of the next chapter.*

2. *This representation theorem will be seen to play an important role in generating random samples of ordered exponentials variates in the Section 5.2, as well as in the continued development in Section 7.2 of extreme value theory.*

3. *Because* $\sum_{j=1}^{N} \frac{1}{j} \approx \ln N$ *as* $N \to \infty$, *we have that for* M *large:*

$$\mu_{(M)} \approx \frac{1}{\lambda} \ln M.$$

More generally, by comparing the series to the integral of $1/x$:

$$\ln N + \frac{1}{N} < \sum_{j=1}^{N} \frac{1}{j} < \ln N + 1.$$

Thus:

$$\frac{1}{\lambda}\left[\ln M + \frac{1}{M}\right] < \mu_{(M)} < \frac{1}{\lambda}\left[\ln M + 1\right],$$

and for $k < M$:

$$\frac{1}{\lambda}\left[\ln\left(\frac{M}{M-k}\right) - \frac{M-1}{M}\right] < \mu_{(k)} < \frac{1}{\lambda}\left[\ln\left(\frac{M}{M-k}\right) + \frac{M-k-1}{M-k}\right].$$

4

Expectations of Random Variables 1

In this chapter we begin the study of "expectations" of random variables and introduce some of their important properties. These results can be largely appreciated with the current state of our theoretical development in the special cases of continuously differentiable and discrete distribution functions. But as will be summarized below, even this development requires a leap of faith regarding the fundamental definitions.

This definitional ambiguity will be investigated here, and the framework for a resolution will be outlined. But these matters can only be finally resolved in Book VI with the more advanced integration theory of Book V.

4.1 General Definitions

We begin by introducing the definition of the expectation of a random variable or function of this random variable in the general case. The special cases of continuously differentiable and discrete distribution functions will then likely look familiar. But it will be both necessary and important to identify an inherent ambiguity in this definition, and review the forthcoming mathematical tools of Book V that will ultimately be used to put this definition on a solid, unambiguous footing in Book VI.

For the following definition, recall the development of the Riemann-Stieltjes integral in Chapter III.4.

Definition 4.1 (Expectation) *Let $X : \mathcal{S} \to \mathbb{R}$ be a random variable defined on a probability space $(\mathcal{S}, \mathcal{E}, \lambda)$ with distribution function $F(x)$. If $g(x)$ is a Borel measurable function, the* **expectation of** *$g(X)$, denoted $E([g(X)]$, is defined by the* **Riemann-Stieltjes integral:**

$$E[g(X)] = \int_{-\infty}^{\infty} g(x)dF, \tag{4.1}$$

when:

$$\int_{-\infty}^{\infty} |g(x)|\, dF < \infty. \tag{4.2}$$

Remark 4.2 *Two comments on this definition:*

1. **Existence:** *Since $F(x)$ is increasing and bounded, Proposition III.4.21 assures that this integral exists for $g(x)$ continuous and bounded. The existence theory also applies to $|g(x)|$. Of course boundedness of $g(x)$ is a big restriction, but it will be seen that at least in the special cases of integrators addressed in Proposition III.4.28, that this integral also exists for certain unbounded integrands.*

2. **Random Vectors:** *The above definition is equally applicable when $X : \mathcal{S} \to \mathbb{R}^n$ is a random vector defined on a probability space $(\mathcal{S}, \mathcal{E}, \lambda)$ with joint distribution function*

DOI: 10.1201/9781003264583-4

$F(x)$, and $g : \mathbb{R}^n \to \mathbb{R}$ *a Borel measurable function. This then uses the Riemann-Stieltjes theory and results from Section III.4.4.*

The following result is an immediate application of Proposition III.4.24.

Proposition 4.3 (E is linear) *If $X : S \to \mathbb{R}$ is a random variable defined on a probability space $(S, \mathcal{E}, \lambda)$ with distribution function $F(x)$, and g and h are Borel measurable functions for which $E[g(x)]$ and $E[h(x)]$ are well defined in the sense of (4.2), then for any real constants a, b, $E[ag(x) + bh(x)]$ exists and:*

$$E[ag(x) + bh(x)] = aE[g(x)] + bE[h(x)]. \tag{4.3}$$

Proof. *Once we prove that $|ag(x) + bh(x)|$ is integrable and thus $E[ag(x) + bh(x)]$ exists, the equality in (4.3) follows from Proposition III.4.24.*
By the triangle inequality:

$$|ag(x) + bh(x)| \leq |a|\,|g(x)| + |b|\,|h(x)|,$$

and hence linearity of the integral and integrability of $|g(x)|$ and $|h(x)|$ obtains the result. ■

For distribution functions of random variables, any such function can be decomposed as in (1.24) of Proposition 1.18:

$$F(x) = \alpha F_{SLT}(x) + \beta F_{AC}(x) + \gamma F_{SN}(x).$$

At least in the case of continuous and bounded $g(x)$, the existence theory of Proposition III.4.21 applies to the Riemann-Stieltjes integrals defined with respect to each of the three component functions, and then Proposition III.4.24 assures that for such $g(x)$:

$$\int g(x)dF = \alpha \int g(x)dF_{SLT} + \beta \int g(x)dF_{AC} + \gamma \int g(x)dF_{SN}.$$

By Proposition 1.18, with $f_{SLT}(x_n) \equiv F(x_n) - F(x_n^-)$ defined on the at most countably many discontinuities $\{x_n\}$ of $F(x)$:

$$F_{SLT}(x) \equiv \sum_{x_n \leq x} f_{SLT}(x_n).$$

When $F_{AC}(x)$ is continuously differentiable with $f_{AC}(x) \equiv F'_{AC}(x)$, it follows from Proposition III.1.32 that defined as a Riemann integral:

$$F_{AC}(x) = (\mathcal{R}) \int_{-\infty}^{x} f_{AC}(y)dy.$$

Using Lebesgue integration, this last result generalizes to arbitrary absolutely continuous functions by Proposition III.3.62, where now $f_{AC}(x) \equiv F'_{AC}(x)$ a.e., recalling that $F'_{AC}(x)$ exists a.e. by Proposition III.3.59.

For these last two conclusions, each noted proposition stated that for any a, such $F_{AC}(x)$ could be expressed:

$$F_{AC}(x) = F_{AC}(a) + (\mathcal{R}/\mathcal{L}) \int_{a}^{x} f_{AC}(y)dy,$$

where $f_{AC}(x)$ is defined as above. Since $F_{AC}(a) \to 0$ as $a \to -\infty$ by Proposition 1.3, the result follows.

For discrete $F_{SLT}(x)$ and continuously differentiable $F_{AC}(x)$ component functions in the decomposition of $F(x)$, the Riemann-Stieltjes integral can be recast as follows by Proposition

III.4.28. For more general absolutely continuous $F_{AC}(x)$, the above Lebesgue integrals do not fit neatly within the Riemann-Stieltjes framework of this result. But we will see in Book V that these integrals are intimately related to the Lebesgue-Stieltjes framework studied there, and that there will be a parallel result for Lebesgue-Stieltjes integrals that looks much like item 1 of Proposition III.4.28.

Definition 4.4 (Expectations – Special Cases) *Let $X : \mathcal{S} \to \mathbb{R}$ be a random variable defined on a probability space $(\mathcal{S}, \mathcal{E}, \lambda)$ with distribution function $F(x)$ given by:*

$$F(x) = \alpha F_{SLT}(x) + \beta F_{AC}(x),$$

where $f_{AC}(x) \equiv F'_{AC}(x)$ is continuous, and the collection $\{x_n\}$ underlying the definition of $f_{SLT}(x)$ have no accumulation points.

*If $g(x)$ is a continuous function, the **expectation of** $g(X)$, denoted $E([g(X)]$, is defined by:*

$$E[g(X)] = \alpha \sum_n g(x_n) f_{SLT}(x_n) + \beta \ (\mathcal{R}) \int_{-\infty}^{\infty} g(x) f_{AC}(x) dx, \tag{4.4}$$

when:

$$\sum_n |g(x_n)| \, f_{SLT}(x_n) + (\mathcal{R}) \int_{-\infty}^{\infty} |g(x)| \, f_{AC}(x) dx < \infty. \tag{4.5}$$

In many applications using distribution functions as defined above, only one of $F_{SLT}(x)$ or $F_{AC}(x)$ will be present and thus $E[g(X)]$ will be defined in terms of only one of the components in 4.4. These are then the standard applications in the discrete and continuous probability theories.

4.1.1 Is Expectation Well Defined?

The definition of $E[g(x)]$ in the special cases of Definition 4.4 is quite likely familiar to students of probability theory, even if the more unifying Riemann-Stieltjes approach of Definition 4.1 is perhaps new. However, a closer examination of the above definitions reveals a potential ambiguity, and a resultant conclusion that $E[g(X)]$ may not be well defined.

This definitional ambiguity is often not highlighted by authors of introductory texts because as we will see below, the resolution involves advanced notions that would likely be far outside such texts' mathematical tool kits. The ambiguity identified is equally applicable whether X is a random variable or random vector, but equally resolvable with the integration theory of Book V. In this book we focus on results for random variables and defer the more general discussion to Book VI.

Here is the problem. If X is a random variable on $(\mathcal{S}, \mathcal{E}, \lambda)$, then by Proposition I.3.33 so too is $Y \equiv g(X)$ as the composition of measurable $X : \mathcal{S} \to \mathbb{R}$ and Borel measurable $g : \mathbb{R} \to \mathbb{R}$. Let $F_Y(y)$ denote the distribution function of Y:

$$F_Y(y) = \lambda[(g \circ X)^{-1}(-\infty, y]] = \lambda[X^{-1}[g^{-1}(-\infty, y]]].$$

Then $F_Y(y)$ is well defined since $g^{-1}(-\infty, y] \in \mathcal{B}(\mathbb{R})$ assures that $X^{-1}[g^{-1}(-\infty, y]] \in \mathcal{E}$, and $F_Y(y)$ is an increasing and right continuous function as are all distribution functions by Proposition 1.3.

We now have two approaches to the definition of the expectation of $Y \equiv g(X)$:

1. As a function of X :

$$E[g(X)] = \int_{-\infty}^{\infty} g(x) dF_X.$$

2. As the random variable Y :

$$E[Y] = \int_{-\infty}^{\infty} y dF_Y.$$

The outstanding questions are:

1. *Must it be the case that either both integrals exist, or both don't exist?*

2. *When both integrals exist, must:*

$$\int_{-\infty}^{\infty} g(x)dF_X = \int_{-\infty}^{\infty} y dF_Y \ ? \tag{4.6}$$

For the special cases of Definition 4.4, we can derive insights to the affirmative answer to (4.6).

- $F_X(x) = F_{SLT}^X(x)$ **a saltus function, and** $g(x)$ **monotonic.**

  Assume that $F_X(x)$ is defined by $\{x_n\}_{n=-\infty}^{\infty}$ in increasing order, with probabilities $\{f_X(x_n)\}_{n=-\infty}^{\infty}$, and define $y_n \equiv g(x_n)$.

  If $g(x)$ is **increasing,** then $\{y_n\}_{n=-\infty}^{\infty}$ is an increasing sequence and:

  $$F_Y(y_n) \equiv \Pr[g(X) \le g(x_n)] = F_X(x_n).$$

Thus by (1.3):

$$f_Y(y_n) \equiv F_Y(y_n) - F_Y(y_n^-) = F_X(x_n) - F_Y(x_n^-) = f_X(x_n).$$

When $g(x)$ is **decreasing,** then $\{y_n\}_{n=-\infty}^{\infty}$ is a decreasing sequence and so:

$$F_Y(y_n) \equiv \Pr[g(X) \le g(x_n)] = \Pr[X \ge x_n] = 1 - F_X(x_n^-),$$

$$F_Y(y_n^-) \equiv \Pr[g(X) < g(x_n)] = \Pr[X > x_n] = 1 - F_X(x_n).$$

Thus:

$$f_X(x_n) = F_X(x_n) - F_X(x_{n-1}^-) = F_Y(y_n) - F_Y(y_{n+1}) = f_Y(y_n).$$

By (4.4), it then follows that for $g(x)$ monotonic:

$$\int_{-\infty}^{\infty} g(x)dF_X = \sum_n g(x_n)f_X(x_n) = \sum_n y_n f_Y(y_n) = \int_{-\infty}^{\infty} y dF_Y,$$

and (4.6) is validated for this case.

- $F_X(x) = F_{AC}^X(x)$ **a continuously differentiable function, and** $g(x)$ **monotonic and continuously differentiable with** $g'(x) \ne 0$ **for all** x.

  Both cases can be accommodated with the aid of (2.4), and this obtains:

  $$f_Y(y) = f_X(g^{-1}(y)) \left| \frac{dg^{-1}(y)}{dy} \right|.$$

As Riemann integrals with substitution $x = g^{-1}(y)$:

$$
\begin{aligned}
E[g(X)] &= \int_{-\infty}^{\infty} g(x)f_X(x)dx \\
&= \int_{-\infty}^{\infty} y f_X(g^{-1}(y)) \left| \frac{dg^{-1}(y)}{dy} \right| dy \\
&= \int_{-\infty}^{\infty} y f_Y(y)dy = E[Y].
\end{aligned}
$$

Exercise 4.5 *Show that with $F_X(x)$ a general distribution function, and $g(x)$ monotonically increasing and continuously differentiable with $\left|\frac{\partial g^{-1}(y)}{\partial y}\right| \leq M$ for all x, that if $E[X]$ exists then (4.6) is satisfied with $Y \equiv g(X)$. Hint: From (2.2), $F_Y(y) = F_X(g^{-1}(y))$. Investigate the Riemann-Stieltjes summations for $\int y dF_Y$, approximating ΔF_Y with ΔF_X, and noting that if $\{y_j\}$ is a partition for the dF_Y-integral, then by continuity and monotonicity of $g^{-1}(y)$ it follows that $\{g^{-1}(y_j)\}$ is a partition for the dF_X-integral.*

4.1.2 Formal Resolution of Well-Definedness

While the well-definedness question on expectations has been raised and at best partially answered by the results of the prior section, it is clear that we are still a long way from a statement for general $F_X(x)$ and general Borel measurable $g(x)$. This result is needed for all such $g(x)$ since Borel measurability ensures that $Y = g(X)$ is a random variable on $(\mathcal{S}, \mathcal{E}, \lambda)$.

The resolution will be formalized in Book VI using the following results that will be developed in Book V. This development will accommodate both random variables and random vectors, but here we summarize the former case.

1. General Definition of $E[X]$

If $X : \mathcal{S} \to \mathbb{R}$ is a random variable defined on a probability space $(\mathcal{S}, \mathcal{E}, \lambda)$, we will formally define:

$$E[X] = \int_{\mathcal{S}} X(s) d\lambda(s), \tag{4.7}$$

when

$$\int_{\mathcal{S}} |X(s)| \, d\lambda(s) < \infty. \tag{4.8}$$

This definition requires the development of an integration theory on the measure space $(\mathcal{S}, \mathcal{E}, \lambda)$ which will be addressed in Book V. But for now, we mention that such integrals possess many of the familiar properties seen in the development of the Riemann (Proposition III.1.72), Lebesgue (Proposition III.2.49), and Riemann-Stieltjes integrals (Proposition III.4.24).

Summary 4.6 (On $\int_{\mathcal{S}} X(s) d\lambda(s)$) *Three important and likely familiar properties of integrals from Book III to be seen in Book V are summarized below. But before continuing, the reader may want to look back to Definitions III.2.2 and III.2.4 to recall how we initiated the development of a Lebesgue integral, and then to Remark III.2.6 which attempted to set the stage for future generalizations, including the current one.*

(a) **Characteristic functions:** *Given $A \in \mathcal{E}$, let $X \equiv \chi_A$, the characteristic function of A, defined to equal 1 on A and 0 elsewhere. As was seen in the Lebesgue theory in Definition III.2.4:*

$$\int_{\mathcal{S}} \chi_A(s) d\lambda(s) \equiv \lambda(A). \tag{4.9}$$

Recalling Definition III.2.9, this integral is also defined:

$$\int_{\mathcal{S}} \chi_A(s) d\lambda(s) = \int_A d\lambda(s).$$

(b) **Linearity:** *If X and Y are integrable, which means that (4.8) is satisfied for each, then so too is $aX + bY$ for all $a, b \in \mathbb{R}$, and:*

$$\int_{\mathcal{S}} [aX(s) + bY(s)] \, d\lambda(s) = a \int_{\mathcal{S}} X(s) d\lambda(s) + b \int_{\mathcal{S}} Y(s) d\lambda(s). \tag{4.10}$$

(c) **Monotonicity:** *If* $X \leq Y$, λ-*a.e., then:*

$$\int_S X(s)d\lambda(s) \leq \int_S Y(s)d\lambda(s). \qquad (4.11)$$

A consequence of this and (4.10) is that if $X \geq 0$, λ-*a.e., then:*

$$\int_S X(s)d\lambda(s) \geq 0.$$

Exercise 4.7 ($X \geq 0$ **and** $E[X] = 0$ **imply** $X = 0$, λ-**a.e.**) *Prove that (4.9) and (4.11) imply that if* $X \geq 0$ *and* $E[X] = 0$, *then* $X = 0$, λ-*a.e., meaning outside a set of* λ-*measure 0. Hint: Let* $A_n = \{X > 1/n\}$, *and prove that* $\lambda(A_n) = 0$. *Now apply continuity from above of Proposition I.2.45 to* $A = \{X > 0\}$.

Exercise 4.8 ($X = 0$, λ-**a.e. implies** $E[X] = 0$) *Prove that (4.11) and (4.10) imply that if* $X = 0$, λ-*a.e., then* $E[X] = 0$. *Hint: Both* $0 \leq X$, λ-*a.e. and* $0 \leq -X$, λ-*a.e. are true.*

Exercise 4.9 (Triangle inequality: $|E[X]| \leq E[|X|]$**)** *Prove the triangle inequality, that:*

$$|E[X]| \leq E[|X|], \qquad (4.12)$$

noting that $|X|$ *is integrable if* X *is integrable by (4.8).*

Once $E[X]$ is defined as in (4.7), the ambiguities of the previous section disappear. Defining a new random variable $Y \equiv g(X)$ with Borel measurable $g(x)$, then:

$$E[Y] \equiv \int_S Y(s)d\lambda(s)$$
$$\equiv \int_S g(X(s))d\lambda(s) \equiv E[g(X)].$$

Since the general definition does not even mention the distribution function of the variable we are integrating, it matters not whether we consider Y as the random variable, or X as the random variable which is then composed with Borel measurable $g(x)$. As measurable functions on S, $Y(s) \equiv g(X(s))$ for all s by definition.

2. Change of Variables I - Transformation from S to $\mathbb{R}$

While this approach circumvents the apparent definitional problem, it raises the question of how in any given application one actually evaluates such an integral on S. If $F(x)$ is the distribution function of X, and λ_{F_X} the associated Borel measure on $\mathbb{R}$ as summarized in Proposition 1.5, it will be proved that for any Borel measurable function g :

$$\int_S g(X(s))d\lambda(s) = \int_{\mathbb{R}} g(x)d\lambda_{F_X}. \qquad (4.13)$$

The integral on the right is a **Lebesgue-Stieltjes integral,** and was briefly introduced in Section III.4.1.2. It is named for **Henri Lebesgue** (1875–1941) and **Thomas Stieltjes** (1856–1894).

The original Stieltjes integral, which modified the Riemann approach and is now known as the Riemann-Stieltjes integral, was the subject of most of Chapter III.4. As the Lebesgue integral was introduced in 1904, it is apparent from the above timelines that Stieltjes did develop the modification that is now known as the Lebesgue-Stieltjes integral. However, this integral adapts the original Lebesgue idea in much the same way as the Stieltjes integral originally modified the Riemann idea, and thus this name was born.

The Lebesgue-Stieltjes integral is also known as the **Lebesgue-Radon integral**, or just the **Radon integral**, after **Johann Radon** (1887–1956) who is credited with many of the ideas in its development. Beyond probability measures on $\mathbb{R}$ induced by distribution functions, it will be seen in Book V that such integrals are well defined on any measure space. Indeed, this is the same integration theory as that needed for integrals on $\mathcal{S}$.

Applying (4.13) to the Y-integral obtains:

$$\int_{\mathcal{S}} Y(s)d\lambda(s) = \int_{-\infty}^{\infty} yd\lambda_{F_Y},$$

and thus as integrals on $\mathbb{R}$, this implies a change of variables result that for $y = g(x)$;

$$\int_{-\infty}^{\infty} g(x)d\lambda_{F_X} = \int_{-\infty}^{\infty} yd\lambda_{F_Y}.$$

When $g(x)$ is continuous it will turn out that Lebesgue-Stieltjes and Riemann-Stieltjes integrals agree, and thus for example:

$$\int_{-\infty}^{\infty} g(x)d\lambda_F = \int_{-\infty}^{\infty} g(x)dF.$$

3. **Change of Variables II - Transformation from $d\lambda_F$ to dx or to a summation for Special Distributions**

In this last step we derive the above formulas for $E[g(x)]$ in the special case where $F(x)$ has no singular component $F_{SN}(x)$.

(a) In the special case where $F(x) = F_{AC}(x)$ is absolutely continuous, then with $f_{AC}(x) \equiv F'_{AC}(x)$ defined almost everywhere and Lebesgue measurable (Proposition III.3.59), and Lebesgue measurable $g(x)$:

$$\int_{-\infty}^{\infty} g(x)d\lambda_F = (\mathcal{L}) \int_{-\infty}^{\infty} g(x)f_{AC}(x)dx, \tag{4.14}$$

defined as a Lebesgue integral.

If $f_{AC}(x)$ and $g(x)$ are continuous, the usual set-up in continuous probability theory, then the integral on the right exists as a Riemann integral over every bounded interval $[a,b]$ by Proposition III.1.15, and equals the corresponding Lebesgue integral by Proposition III.2.18. Letting $a \to -\infty$ and $b \to \infty$, it follows that when the Lebesgue integral exists, so too does the Riemann integral, and these integrals agree.

In this case of continuous $f_{AC}(x)$ and $g(x)$, the integral on the left equals also the **Riemann-Stieltjes integral** $\int_{\mathbb{R}} g(x)dF$ as noted above, and as in Proposition III.4.28:

$$\int_{-\infty}^{\infty} g(x)dF = (\mathcal{R}) \int_{-\infty}^{\infty} g(x)f_{AC}(x)dx. \tag{4.15}$$

(b) In the special case where $F(x) = F_{SLT}(x)$ is discrete with discontinuity set $\{x_n\}$ having no accumulation points, then with $f_{SLT}(x_n) \equiv F(x_n) - F(x_n^-)$ and continuous $g(x)$:

$$\int_{-\infty}^{\infty} g(x)d\lambda_F = \int_{-\infty}^{\infty} g(x)dF = \sum_n g(x_n)f_{SLT}(x_n), \tag{4.16}$$

as in Proposition III.4.28.

Remark 4.10 (Random vectors) *The above program of study will also apply when* $X :$ $S \to \mathbb{R}^n$ *is a random vector defined on a probability space* $(S, \mathcal{E}, \lambda)$, $F(x)$ *is the associated joint distribution function, and* $g : \mathbb{R}^n \to \mathbb{R}$ *a Borel measurable function. We will review the multivariate version of steps 2 and 3 in Section 4.2.3 on moments of sums of random variables.*

4.2 Moments of Distributions

As may be recalled from past experiences in probability theory, there are a number of special expectation values which are commonly defined relative to the functions $g(x) = x^n$, and $g(x) = (x - a)^n$ for positive integer n and for a given value of a. It is common to refer to these expectations as **the moments of the distribution,** and sometimes, **the moments of the random variable.** Such functions are not bounded unless the range of X is bounded, so there is, in general, no applicable existence theory from Book III. Thus, it is important to note that such moments need not exist.

We will present these definitions in the general form of the above Riemann-Stieltjes integral in (4.1) to simplify notation, and note that in the special but common cases of continuously differentiable and/or discrete distribution functions, these definitions transform to the familiar results involving Riemann integrals and/or summations as in (4.4).

4.2.1 Common Types of Moments

There are three types of moments commonly defined. The first two are likely familiar to the reader and have well-established notational conventions, while the third is more specialized and used primarily for moment inequalities.

1. Moments about the Origin

Sometimes referred to as the **raw moments** or simply the **moments,** these are the expectations defined relative to the function $g(x) = x^n$.

Definition 4.11 (Moments; mean) *Let* $X : S \to \mathbb{R}$ *be a random variable defined on a probability space* $(S, \mathcal{E}, \lambda)$ *with distribution function* $F(x)$.

The nth **moment** *of* X, *denoted* μ'_n, *is defined by:*

$$\mu'_n \equiv \int_{-\infty}^{\infty} x^n dF, \tag{4.17}$$

when (4.2) is satisfied, and is undefined otherwise.

When $n = 1$, μ'_1 *is called the* **mean of** X, *or of the distribution* F, *and denoted by* μ :

$$\mu \equiv \mu'_1. \tag{4.18}$$

When $F(x)$ *is continuously differentiable with a continuous density function* $f(x)$, *the Riemann-Stieltjes integral in (4.17) can be defined as in (4.4):*

$$\mu'_n \equiv (\mathcal{R}) \int_{-\infty}^{\infty} x^n f(x) dx, \\ \mu \equiv (\mathcal{R}) \int_{-\infty}^{\infty} x f(x) dx. \tag{4.19}$$

When $F(x)$ *is a discrete distribution function with discontinuities on* $\{x_i\}_{i=-\infty}^{\infty}$, *then when these points have no accumulation points, the Riemann-Stieltjes integral in (4.17) can defined as in (4.4):*

$$\mu'_n \equiv \sum_{i=-\infty}^{\infty} x_i^n f(x_i), \\ \mu \equiv \sum_{i=-\infty}^{\infty} x_i f(x_i). \tag{4.20}$$

2. Central Moments

The **central moments** are defined with $g(x) = (x - \mu)^n$, where μ denotes the mean of the distribution.

Definition 4.12 (Central moments; variance) *Let $X : S \to \mathbb{R}$ be a random variable defined on a probability space $(S, \mathcal{E}, \lambda)$ with distribution function $F(x)$.*

*The nth **central moment of X**, denoted μ_n, is defined by:*

$$\mu_n \equiv \int_{-\infty}^{\infty} (x - \mu)^n dF, \tag{4.21}$$

when (4.2) is satisfied, and undefined otherwise.

*When $n = 2$, μ_2 is called the **variance of X, or of the distribution** F, and denoted by σ^2:*

$$\sigma^2 \equiv \mu_2, \tag{4.22}$$

and the positive square root:

$$\sigma \equiv \sqrt{\mu_2}, \tag{4.23}$$

*is called the **standard deviation of X, or of the distribution** F.*

When $F(x)$ is continuously differentiable with a continuous density function $f(x)$, the Riemann-Stieltjes integral in (4.21) can be defined as in (4.4):

$$\mu_n \equiv (\mathcal{R}) \int_{-\infty}^{\infty} (x - \mu)^n f(x) dx,$$
$$\sigma^2 \equiv (\mathcal{R}) \int_{-\infty}^{\infty} (x - \mu)^2 f(x) dx. \tag{4.24}$$

When $F(x)$ is a discrete distribution function with discontinuities on $\{x_i\}_{i=-\infty}^{\infty}$, then when these points have no accumulation points, the Riemann-Stieltjes integral in (4.21) can defined as in (4.4):

$$\mu_n \equiv \sum_{i=-\infty}^{\infty} (x_i - \mu)^n f(x_i),$$
$$\sigma^2 \equiv \sum_{i=-\infty}^{\infty} (x_i - \mu)^2 f(x_i). \tag{4.25}$$

3. Absolute Moments

There are both **absolute moments** and **absolute central moments** defined respectively in terms of $g(x) = |x|^n$ and $g(x) = |x - \mu|^n$. Of course, the absolute value is redundant when n is an even integer. By definition, these moments exist whenever the associated moments and central moments exist due to the constraint in (4.2).

There is no standard notation for these moments, but $\mu'_{|n|}$ and $\mu_{|n|}$ seem self-explanatory and will be used in this text.

4.2.2 Moment Generating Function

In contrast to the above moment definitions which produce numerical values, the moment generating function is defined as an expectation of a parametrized exponential function. Specifically, $g(x)$ is defined by $g(x) = e^{tx}$. Thus the moment generating function is an expectation parametrized by t, and is truly a function of t.

Since exponential functions are nonnegative, there is no explicit mention of the constraint in (4.2).

Definition 4.13 (Moment generating function) *Let $X : S \to \mathbb{R}$ be a random variable defined on a probability space $(S, \mathcal{E}, \lambda)$ with distribution function $F(x)$.*

*The **moment generating function of** X, denoted $M_X(t)$, is defined by the **Riemann-Stieltjes integral**:*

$$M_X(t) \equiv \int_{-\infty}^{\infty} e^{tx} dF(x), \tag{4.26}$$

for all t for which the integral is finite.

When $F(x)$ is continuously differentiable with a continuous density function $f(x)$, the Riemann-Stieltjes integral in (4.26) can be defined as in (4.4):

$$M_X(t) \equiv (\mathcal{R}) \int_{-\infty}^{\infty} e^{tx} f(x) dx. \tag{4.27}$$

When $F(x)$ is a discrete distribution function with discontinuities on $\{x_i\}_{i=-\infty}^{\infty}$, then when these points have no accumulation points, the Riemann-Stieltjes integral in (4.26) can defined as in (4.4):

$$M_X(t) \equiv \sum_{i=-\infty}^{\infty} e^{tx_i} f(x_i). \tag{4.28}$$

Notation 4.14 *It is common is probability theory to abbreviate moment generating function by m.g.f. in the same way as p.d.f. and d.f. are used as abbreviations for the probability density function and distribution function.*

It is also common to denote $M_X(t)$ by $M(t)$ when the random variable is obvious from the context, or by $M_F(t)$ when one wants to highlight the underlying distribution function.

Remark 4.15 (Laplace transform) *When written as a **Lebesgue-Stieltjes integral** as in (4.13):*

$$M_X(t) \equiv \int_{-\infty}^{\infty} e^{tx} d\lambda_F(x),$$

*the integral in (4.26) is related to the **bilateral** or **two-sided Laplace transform of the Borel measure** λ_F, and named for **Pierre-Simon Laplace** (1749–1827). However, it is then conventional to use the exponential e^{-tx} in this definition, and to define this function on complex $t = a + ib$.*

Thus the moment generating function is the two-sided Laplace transform of the Borel measure λ_F restricted to the real numbers, and with reversed orientation.

*In Book VI, we will introduce the **characteristic function** of a distribution, and this will be explicitly defined in terms of the **Fourier transform** of this measure, again restricted to the real line. This transform is named for **Jean-Baptiste Joseph Fourier** (1768–1830) and will be studied in Book V.*

Splitting the integral in (4.26):

$$M_X(t) = \int_{-\infty}^{0} e^{tx} dF(x) + \int_{0}^{\infty} e^{tx} dF(x), \tag{4.29}$$

it is apparent that the first integral exists for all $t \geq 0$ since $\int_{-\infty}^{0} dF(x) = F(0)$ and $0 < e^{tx} < 1$. Similarly the second integral exists for all $t \leq 0$. But for the general case, it is not at all obvious that $M_X(t)$ exists for some interval about the origin, $(-t_0, t_0)$, as will be essential for important results in the next section. While $M_X(0)$ exists for any $F(x)$:

$$M_X(0) = \int_{-\infty}^{\infty} dF(x) = 1,$$

it is possible that $M_X(t)$ exists only for $t = 0$.

Exercise 4.16 (Nonexistence of $M_X(t)$, $t \neq 0$) *Using (4.27) and the continuous density function for the lognormal distribution in (1.67) defined on $[0, \infty)$, investigate the conclusion that $M_{LN}(t)$ exists only for $t = 0$. The solution to this will be found in Example 4.32.*

Based on the above splitting of the integral defining $M_X(t)$ into negative and positive domains of integration, we have the simple result:

Proposition 4.17 (Simple existence criterion) *If the first integral in (4.29) exists for some $t'_0 < 0$, and the second exists for some $t''_0 > 0$, then $M_X(t)$ is well defined on $(-t_0, t_0)$ for $t_0 = \min[-t'_0, t''_0]$.*

Proof. *If $\int_{-\infty}^0 e^{t'_0 x} dF(x) < \infty$, then $e^{tx} \leq e^{t'_0 x}$ for $x \leq 0$ obtains that $\int_{-\infty}^0 e^{tx} dF(x) < \infty$ for $t'_0 < t \leq 0$ by Proposition III.4.24. As the second integral automatically exists for all $t \leq 0$ as noted above, $M_X(t)$ exists on $(t'_0, 0]$.*

Similarly, if the second integral exists for some $t''_0 > 0$, then $M_X(t)$ exists on $[0, t''_0)$, and the result follows. ∎

Finally, we note a simple but useful result which is an application of (4.3).

Proposition 4.18 ($M_{aX+b}(t)$) *Assume that $M_X(t)$ exists for $t \in (-t_0, t_0)$ with $t_0 > 0$, and define $Y \equiv aX + b$ for $a, b \in \mathbb{R}$. Then $M_Y(t)$ exists for $t \in (-t_0/|a|, t_0/|a|)$, and:*

$$M_Y(t) = e^{bt} M_X(at). \tag{4.30}$$

Proof. *By definition and (4.3):*

$$M_Y(t) = E\left[e^{(aX+b)t}\right] = e^{bt} E\left[e^{Xat}\right],$$

and the result follows. ∎

4.2.3 Moments of Sums – Theory

If $\{X_i\}_{i=1}^n$ are random variables defined on a probability space $(\mathcal{S}, \mathcal{E}, \lambda)$, so $X_i : \mathcal{S} \to \mathbb{R}$ for $i = 1, 2, ..., n$, we often need to calculate the moments of $\sum_{i=1}^n X_i$. More generally, if $g : \mathbb{R}^n \to \mathbb{R}$ is a Borel measurable function, we are interested in the expectation of $g(X_1, X_2, ..., X_n)$. For such general situations, we again require the tools of Book V and specifically, the multivariate version of the development of $E[g(X)]$ in Section 4.1.

We summarize the needed results here to provide a formal framework for the calculations below, and supplement the reader's prior experience.

1. Definition of $E[g(X)]$

There is no change to the general definition given in Section 4.1. If the **random vector** $X : \mathcal{S} \longrightarrow \mathbb{R}^n$ is defined on the probability space $(\mathcal{S}, \mathcal{E}, \lambda)$ by:

$$X(s) = (X_1(s), X_2(s), ..., X_n(s)),$$

then $g(X) : \mathcal{S} \longrightarrow \mathbb{R}$ is a random variable for Borel measurable $g : \mathbb{R}^n \to \mathbb{R}$. Thus by (4.4):

$$E[g(X)] \equiv \int_{\mathcal{S}} g(X(s)) d\lambda,$$

when

$$\int_{\mathcal{S}} |g(X(s))| \, d\lambda < \infty,$$

and $E[g(X)]$ is undefined otherwise.

Example 4.19 $(X = \sum_{i=1}^{n} g_i(X_i))$ *When X is a sum of Borel functions of random variables:*

$$X = \sum_{i=1}^{n} g_i(X_i),$$

the existence of $E[g_i(X_i)]$ for all i assures the existence of $E[X]$ by the triangle inequality:

$$|X| \leq \sum_{i=1}^{n} |g_i(X_i)|,$$

and (4.11). The linearity property of integrals in (4.10) then obtains:

$$E\left[\sum_{i=1}^{n} g_i(X_i)\right] = \sum_{i=1}^{n} E[g_i(X_i)]. \tag{4.31}$$

As a special case, it follows that if $E[X_i]$ exists for all i, then $E[\sum_{i=1}^{n} X_i]$ exists with:

$$E\left[\sum_{i=1}^{n} X_i\right] = \sum_{i=1}^{n} E[X_i]. \tag{4.32}$$

2. **Change of Variables I – Transformation from $d\lambda$ on $\mathcal{S}$ to $d\lambda_F$ on $\mathbb{R}^n$**

In order to evaluate expectations it is necessary to move this integral to $\mathbb{R}^n$. Using the same mathematical result as in the one dimensional case, it will turn out that with $F(x_1, x_2, ..., x_n)$ defined as the joint distribution function of $(X_1, X_2, ..., X_n)$, that:

$$\int_{\mathcal{S}} g(X(s)) d\lambda = \int_{\mathbb{R}^n} g(x_1, x_2, ..., x_n) d\lambda_F, \tag{4.33}$$

where λ_F is the Borel measure on $\mathbb{R}^n$ induced by F and introduced in Proposition 1.9. This integral over $\mathbb{R}^n$ is again a **Lebesgue-Stieltjes integral** as was the case for the one dimensional result.

The measure λ_F is defined on rectangles of the form $\prod_{i=1}^{n}(-\infty, x_i]$ by:

$$\lambda_F\left[\prod_{i=1}^{n}(-\infty, x_i]\right] = F(x_1, x_2, ..., x_n),$$

while the measure of a bounded measurable rectangle is given by:

$$\lambda_F\left[\prod_{i=1}^{n}(a_i, b_i]\right] = \sum_x sgn(x) F(x). \tag{4.34}$$

Each $x = (x_1, ..., x_n)$ in the summation is one of the 2^n vertices of $\prod_{i=1}^{n}(a_i, b_i]$, so each $x_i = a_i$ or $x_i = b_i$, and $sgn(x)$ is defined as -1 if the number of a_i-components of x is odd and $+1$ otherwise.

Recalling the development of Chapter I.8, given a distribution function $F(x)$ and set function λ_F so defined on rectangles, this set function can be extended to the measure λ_F on the Borel sigma algebra $\mathcal{B}(\mathbb{R}^n)$, as well as to a complete sigma algebra there denoted $\mathcal{M}_F(\mathbb{R}^n)$.

3. **Change of Variables II – Transformation from $d\lambda_F$ to dF or dx for Special Distributions**

Similar to (4.15), in the special case of absolutely continuous $F(x)$, there exists a measurable density function $f(x)$ so that for all $R_x \equiv \prod_{i=1}^{n}(-\infty, x_i]$:

$$F(x_1, x_2, ..., x_n) = (\mathcal{L}) \int_{R_x} f(y) dy,$$

where $y = (y_1, ..., y_n)$ and dy denotes the Lebesgue product measure on $\mathbb{R}^n$. It will then follow that:

$$\int_{\mathbb{R}^n} g(x_1, x_2, ..., x_n) d\lambda_F = (\mathcal{L}) \int_{\mathbb{R}^n} g(x_1, x_2, ..., x_n) f(x_1, x_2, ..., x_n) dx. \qquad (4.35)$$

A similar transformation is valid when $F(x)$ is a discrete multivariate distribution function, replacing the Lebesgue integral with a summation.

When $f(x)$ and $g(x)$ are continuous, the integral on the right in (4.35) is definable as a Riemann integral over every bounded rectangle $R = \prod_{j=1}^{n} (a_i, b_i]$ by Proposition III.1.63, and such integrals equal the respective Lebesgue integrals by Proposition III.2.18. Letting all $a_i \to -\infty$ and $b_i \to \infty$ obtain that the improper Riemann integral exists if the Lebesgue integral exists.

In this case, $F(x_1, x_2, ..., x_n)$ is also definable as a Riemann integral, and so by Proposition III.4.97, when $g(x)$ is continuous:

$$\int_{\mathbb{R}^n} g(x_1, x_2, ..., x_n) d\lambda_F = \int_{\mathbb{R}^n} g(x_1, x_2, ..., x_n) dF,$$

defined as a **Riemann-Stieltjes integral.**

4. Iterated Integrals and Independent Random Variables: From dx to $dx_1 dx_2 ... dx_n$

There is one additional step needed to evaluate multivariate integrals. The Lebesgue integral in (4.35) can be expressed as an iterated integral:

$$(\mathcal{L}) \int_{\mathbb{R}^n} g(x_1, x_2, ..., x_n) f(x_1, x_2, ..., x_n) dx \qquad (4.36)$$

$$= (\mathcal{L}) \int_{-\infty}^{\infty} \cdots \int_{-\infty}^{\infty} g(x_1, x_2, ..., x_n) f(x_1, x_2, ..., x_n) dx_1 dx_2 ... dx_n.$$

This important result from Book V is **Fubini's theorem,** named for **Guido Fubini** (1879–1943). It states that given the constraint in (4.5), this integral over $\mathbb{R}^n$ can be evaluated in an iterated fashion, one variable at a time, and in any order.

When f and g are continuous, the integral in (4.36) is definable as a Riemann integral as in item 3, and this transformation to iterated integrals was proved as Corollary III.1.77.

In the case of **independent random variables**, Proposition 1.14 states that the distribution function of independent $\{X_i\}_{i=1}^{n}$ is given by:

$$F(x_1, x_2, ..., x_n) = \prod_{j=1}^{n} F_j(x_j).$$

In the case of discrete or absolutely continuous distribution functions, this identity extends to an identity in probability density functions:

$$f(x_1, x_2, ..., x_n) = \prod_{j=1}^{n} f_j(x_j).$$

This follows for continuous densities by Corollary III.1.77, and in the general case by Tonelli's theorem in Book V. See also Section 5.1.3 for more detail.

Thus (4.36) can be expressed:

$$\int_{\mathbb{R}^n} g(x_1, x_2, ..., x_n) d\lambda_F = (\mathcal{L}) \int_{-\infty}^{\infty} \cdots \int_{-\infty}^{\infty} g(x_1, x_2, ..., x_n) \prod_{j=1}^{n} f_j(x_j) dx_1 dx_2 ... dx_n.$$

$$(4.37)$$

When $g(x_1, x_2, ..., x_n)$ and all $f_j(x_j)$ are continuous, the integral in (4.37) can be interpreted as a **Riemann integral**, as noted in item 3.

Example 4.20 ($X = \sum_{i=1}^{n} g_i(X_i)$; $X = \prod_{i=1}^{n} g_i(X_i)$ **for independent** X_i) *The above development is somewhat abstract but was needed to connect familiar mathematical manipulations to a rigorous framework.*

As an application, it is worthwhile to revisit the integrals of sums from Example 4.19 in the context of (4.36), and then consider product functions of independent random variables.

(a) $X = \sum_{i=1}^{n} g_i(X_i)$, *general* $\{X_i\}_{i=1}^{n}$
When $\{X_i\}_{i=1}^{n}$ *are independent, then (4.31) follows from (4.37) and linearity of the Lebesgue integral by Proposition III.2.49. In detail:*

$$\int_{\mathbb{R}^n} X d\lambda_F = (\mathcal{L}) \sum_{i=1}^{n} \int_{-\infty}^{\infty} \cdots \int_{-\infty}^{\infty} g_i(x_i) \prod_{j=1}^{n} f_j(x_j) dx_1 dx_2 ... dx_n$$

$$= (\mathcal{L}) \sum_{i=1}^{n} \int_{-\infty}^{\infty} g_i(x_i) f_i(x_i) dx_i,$$

since:

$$\int_{-\infty}^{\infty} \cdots \int_{-\infty}^{\infty} \prod_{j \neq i} f_j(x_j) dx_j = 1.$$

But independence is not needed for this result. In the general case:

$$\int_{\mathbb{R}^n} X d\lambda_F = (\mathcal{L}) \sum_{i=1}^{n} \int_{-\infty}^{\infty} \cdots \int_{-\infty}^{\infty} g_i(x_i) f(x_1, x_2, ..., x_n) dx_1 dx_2 ... dx_n$$

$$= (\mathcal{L}) \sum_{i=1}^{n} \int_{-\infty}^{\infty} g_i(x_i) f_i(x_i) dx_i,$$

since recalling the marginal density function of (3.17):

$$\int_{-\infty}^{\infty} \cdots \int_{-\infty}^{\infty} f(x_1, x_2, ..., x_n) \prod_{j \neq i} dx_j \equiv f(x_i).$$

(b) $X = \prod_{i=1}^{n} g_i(X_i)$, *independent* $\{X_i\}_{i=1}^{n}$
In this case:

$$\int_{\mathbb{R}^n} X d\lambda_F = (\mathcal{L}) \int_{-\infty}^{\infty} \cdots \int_{-\infty}^{\infty} \prod_{j=1}^{n} f_i(x_i) g_i(x_i) dx_1 dx_2 ... dx_n$$

$$= (\mathcal{L}) \prod_{i=1}^{n} \int_{-\infty}^{\infty} g_i(x_i) f_i(x_i) dx_i.$$

In other words, for independent $\{X_i\}_{i=1}^{n}$ *and Borel measurable* $\{g_i\}_{i=1}^{n}$, *existence of* $E[g_i(X_i)]$ *for all* i *assure the existence of* $X = \prod_{i=1}^{n} g_i(X_i)$, *and:*

$$E\left[\prod_{i=1}^{n} g_i(X_i)\right] = \prod_{i=1}^{n} E[g_i(X_i)]. \tag{4.38}$$

4.2.4   Moments of Sums – Applications

For independent random variables, the evaluation of moments is relatively simple in theory using (4.32), or more generally (4.37), though some calculations may be complicated. In this section, we develop the mean and variance of a sum, and the general moments and moment generating function of an independent sum.

Notation 4.21 (On $E^{(n)}$; $(\mathcal{L})/(\mathcal{R})$) *In an attempt at notational clarity, we denote by $E^{(n)}$ the expectation of a multivariate function relative to the joint distribution function of $(X_1, ..., X_n)$ as defined in (4.36) or its discrete counterpart, and by E an expectation of any random variable as in (4.14) or its discrete counterpart. This notation is not standard and is used here to avoid ambiguity.*

For the integrals below, we do not specify if Riemann or Lebesgue. As we will be implementing (4.14) and (4.36), our minimal requirement on every distribution function is absolute continuity, which assures the existence of an integrable density function. This was proved in Proposition III.3.62 in the 1-dimensional case, and will be generalized in Book V.

Thus if these density functions are merely measurable, the integrals below will be Lebesgue, while if continuous, they can also be understood as Riemann integrals as noted above.

1. **Mean:**

   Let $X = \sum_{i=1}^n X_i$ where $\{X_i\}_{i=1}^n$ are random variables defined on $(\mathcal{S}, \mathcal{E}, \lambda)$. Using (4.32) obtains that

   $$E^{(n)}[X] = \sum_{i=1}^n E^{(n)}[X_i].$$

   In the case of independent variates, (4.37) and $\int_{\mathbb{R}} f(x_j)dx_j = 1$ for all j obtains:

   $$\begin{aligned} E^{(n)}[X_i] &= \int_{-\infty}^{\infty} x_i f(x_i) dx_i \prod_{j \neq i} \int_{-\infty}^{\infty} f(x_j) dx_j \\ &= \mu_i, \end{aligned}$$

   where $\mu_i = E[X_i]$ denotes the mean of X_i as a random variable on $\mathcal{S}$.

   This same formula applies for variates that are not independent, but requires the notion of a marginal density function, as in Example 4.20. As noted there:

   $$\int_{-\infty}^{\infty} \cdots \int_{-\infty}^{\infty} f(x_1, x_2, ..., x_n) \prod_{j \neq i} dx_j \equiv f(x_i),$$

   and applying this in (4.36) obtains:

   $$\begin{aligned} E^{(n)}[X_i] &= \int_{-\infty}^{\infty} \cdots \int_{-\infty}^{\infty} x_i f(x_1, x_2, ..., x_n) dx_1 dx_2 ... dx_n \\ &= \int_{-\infty}^{\infty} x_i f(x_i) dx_i = \mu_i, \end{aligned}$$

   Hence, for **any collection of random variables** $\{X_i\}_{i=1}^n$:

   $$E^{(n)}\left[\sum_{i=1}^n X_i\right] = \sum_{i=1}^n E[X_i] = \sum_{i=1}^n \mu_i. \tag{4.39}$$

2. **mth Moment:**

   Using the **multinomial theorem**, which is to be proved in Exercise 4.22:

   $$\left(\sum_{i=1}^n X_i\right)^m = \sum_{m_1, m_2, .. m_n} \frac{m!}{m_1! m_2! ... m_n!} X_1^{m_1} X_2^{m_2} ... X_n^{m_n}, \tag{4.40}$$

   where this summation is over all distinct n-tuples $(m_1, m_2, .. m_n)$ with $m_j \geq 0$ and $\sum_{j=1}^n m_j = m$.

Using the same approach as for the mean, it follows that for **independent variates**:

$$
\begin{aligned}
E^{(n)}[X_1^{m_1} X_2^{m_2} ... X_n^{m_n}] &= \int_{-\infty}^{\infty} \cdots \int_{-\infty}^{\infty} x_1^{m_1} x_2^{m_2} ... x_n^{m_n} f(x_1, x_2, ..., x_n) dx_1 dx_2 ... dx_n \\
&= \mu_{m_1}^{(1)\prime} \mu_{m_2}^{(2)\prime} \cdots \mu_{m_n}^{(n)\prime},
\end{aligned}
$$

where $\mu_{m_i}^{(i)\prime}$ is the $m_i th$ moment of X_i, and so $\mu_0^{(i)\prime} = 1$ and $\mu_1^{(i)\prime} = \mu_i$.

Thus by linearity in (4.10), for **any collection of independent random variables** $\{X_i\}_{i=1}^n$:

$$
E^{(n)}\left[\left(\sum_{i=1}^n X_i\right)^m\right] = \sum_{m_1, m_2, .. m_n} \frac{m!}{m_1! m_2! ... m_n!} \mu_{m_1}^{(1)\prime} \mu_{m_2}^{(2)\prime} \cdots \mu_{m_n}^{(n)\prime}, \qquad (4.41)
$$

Exercise 4.22 (Multinomial theorem) *Prove the formula in (4.40) using induction on m.*

3. **mth Central Moment:**

As for the mth moments, it follows that for **any collection of independent random variables** $\{X_i\}_{i=1}^n$:

$$
E^{(n)}\left[\left(\sum_{i=1}^n (X_i - \mu_i)\right)^m\right] = \sum_{m_1, m_2, .. m_n} \frac{m!}{m_1! m_2! ... m_n!} \mu_{m_1}^{(1)} \mu_{m_2}^{(2)} \cdots \mu_{m_n}^{(n)}, \qquad (4.42)
$$

where this summation is over all distinct n-tuples $(m_1, m_2, .. m_n)$ with $m_j \geq 0$ and $\sum_{j=1}^n m_j = m$. Here $\mu_{m_i}^{(i)}$ is the $m_i th$ central moment of X_i, and so $\mu_0^{(i)} = 1$ and $\mu_1^{(i)} = 0$.

4. **Variance for Independent Variates:**

With $X = \sum_{i=1}^n X_i$, it follows from (4.39) that $X - E^{(n)}[X] = \sum_{i=1}^n (X_i - \mu_i)$, and so:

$$
\left(X - E^{(n)}[X]\right)^2 = \sum_{i=1}^n (X_i - \mu_i)^2 + 2 \sum_{j<i} (X_j - \mu_j)(X_i - \mu_i). \qquad (1)
$$

The second summation in (1) is sometimes expressed as $\sum_{j \neq i}$ without the coefficient 2.

With (4.37) applied to independent random variables:

$$
\begin{aligned}
E^{(n)}\left[(X_i - \mu_i)^2\right] &= \int_{-\infty}^{\infty} \cdots \int_{-\infty}^{\infty} (X_i - \mu_i)^2 f(x_1, x_2, ..., x_n) dx_1 dx_2 ... dx_n \\
&= \int_{-\infty}^{\infty} (X_i - \mu_i)^2 f(x_i) dx_i = \sigma_i^2,
\end{aligned}
$$

where σ_i^2 denotes the variance of X_i.

Similarly for $i \neq j$:

$$
\begin{aligned}
E^{(n)}\left[(X_j - \mu_j)(X_i - \mu_i)\right] &= \int_{-\infty}^{\infty} \cdots \int_{-\infty}^{\infty} (X_j - \mu_j)(X_i - \mu_i) f(x_1, x_2, ..., x_n) dx_1 dx_2 ... dx_n \\
&= \int_{-\infty}^{\infty} (X_j - \mu_j) f(x_j) dx_j \int_{-\infty}^{\infty} (X_i - \mu_i) f(x_i) dx_i = 0.
\end{aligned}
$$

As it is common in probability to use Var to denote the variance of a complicated expression, it follows from (1) that the variance of a sum of **independent random variables** is given:

$$
Var\left[\sum_{i=1}^n X_i\right] = \sum_{i=1}^n \sigma_i^2. \qquad (4.43)
$$

5. Variance for Dependent Variates:

When $\{X_i\}_{i=1}^n$ are not independent, we obtain from (1) and the prior result:

$$Var\left[\sum_{i=1}^n X_i\right] = \sum_{i=1}^n \sigma_i^2 + 2\sum_{j<i} E^{(n)}\left[(X_j - \mu_j)(X_i - \mu_i)\right].$$

The terms of this summation again involve a marginal density function of (3.17):

$$f(x_i, x_j) = \int_{-\infty}^{\infty} \cdots \int_{-\infty}^{\infty} f(x_1, x_2, ..., x_n) \prod_{k \neq i,j} dx_k,$$

and so:

$$\begin{aligned} E^{(n)}\left[(X_j - \mu_j)(X_i - \mu_i)\right] &= \int_{-\infty}^{\infty}\int_{-\infty}^{\infty} (X_j - \mu_j)(X_i - \mu_i) f(x_i, x_j) dx_i dx_j \\ &\equiv E^{(2)}\left[(X_j - \mu_j)(X_i - \mu_i)\right]. \end{aligned}$$

Definition 4.23 (Covariance; correlation) *Given random variables X, Y on a probability space $(\mathcal{S}, \mathcal{E}, \lambda)$, the **covariance of X and Y**, denoted $cov(X,Y)$, is defined:*

$$cov(X,Y) \equiv E^{(2)}[(X - \mu_X)(Y - \mu_Y)] = E^{(2)}[XY] - \mu_X\mu_Y. \tag{4.44}$$

*The **correlation between X and Y**, denoted $corr(X,Y)$ and often $\rho(X,Y)$ or ρ_{XY}, is defined:*

$$corr(X,Y) \equiv \frac{cov(X,Y)}{\sigma_X\sigma_Y}, \tag{4.45}$$

where σ_X, σ_Y are the standard deviations of these variates.

Hence, the general formula for the variance of a summation can be expressed:

$$Var\left[\sum_{i=1}^n X_i\right] = \sum_{i=1}^n \sigma_i^2 + 2\sum_{j<i} cov(X_i, X_j), \tag{4.46}$$

or equivalently:

$$Var\left[\sum_{i=1}^n X_i\right] = \sum_{i=1}^n \sigma_i^2 + 2\sum_{j<i} \rho_{ij}\sigma_i\sigma_j. \tag{4.47}$$

6. Moment Generating Function:

By the definition in (4.26) applied to a sum of **independent variates** $X = \sum_{i=1}^n X_i$, and using the same manipulations:

$$\begin{aligned} M_X(t) &= E^{(n)}\left[\exp\left(\sum_{i=1}^n tX_i\right)\right] \\ &= \prod_{i=1}^n \int_{-\infty}^{\infty} \exp(tx_i) f(x_i) dx_i, \end{aligned}$$

and so:

$$M_X(t) = \prod_{i=1}^n M_{X_i}(t). \tag{4.48}$$

Thus if $M_{X_i}(t)$ exists for $t \in (-t_0^{(i)}, t_0^{(i)})$, then $M_X(t)$ exists on $(-t_0, t_0) \equiv \bigcap_{i=1}^n (-t_0^{(i)}, t_0^{(i)})$, and on this interval (4.48) is satisfied.

This result is a special case of (4.38) since $\exp\left(\sum_{i=1}^n tX_i\right) = \prod_{i=1}^n \exp(tX_i)$.

4.2.5   Properties of Moments

In this section, we investigate various properties of moments, and the connection between moments and the moment generating function. See also Propositions 4.26 and 4.27 for other results on this connection.

As an example of one property, note that for $m < n$:

$$(x - a)^m \leq (x - a)^n + 1.$$

Thus by (4.11) and setting $a = 0$, the existence of an nth moment assures the existence of the mth moments for all $m < n$. Taking $a = \mu$, the same result applies to central moments.

In applications, one is often interested in whether moments exist for all n, and if not, determining the largest n for which various moments exist. The following result assures that if the nth moment exists, so too does the nth central moment, and conversely. The absolute versions of these moments then always exist by definition of expectation.

Proposition 4.24 ($\mu_n \Leftrightarrow \mu_n'$) *For any n, μ_n exists if and only if μ_n' exists, and:*

$$\mu_n = \sum\nolimits_{j=0}^{n} (-1)^{n-j} \binom{n}{j} \mu_j' \mu^{n-j}, \qquad \mu_n' = \sum\nolimits_{j=0}^{n} \binom{n}{j} \mu_j \mu^{n-j}. \qquad (4.49)$$

Thus for $n = 2$:

$$\sigma^2 = \mu_2' - \mu^2. \qquad (4.50)$$

Proof. *Left as an exercise. Hint: Recall the binomial theorem in (1.42).* ∎

The next result states that in order for $M_X(t)$ to exist for some interval $(-t_0, t_0)$ with $t_0 > 0$, it is necessary that μ_n' and hence μ_n exist for all n. In other words, the existence of the moment generating function is at least as restrictive on a distribution function as is the existence of all finite moments.

As will be seen in Examples 4.32 and 4.58, the existence of $M_X(t)$ is even more restrictive than the existence of all moments. Specifically, there are infinitely many distributions for which μ_n' exists for all n, but for which $M_X(t)$ exists only for $t = 0$.

Proposition 4.25 ($M_X(t)$ **implies** μ_n', μ_n **all** n) *If $M_X(t)$ exists for some interval $(-t_0, t_0)$ with $t_0 > 0$, then μ_n' and thus μ_n exist for all n.*
Proof. *Choose t with $0 < t < t_0$. Since $e^{t|x|} \leq e^{tx} + e^{-tx}$, the existence of $M_X(t)$ and $M_X(-t)$ implies by items 1 and 2 of Proposition III.4.24:*

$$\int_{-\infty}^{\infty} e^{t|x|} dF \leq M_X(t) + M_X(-t) < \infty. \qquad (1)$$

Given n, $e^{t|x|} \geq |x|^n$ for $|x| / \ln |x| \geq n/t$. Since $|x| / \ln |x|$ is increasing and unbounded, choose $\{x_n\}_{n=1}^{\infty}$ with $x_n > 0$ and $|x_n| / \ln |x_n| \geq n/t$. Then by (1):

$$\int_{-\infty}^{\infty} |x|^n \, dF \;=\; \int_{|x|<x_n} |x|^n \, dF + \int_{|x|\geq x_n} |x|^n \, dF$$

$$\leq\; c|x_n|^n + \int_{|x|\geq x_n} e^{t|x|} dF$$

$$\leq\; c|x_n|^n + M_X(t) + M_X(-t),$$

where:

$$c \;=\; \int_{|x|<x_n} dF$$

$$=\; F\left(x_n^-\right) - F(-x_n) \leq 1.$$

Thus μ_n' exists for every n, and the existence of μ_n for all n is Proposition 4.24. ∎

The name "moment generating" function for $M_X(t)$ is justified in the next two propositions. When $M_X(t)$ exists, it not only assures the existence of all moments by the prior result, but this function can also be used to generate these moments.

To make these more general proofs rigorous requires the **Lebesgue dominated convergence theorem** for Lebesgue-Stieltjes integrals. This result generalizes the result for the Lebesgue integral of Proposition III.2.52, and will be proved in Book V. As noted above, it will also be proved in Book V that Lebesgue-Stieltjes and Riemann-Stieltjes integrals agree for continuous integrands, and thus this result obtains the Lebesgue dominated convergence theorem for Riemann-Stieltjes integrals used below.

To make this result more accessible, we note some important special cases which require "mostly" prior results.

1. As noted in the proof, if $F(x) = 1$ for $x \geq b$ and $F(x) = 0$ for $x \leq a$, then Proposition III.4.27 obtains the needed convergence result.

2. When the distribution function $F(x)$ is discrete or continuously differentiable, the respective Riemann-Stieltjes integrals reduce to summations or Riemann integrals by Proposition III.4.28.

   (a) For discrete distributions, we can appeal to standard results on absolutely convergent series.

   (b) In the continuously differentiable case, these Riemann integrals agree with Lebesgue integrals over every bounded interval $[a, b]$ by Proposition III.1.15, and such integrals equal the respective Lebesgue integrals by Proposition III.2.18. Letting $a \to -\infty$ and $b \to \infty$ obtains that the improper Riemann integral exists if the Lebesgue integral exists. Thus the Lebesgue dominated convergence result of Proposition III.2.52 suffices.

3. If $F(x)$ is absolutely continuous and thus has Lebesgue measurable density $f(x)$ by Proposition III.3.62, the Riemann-Stieltjes integrals below equal Lebesgue-Stieltjes integrals (Book V). Then as in (4.14), this integral reduces to a Lebesgue integral, and again the Book III dominated convergence theorem suffices.

Proposition 4.26 ($M_X(t)$ **implies** $\mu'_n = M_X^{(n)}(0)$) *Assume that $M_X(t)$ exists for some interval $(-t_0, t_0)$ with $t_0 > 0$.*

Then $M_X(t)$ is infinitely differentiable on this interval, and for all n, the nth derivative of $M_X(t)$ is given by:

$$M_X^{(n)}(t) = \int_{-\infty}^{\infty} y^n e^{ty} dF. \tag{4.51}$$

Thus the moments of X are given by:

$$\mu'_n = M_X^{(n)}(0). \tag{4.52}$$

Proof. By (4.26):

$$M_X(t) \equiv \int_{-\infty}^{\infty} e^{tx} dF(x).$$

This result would be obvious if we could assume a generalization of the Leibniz rule of Proposition III.1.40 to differentiate under the integral sign, but a proof is required to justify this.

Proceeding by induction, since $M_X^{(0)}(t) \equiv M_X(t)$ by definition, we prove that for $n \geq 0$:

$$\text{If } M_X^{(n)}(t) = \int_{-\infty}^{\infty} x^n e^{tx} dF, \text{ then, } M_X^{(n+1)}(t) = \int_{-\infty}^{\infty} x^{n+1} e^{tx} dF. \tag{1}$$

Let:

$$f_m^+(x) = x^n e^{tx} m \left[e^{x/m} - 1 \right], \qquad f_m^-(x) = x^n e^{tx} m \left[1 - e^{-x/m} \right].$$

The assumed existence of $M_X^{(n)}(t)$ for $t \in (-t_0, t_0)$ assures that for m large:

$$\begin{aligned}
m \left[M_X^{(n)}(t + 1/m) - M_X(t) \right] &= \int_{-\infty}^{\infty} f_m^+(x) dF, \\
m \left[M_X^{(n)}(t) - M_X(t - 1/m) \right] &= \int_{-\infty}^{\infty} f_m^-(x) dF.
\end{aligned} \tag{2}$$

Both integrals on the right are well-defined and bounded if $t \pm 1/m \in (-t_0, t_0)$.
    Now:

$$m \left[1 - e^{-x/m} \right] \leq x \leq m \left[e^{x/m} - 1 \right],$$

and this obtains:

$$\min \left[\left| f_m^+(x) \right|, \left| f_m^-(x) \right| \right] \leq \left| x^{n+1} e^{tx} \right| \leq \max \left[\left| f_m^+(x) \right|, \left| f_m^-(y) \right| \right].$$

Since all bounding functions are integrable for $t \pm 1/m \in (-t_0, t_0)$, item 2 of Proposition III.4.24 assures that:

$$\int_{-\infty}^{\infty} \left| x^{n+1} e^{tx} \right| dF < \infty, \tag{3}$$

for all $t \in (-t_0, t_0)$.
    Thus by (3) and another application of this item 2:

$$\int_{-\infty}^{\infty} x^{n+1} e^{tx} dF,$$

is well defined for all $t \in (-t_0, t_0)$.
    To prove that this integral equals $M_X^{(n+1)}(t)$ and thus prove (1), first note that $f_m^{\pm}(x) \to x^{n+1} e^{tx}$ as $m \to \infty$. This convergence is not only pointwise, but also uniform on bounded intervals using a Taylor series approximation. Thus in the special case where $F(x) = 1$ for $x \geq b$ and $F(x) = 0$ for $x \leq a$, Proposition III.4.27 obtains:

$$\int_{-\infty}^{\infty} f_m^{\pm}(x) dF = \int_a^b f_m^{\pm}(x) dF \to \int_a^b x^{n+1} e^{tx} dF = \int_{-\infty}^{\infty} x^{n+1} e^{tx} dF.$$

Then by (2), the proof of (1) is complete.
    In the general case, $g_m(x) \equiv \min \left[f_m^+(x), f_m^-(x) \right] \to x^{n+1} e^{tx}$ as $m \to \infty$. Since $|g_m(x)| \leq \left| x^{n+1} e^{tx} \right|$ and this bounding function is integrable, the general version of Lebesgue's dominated convergence theorem of Book V obtains:

$$\int_{-\infty}^{\infty} g_m(x) dF \to \int_{-\infty}^{\infty} x^{n+1} e^{tx} dF. \tag{4}$$

1. *If n is even, then $f_m^-(x) \leq f_m^+(x)$ since $e^{-x/m} + e^{x/m} \geq 2$ by concavity of the exponential function. Thus $g_m(x) = f_m^-(x)$ and this obtains by (2) and (4) that $M_X^{(n+1)}(t) = \int_{-\infty}^{\infty} x^{n+1} e^{tx} dF$.*

2. *If n is odd, then $f_m^-(x) \leq f_m^+(x)$ for $x \geq 0$ and $f_m^-(x) \geq f_m^+(x)$ for $x \leq 0$, but the same conclusion will follow if it can be proved that as $m \to \infty$:*

$$\int_0^{\infty} \left| f_m^+(x) - f_m^-(x) \right| dF \to 0.$$

Since $f_m^+(x) - f_m^-(x) \to 0$ pointwise, this result will again follow by Lebesgue's dominated convergence theorem if the integrand is appropriately dominated.

A Taylor series analysis shows that if $t + 1/m_0 \in (-t_0, t_0)$ and $m \geq m_0$, then for $x \geq 0$:

$$0 \leq f_m^+(x) - f_m^-(x) \leq 2x^{n+1} e^{(t+1/m)x} \leq 2x^{n+1} e^{(t+1/m_0)x},$$

and this bounding function is integrable, as noted above.

■

The next result provides a second insight to the name moment generating function by way of a power series representation using the moments $\{\mu_n'\}_{n=1}^{\infty}$. This proof again requires the Book V generalization of the Lebesgue dominated convergence theorem, and more specifically, the corollary to this result that generalizes Corollary III.2.53. This needed result addresses the question of when the integral of a infinite sum of functions equals the sum of the associated integrals.

Note that the power series representation in (4.53) provides an alternative proof of the infinite differentiability of $M_X(t)$ and derivation of (4.52). See for example Proposition 9.111 in **Reitano** (2010).

Proposition 4.27 (Power series for $M_X(t)$) *If $M_X(t)$ exists for some interval $(-t_0, t_0)$ with $t_0 > 0$, then on this interval:*

$$M_X(t) = \sum_{n=0}^{\infty} \frac{\mu_n'}{n!} t^n, \tag{4.53}$$

where $\{\mu_n'\}_{n=1}^{\infty}$ are the moments of X.

Proof. *Because $e^{|tx|} \leq e^{tx} + e^{-tx}$, the existence of $M_X(t)$ for $|t| < t_0$ assures that $\int e^{|tx|} dF < \infty$.*

The exponential Taylor series:

$$e^{tx} \equiv \sum_{n=0}^{\infty} \frac{(tx)^n}{n!}, \tag{4.54}$$

is absolutely convergent in x for all t. Hence the partial sums:

$$\sum_{n=0}^{N} \frac{|tx|^n}{n!} \leq e^{|tx|},$$

are bounded by an integrable function, and this implies by the triangle inequality that for all N :

$$\left| \sum_{n=0}^{N} \frac{(tx)^n}{n!} \right| \leq e^{|tx|}.$$

Hence as $N \to \infty$:

$$\sum_{n=0}^{N} \frac{(tx)^n}{n!} \to e^{tx},$$

and these partial sums are dominated by an integrable function. The above noted corollary to the Lebesgue dominated convergence theorem of Book V obtains for $|t| < t_0$:

$$
\begin{aligned}
M_X(t) &= \lim_{N \to \infty} \sum_{n=0}^{N} \frac{t^n}{n!} \int x^n dF \\
&= \sum_{n=0}^{\infty} \frac{\mu_n'}{n!} t^n.
\end{aligned}
$$

■

For the purpose of calculating the mean, variance, and third central moment of a distribution, the following corollary is often useful.

Corollary 4.28 (The cumulant generating function: $\ln M_X(t)$) *If $M_X(t)$ exists for some interval $(-t_0, t_0)$ with $t_0 > 0$, then defining the **cumulant generating function** $g_X(t) \equiv \ln M_X(t)$:*

$$\mu = g_X'(0), \qquad \sigma^2 = g_X''(0), \qquad \mu_3 = g_X^{(3)}(0). \tag{4.55}$$

Proof. Left as an exercise. ∎

Remark 4.29 (On cumulants) *If $M_X(t)$ exists for $t \in (-t_0, t_0)$, then $g_X(t) \equiv \ln M_X(t)$ exists as a Taylor series for all such t since $M_X(t) > 0$ by definition.*

*The **cumulants** of X are then defined in terms of this Taylor series:*

$$\ln M_X(t) = \sum_{n=0}^{\infty} \frac{t^n}{n!} \kappa_n,$$

*where κ_n denotes the nth **cumulant of** X. As above, it follows that:*

$$\kappa_n = g_X^{(n)}(0).$$

Corollary 4.30 (Convexity of $M_X(t)$) *If $M_X(t)$ exists for some interval $(-t_0, t_0)$ with $t_0 > 0$, then $M_X(t)$ is **convex** on on this interval.*

That is by Definition 4.39, if $[t_1, t_2] \subset (-t_0, t_0)$, then for any α, $0 \leq \alpha \leq 1$:

$$M_X(\alpha t_1 + (1 - \alpha) t_2) \leq \alpha M_X(t_1) + (1 - \alpha) M_X(t_2). \tag{4.56}$$

Proof. Left as an exercise. Hint: First prove that e^{xt} is convex in t for any x. ∎

4.2.6   Examples–Discrete Distributions

In this section, we present various results related to moments of the discrete distributions presented in Section 1.4.1. The various results, which are largely assigned as exercises, utilize (4.16).

1. **Discrete Rectangular Distribution on $[0, 1]$:**

   Recalling (1.38):

   $$f_R\left(\frac{j}{n}\right) = \frac{1}{n}, \qquad j = 1, 2, .., n,$$

   a calculation obtains:

   $$\mu_R = \frac{n+1}{2n}, \qquad \sigma_R^2 = \frac{n^2 - 1}{12n^2}, \tag{4.57}$$

   and:

   $$M_R(t) = \frac{\exp[(1 + 1/n)t] - \exp[t/n]}{n(\exp[t/n] - 1)}, \qquad t \in \mathbb{R}. \tag{4.58}$$

   These results can be generalized to a discrete rectangular distribution on $[a, b]$ by defining $Y = (b - a)X + a$:

   $$f_{R_{a,b}}\left(\frac{j(b-a)}{n} + a\right) = 1/n, \qquad j = 1, 2, .., n. \tag{4.59}$$

   Utilizing linearity of expectations in (4.3):

   $$\mu_{R_{a,b}} = (b - a)\mu_R + a, \qquad \sigma_{R_{a,b}}^2 = (b - a)^2 \sigma_R^2, \tag{4.60}$$

and by (4.30):

$$M_{R_{a,b}}(t) = e^{at} M_R([b-a]t).$$ (4.61)

2. Binomial Distribution:

Let $f_{B_n}(j)$ denote the probability density function of the sum of n independent standard binomials with $0 < p < 1$ as in (1.40):

$$f_{B_n}(j) = \binom{n}{j} p^j (1-p)^{n-j}, \quad j = 0, 1, .., n.$$

For $n = 1$, the standard binomial random variable X_1^B has moments:

$$\mu_{B_1} = p, \qquad \sigma_{B_1}^2 = p(1-p),$$

and moment generating function:

$$M_{X_1^B}(t) = 1 + p(e^t - 1), \quad t \in \mathbb{R}.$$

Hence from Section 4.2.4:

$$\mu_{B_n} = np, \qquad \sigma_{B_n}^2 = np(1-p),$$ (4.62)

$$M_{B_n}(t) = \left(1 + p(e^t - 1)\right)^n.$$ (4.63)

While the moment generating function of $f_{B_n}(j)$ has a simple form, the mth moment $\mu'_{B_n,m}$ of B_n is a messy polynomial in p of degree m since $\mu'^{(1)}_{m_1} \mu'^{(2)}_{m_2} \cdots \mu'^{(n)}_{m_n} = p^r$ in (4.41), where r equals the number of subscripts with $m_j > 0$. The derivation of $\mu'_{B_n,m}$ is not simplified by using (4.52).

3. Geometric Distribution::

From (1.43):

$$f_G(j) = p(1-p)^j, \; j = 0, 1, 2, ...,$$

and for this distribution, it is easiest to first calculate $M_G(t)$, then evaluate some of the moments using (4.52).

To this end:

$$M_G(t) = p \sum_{j=0}^{\infty} \left[(1-p)e^t\right]^j,$$

and as a geometric summation, this series is convergent if $(1-p)e^t < 1$. Then:

$$M_G(t) = \frac{p}{1 - (1-p)e^t}, \quad t < -\ln(1-p),$$ (4.64)

and derivatives can now be evaluated using (4.55) to produce:

$$\mu_G = \frac{1-p}{p}, \qquad \sigma_G^2 = \frac{1-p}{p^2}.$$ (4.65)

4. Negative Binomial Distribution::

This is another distribution for which it is easiest to evaluate $M_{NB}(t)$ first. Using (1.46):

$$f_{NB}(j) = \binom{j+k-1}{k-1} p^k (1-p)^j, \quad j = 0, 1, 2, ...,$$

and it again follows that the resulting series is convergent for $(1 - p)e^t < 1$, and:

$$M_{NB}(t) = \left(\frac{p}{1 - (1 - p)e^t}\right)^k, \quad t < -\ln(1 - p). \tag{4.66}$$

Calculating two derivatives with (4.55) produces:

$$\mu_{NB} = \frac{k(1 - p)}{p}, \qquad \sigma^2_{NB} = \frac{k(1 - p)}{p^2}. \tag{4.67}$$

Remark 4.31 (Independent geometric variables) *By (4.48), $M_{NB}(t)$ is also the moment generating function of the sum of k independent geometric variables, thereby raising the question of uniqueness of this function. Using the tools of Section 4.4, we will conclude that the negative binomial is uniquely defined by its moments and moment generating function, and hence will conclude that a negative binomial variate equals the sum of k independent geometric variates in distribution. See item 2 of Section 4.4.1.*

5. **Poisson Distribution:**

   Recalling (1.47):

   $$f_P(j) = \frac{e^{-\lambda}\lambda^j}{j!}, \quad j = 0, 1, 2, ..., $$

   and the moment generating function is calculated:

   $$M_P(t) = \exp[\lambda(e^t - 1)], \quad t \in \mathbb{R}. \tag{4.68}$$

   The mean and variance follow from (4.55):

   $$\mu_P = \lambda, \qquad \sigma^2_P = \lambda. \tag{4.69}$$

4.2.7   Examples–Continuous Distributions

In this section, we identify various results related to moments of the continuous distributions presented in Section 1.4.2. For the various results, which are largely assigned as exercises, we utilize (4.15).

1. **Continuous Uniform Distribution:**

   From (1.51):

   $$f_U(x) = 1/(b - a), \quad x \in [a, b],$$

   and $f_U(x) = 0$ otherwise. It follows that:

   $$\mu_U = \frac{b + a}{2}, \qquad \sigma^2_U = \frac{(b - a)^2}{12}, \tag{4.70}$$

   and:

   $$M_U(t) = \frac{e^{bt} - e^{at}}{t(b - a)}, \quad t \in \mathbb{R}. \tag{4.71}$$

   Note that $M_U(t)$ is well defined at $t = 0$ despite the apparent singularity, as justified using the exponential Taylor series in (4.54).

2. **Exponential Distribution** and **Gamma Distribution**:

The exponential density is defined with a single scale parameter $\lambda > 0$ in (1.53):

$$f_E(x) = \lambda e^{-\lambda x}, \quad x \geq 0.$$

This distribution is a special case of the more general gamma density defined with a shape parameter α and scale parameter $\lambda > 0$ in (1.55) by:

$$f_\Gamma(x) = \frac{1}{\Gamma(\alpha)} \lambda^\alpha x^{\alpha-1} e^{-\lambda x}, \quad x \geq 0,$$

with the **gamma function** $\Gamma(\alpha)$ defined by:

$$\Gamma(\alpha) = \int_0^\infty x^{\alpha-1} e^{-x} dx.$$

Moments of the gamma are derived by integration by parts and (1.57), that for $\alpha > 1$:

$$\Gamma(\alpha + 1) = \alpha \Gamma(\alpha),$$

and thus $\Gamma(n) = (n-1)!$ for integer n as in (1.58). This obtains:

$$\mu'_{\Gamma,n} = \frac{\alpha}{\lambda^n} \left[\prod_{j=1}^{n-1} (\alpha + j) \right], \qquad \mu_\Gamma = \frac{\alpha}{\lambda}, \qquad \sigma_\Gamma^2 = \frac{\alpha}{\lambda^2}. \tag{4.72}$$

The moment generating function can also be calculated:

$$M_\Gamma(t) = \left(1 - \frac{t}{\lambda} \right)^{-\alpha}, \qquad t < \lambda. \tag{4.73}$$

These formulas produce results for:

(a) The **exponential distribution** with parameter λ, by setting $\alpha = 1$.

(b) The **chi-squared distribution** with **n-degrees of freedom** by Remark 1.30, by setting $\lambda = 1/2$ and $\alpha = n/2$.

3. **Beta Distribution**

The **beta density** contains two shape parameters, $v > 0, w > 0$, and is defined on the interval $[0, 1]$ by the density function in (1.61):

$$f_\beta(x) = \frac{1}{B(v, w)} x^{v-1} (1 - x)^{w-1},$$

where the **beta function** $B(v, w)$ is defined by a definite integral in (1.62):

$$B(v, w) = \int_0^1 y^{v-1} (1 - y)^{w-1} dy.$$

For any positive integer n, it follows by definition that:

$$E[x^n] = \frac{B(v + n, w)}{B(v, w)}.$$

The beta function $B(v, w)$ satisfies an important identity which is useful in evaluating these moments:

$$B(v + 1, w) = \frac{v}{v + w} B(v, w), \tag{4.74}$$

as can be derived using (1.63) and (1.57).

Applying the iterative formula in (4.74) and noting that $B(1,1) = 1$ produces:

$$\mu_\beta = \frac{v}{v+w}, \qquad \mu'_{\beta,n} = \prod_{j=0}^{n-1}\left(\frac{v+j}{v+w+j}\right), \tag{4.75}$$

$$\sigma_\beta^2 = \frac{vw}{(v+w)^2\,(v+w+1)}.$$

Thus $\mu'_{\beta,n} \to 0$ as $n \to \infty$ since $w > 0$.

The moment generating function $M_\beta(t)$ exists for all t since $e^{xt} \leq \max[1, e^t]$ for $x \in [0,1]$, and so:

$$|M_\beta(t)| \leq \max[1, e^t].$$

Thus $M_\beta(t)$ can be expressed in terms of its moments by (4.53):

$$M_\beta(t) = 1 + \sum_{n=1}^{\infty}\prod_{i=0}^{n-1}\left(\frac{v+i}{v+w+i}\right)\frac{t^n}{n!}. \tag{4.76}$$

4. Cauchy Distribution

The **Cauchy distribution** of (1.68) is of interest as an example of a distribution that has no finite moments. This density function is defined on $\mathbb{R}$ as a function of a location parameter $x_0 \in \mathbb{R}$ and a scale parameter $\gamma > 0$ by:

$$f_C(x) = \frac{1}{\pi\gamma}\frac{1}{1 + ([x-x_0]/\gamma)^2},$$

while the **standard Cauchy distribution** is parameterized with $x_0 = 0$ and $\gamma = 1$ to:

$$f_C(x) = \frac{1}{\pi}\frac{1}{1+x^2}.$$

While one might attempt to derive $E[X]$ by a symmetry argument that:

$$\frac{1}{\pi}\int_{-\infty}^{\infty}\frac{x\,dx}{1+x^2} = 0,$$

it is not the case that $E[X] = 0$ since $E[|X|]$ is unbounded:

$$\frac{1}{\pi}\int_{-N}^{N}\frac{|x|\,dx}{1+x^2} = \frac{1}{\pi}\int_{0}^{N}\frac{2x\,dx}{1+x^2} = \frac{1}{\pi}\ln N.$$

Hence the Cauchy distribution has no mean, and thus by the introduction to Section 4.2.5, it has no finite moments.

5. Normal Distribution

The **normal density** is defined on $(-\infty, \infty)$, depends on a location parameter $\mu \in \mathbb{R}$ and a scale parameter $\sigma > 0$, and is defined in (1.65) by:

$$f_N(x) = \frac{1}{\sigma\sqrt{2\pi}}\exp\left(-\frac{(x-\mu)^2}{2\sigma^2}\right).$$

The associated **unit** or **standard normal distribution** $\phi(x)$ is defined in (1.66) with $\mu = 0$ and $\sigma = 1$:

$$\phi(x) = \frac{1}{\sqrt{2\pi}}\exp\left(-\frac{1}{2}x^2\right).$$

As noted earlier, there is no elementary derivation of the fact that $\phi(x)$, and hence $f_N(x)$, integrate to 1. However, all central moments exist because $\exp\left(-x^2/2\right) < x^{-n}$ for all n as $x \to \infty$. In addition, all odd central moments are 0 by symmetry, and for even moments an integration by parts obtains:

$$\int_{-\infty}^{\infty} x^{2m}\phi(x)dx = (2m-1)\int_{-\infty}^{\infty} x^{2m-2}\phi(x)dx.$$

This plus mathematical induction yields:

$$\int_{-\infty}^{\infty} x^{2m}\phi(x)dy = \frac{(2m)!}{2^m m!}.$$

Hence, justifying the notational convention of parameterizing the normal with μ and σ^2, we have that:

$$\mu'_{N,1} = \mu, \qquad \mu_{N,2} = \sigma^2, \qquad \mu_{N,2m} = \frac{\sigma^{2m}(2m)!}{2^m m!}, \qquad \mu_{N,2m+1} = 0. \qquad (4.77)$$

The moment generating function is derived by completing the square in the exponential function, and then a substitution, to produce:

$$M_N(t) = \exp\left(\mu t + \frac{1}{2}\sigma^2 t^2\right), \qquad t \in \mathbb{R}, \qquad (4.78)$$

which obtains for the unit normal:

$$M_\Phi(t) = \exp\left(\frac{1}{2}t^2\right). \qquad (4.79)$$

6. **Lognormal Distribution:**

The **lognormal distribution** is defined on $[0,\infty)$, depends on a location parameter $\mu \in \mathbb{R}$ and a shape parameter $\sigma > 0$, and has probability density function given in (1.67):

$$f_L(x) = \frac{1}{\sigma x\sqrt{2\pi}}\exp\left(-\frac{(\ln x - \mu)^2}{2\sigma^2}\right).$$

The substitution $y = (\ln x - \mu)/\sigma$ into the integral of $f_L(x)$ produces the integral of the unit normal $\phi(y)$, and moments of all orders exist for the lognormal and are calculated using the same substitution:

$$\mu'_{L,n} = e^{n\mu}M_\Phi(n\sigma).$$

In other words, the moments of the lognormal can be calculated from the moment generating function of the unit normal.

Specifically, using (4.79) obtains:

$$\mu'_{L,n} = e^{n\mu+(n\sigma)^2/2}, \qquad \mu_L = e^{\mu+\sigma^2/2}, \qquad \sigma_L^2 = e^{2\mu+\sigma^2}\left(e^{\sigma^2}-1\right). \qquad (4.80)$$

Example 4.32 (Nonexistence of $M_{LN}(t)$ for $t \neq 0$) *As noted in the introduction to Proposition 4.25, while the existence of $M_X(t)$ on an open interval $(-t_0, t_0)$ assures that all moments of the distribution function exists, it is not the case that the existence of all* *moments assures the existence of $M_X(t)$ on an open interval $(-t_0, t_0)$. The lognormal distribution provides the classical counterexample, in that while μ'_{nL} exists for all n by (4.80), the series:*

$$\sum_{n=0}^{\infty} \frac{t^n \mu'_{nL}}{n!} = \sum_{n=0}^{\infty} \frac{t^n}{n!} e^{n\mu + (n\sigma)^2/2},$$

cannot converge for any $t \neq 0$, and so $M_{LN}(t)$ cannot exist except for $t = 0$.

To see this, recalling Section I.3.4.2 for definitions of limits superior and inferior, the **ratio test** states that for a **positive series** $\sum_{n=0}^{\infty} c_n$:

$$\limsup \frac{c_{n+1}}{c_n} < 1 \Rightarrow \sum_{n=0}^{\infty} c_n < \infty,$$

$$\liminf \frac{c_{n+1}}{c_n} > 1 \Rightarrow \sum_{n=0}^{\infty} c_n = \infty.$$

A calculation shows that for this series and $t > 0$:

$$\frac{c_{n+1}}{c_n} = \frac{t}{n+1} e^{\mu + (n+1/2)\sigma^2},$$

which is unbounded in n for all $t > 0$. Thus $\liminf \frac{c_{n+1}}{c_n} > 1$ and this series diverges.

This series alternates when $t < 0$ and converges by the **alternating series theorem** if and only if as $n \to \infty$:

$$|c_n| \to 0 \text{ as } n \to \infty.$$

By Stirling's formula in (4.105), this series converges if and only if as $n \to \infty$:

$$\left| \frac{t^n e^{n\mu + (n\sigma)^2/2}}{\sqrt{2\pi} n^{n+1/2} e^{-n}} \right| \to 0.$$

As the log of this expression is unbounded for $t \neq 0$, so too is this ratio and thus the alternating series does not converge.

In summary, despite have all finite moments, the moment generating function $M_{LN}(t)$ of the lognormal distribution does not exist except for $t = 0$. By Example 4.58, there are infinitely many distribution functions with this same property.

For results on series, see for example Chapter 6 in **Reitano** (2010).

4.3 Moment Inequalities

In this section, we develop a number of important inequalities related to moments.

Notation 4.33 *If X is a random variable defined on a probability space $(S, \mathcal{E}, \lambda)$ with distribution function F, we use the simplified notation such as $\Pr[|X - \mu| \geq t\sigma]$ as shorthand for:*

$$\Pr[|X - \mu| \geq t\sigma] \equiv \lambda \left[\{s \in \mathcal{S} | X(s) \leq \mu - t\sigma\} \bigcup \{s \in \mathcal{S} | X(s) \geq \mu + t\sigma\} \right]$$

$$= F(\mu - t\sigma) + 1 - F([\mu + t\sigma]^-),$$

where $F([\mu + t\sigma]^-)$ denotes the left limit of $F(x)$ as $x \to \mu + t\sigma$.

4.3.1 Chebyshev's Inequality

Chebyshev's inequality, sometimes spelled as Chebychev or Tchebysheff, applies to any distribution function that has a mean and variance, and hence it is quite generally applicable. It is named for its discoverer, **Pafnuty Chebyshev** (1821–1894), who as a Russian mathematician had several transliterations of his name in English.

This inequality can be stated in many ways, and Chebyshev's inequality is actually a name now given to a family of inequalities as will be seen below. But it is often applied as stated in the first proposition when we are interested in an upper bound for the probability that a random variable is far from its mean, where "far" can be defined in absolute units, or units relative to the variance.

Proposition 4.34 (Chebyshev's inequality) *If $F(x)$ is a distribution function of a random variable X with mean μ and variance σ^2, then for any real number $t > 0$:*

$$\Pr[|X - \mu| \geq t\sigma] \leq \frac{1}{t^2}. \tag{4.81}$$

Equivalently, for any $s > 0$:

$$\Pr[|X - \mu| \geq s] \leq \frac{\sigma^2}{s^2}. \tag{4.82}$$

Proof. *By definition:*

$$
\begin{aligned}
\sigma^2 &\equiv \int_{-\infty}^{\infty} (x - \mu)^2 dF \\
&\geq \int_{|x-\mu| \geq t\sigma} (x - \mu)^2 dF \\
&\geq (t\sigma)^2 \Pr[|X - \mu| \geq t\sigma].
\end{aligned}
$$

The second result is implied by the first with the substitution $t = s/\sigma$. ∎

With the same proof, the above result can be generalized when higher moments exist.

Proposition 4.35 (Generalized Chebyshev inequalities) *If X is a random variable defined on a probability space $(\mathcal{S}, \mathcal{E}, \lambda)$ with distribution function $F(x)$ for which the nth absolute moments exist:*

$$\mu'_{|n|} \equiv E[|X|^n], \quad \mu_{|n|} \equiv E[|X - \mu|^n].$$

Then for any real number $t > 0$:

$$\Pr[|X| \geq t] \leq \frac{\mu'_{|n|}}{t^n}, \tag{4.83}$$

$$\Pr[|X - \mu| \geq t] \leq \frac{\mu_{|n|}}{t^n}. \tag{4.84}$$

Further, these inequalities are well-defined and valid for all real $n > 0$, not just integer n.

If the moment generating function exists for $t' \in (-t_0, t_0)$ with $t_0 > 0$, then:

$$\Pr[|X| \geq t] \leq e^{-t't} \left[M(t') + M(-t') \right]. \tag{4.85}$$

If the moment generating function exists for all t, then:

$$\Pr[X \geq t] \leq M_X(t) e^{-t^2}. \tag{4.86}$$

Proof. *Left as exercises. Hint: Generalize the above proof for the first two and the last bound. For (4.85), consider $t' > 0$ and $t' < 0$ separately.* ∎

The following result is known as **Markov's inequality**, named for **Andrey Markov** (1856–1922), a student of Chebyshev.

Corollary 4.36 (Markov's inequality) *If X is a random variable defined on a probability space $(\mathcal{S}, \mathcal{E}, \lambda)$ with distribution function $F(x)$ for which $E[|X|]$ exists, then:*

$$\Pr[|X| \geq t] \leq \frac{E[|X|]}{t}. \tag{4.87}$$

Proof. *This is a restatement of (4.83) with $n = 1$.* ∎

The final result is intuitively reasonable, but the proof is subtle because we will need to revisit the transformations of Section 4.1.2.

Remark 4.37 (On $\sigma(X)$) *If X is a random variable on $(\mathcal{S}, \mathcal{E}, \lambda)$, then $X^{-1}(B) \in \mathcal{E}$ for every Borel set $B \in \mathcal{B}(\mathbb{R})$ by Exercise II.3.3, and thus $X^{-1}(\mathcal{B}(\mathbb{R})) \subset \mathcal{E}$. By Exercise II.3.44, $X^{-1}(\mathcal{B}(\mathbb{R}))$ is the smallest sigma algebra on $\mathcal{S}$ with respect to which X is measurable, and recalling Definition II.3.43, this sigma algebra is denoted $\sigma(X)$. Thus:*

$$\sigma(X) = X^{-1}(\mathcal{B}(\mathbb{R})). \tag{4.88}$$

Proposition 4.38 (Generalized Chebyshev inequalities) *Let X be a random variable defined on a probability space $(\mathcal{S}, \mathcal{E}, \lambda)$ with distribution function $F(x)$ for which the nth absolute moments exist.*

If $A \in \sigma(X)$, then (4.83) generalizes to:

$$\Pr\left[\{|X| \geq t\} \bigcap A\right] \leq \frac{E[|X|^n \chi_A]}{t^n}, \tag{4.89}$$

where $\chi_A : \mathcal{S} \to \mathbb{R}$ is the characteristic function defined as $\chi_A \equiv 1$ for $s \in A$ and is 0 otherwise.

A similar generalization holds for (4.84).

Proof. *By (4.7):*

$$E[|X|^n \chi_A] \equiv \int_{\mathcal{S}} |X(s)|^n \chi_A(s) d\lambda(s).$$

This expectation is well-defined because (4.8) is automatically satisfied by the existence of $E[|X|^n]$ and (4.11). Recalling Definition III.2.9 for the Lebesgue integral, the integral on the right can be expressed:

$$E[|X|^n \chi_A] = \int_A |X(s)|^n d\lambda(s).$$

The same change of variables result noted in (4.13), but applied to the integral over A, obtains:

$$\int_A |X(s)|^n d\lambda(s) = \int_{X(A)} |x|^n d\lambda_{F_X}$$
$$\equiv \int_{-\infty}^{\infty} |x|^n \chi_{X(A)}(x) d\lambda_{F_X},$$

where $X(A) \in \mathcal{B}(\mathbb{R})$ by definition of A.

Thus:

$$\begin{aligned}
E[|X|^n \chi_A] &\geq \int_{|x| \geq t} |x|^n \chi_{X(A)}(x) d\lambda_{F_X} \\
&\geq t^n \int_{|x| \geq t} \chi_{X(A)}(x) d\lambda_{F_X} \\
&= t^n \int_{-\infty}^{\infty} \chi_{[X(A) \cap \{|x| \geq t\}]}(x) d\lambda_{F_X} \\
&= t^n \Pr\left[\{|X| \geq t\} \bigcap A \right].
\end{aligned}$$

This last step is justified by the observation that by definition of the induced measure λ_{F_X} and (1.7):

$$\begin{aligned}
\int_{-\infty}^{\infty} \chi_{[X(A) \cap \{|x| \geq t\}]}(x) d\lambda_{F_X} &= \lambda_{F_X}\left[X(A) \bigcap \{|x| \geq t\} \right] \\
&= \lambda\left[\{|X| \geq t\} \bigcap A \right].
\end{aligned}$$

The proof for the generalization of (4.84) is similar. ∎

4.3.2 Jensen's Inequality

The notion of convexity was introduced in Corollary 4.30, and in this section, we formalize and investigate this definition. Recall that the **secant line** of a function $g(x)$ over an interval $[x, y]$ is the line segment joining the points $(x, g(x))$ and $(y, g(y))$:

- A **convex function** is one for which the secant line is above the graph of the function over this interval,

- A **concave function** is one for which the secant line is below the graph of the function on this interval.

More formally:

Definition 4.39 (Concave, convex functions) *A function $g(x)$ is **concave** on an interval $I = (a, b)$, if for any $x, y \in I$:*

$$g(tx + (1-t)y) \geq tg(x) + (1-t)g(y) \quad \text{for } t \in [0, 1], \tag{4.90}$$

*and is **convex**on I if for any $x, y \in I$:*

$$g(tx + (1-t)y) \leq tg(x) + (1-t)g(y) \quad \text{for } t \in [0, 1]. \tag{4.91}$$

*When the inequalities are strict for $t \in (0, 1)$, such functions are referred to as **strictly concave** and **strictly convex**, respectively.*

An important result from calculus, which we do not prove but can be found in Section 9.6 of **Reitano** (2010) and elsewhere, is the following.

Proposition 4.40 (Characterizing concavity/convexity) *If $g(x)$ is differentiable, then:*

1. $g(x)$ is concave if and only if $g'(x)$ is a decreasing function on I.

2. $g(x)$ is convex if and only if $g'(x)$ is an increasing function on I.

3. $g(x)$ is strictly concave if and only if $g'(x)$ is strictly decreasing on I.

4. $g(x)$ is strictly convex if and only if $g'(x)$ is strictly increasing on I.

The next proposition states that the graph of a tangent line dominates the graph of a concave function from above, and supports the graph of a convex function from below. To simplify the proof, we assume differentiability.

This result is true without this assumption, but where $g'(a)$ is replaced by a one-sided derivative related to the **Dini derivates** introduced in Definition III.3.4 and named for **Ulisse Dini** (1845–1918). When the right upper and lower Dini derivates agree, this common value is called the right derivative, and similarly for the definition of left derivative. For concave and convex functions, both of these one-sided derivatives exist at **every point**, and they agree except on an at most countable collection of points. Thus a concave or convex function $g(x)$ is differentiable, with at most countably many exceptions.

Proposition 4.41 (Tangents and concavity/convexity) *If $g(x)$ is differentiable, then for any a:*

1. *If $g(x)$ is concave on I, then for any $a \in I$:*
$$g(x) \le g(a) + g'(a)(x - a), \quad x \in I.$$

2. *If $g(x)$ is convex on I, then for any $a \in I$:*
$$g(x) \ge g(a) + g'(a)(x - a), \quad x \in I.$$

If $g(x)$ is strictly concave or strictly convex, then these inequalities are strict.

Proof. *By the mean value theorem of calculus, for any a:*
$$g(x) = g(a) + g'(\theta_x)(x - a), \tag{1}$$

where either $x < \theta_x < a$ or $a < \theta_x < x$. If $g(x)$ is concave then $g'(x)$ is a decreasing function by Proposition 4.40, and hence, $g'(\theta_x) \le g'(a)$ if $x > a$, and $g'(\theta_x) \ge g'(a)$ if $x < a$.

Thus for concave $g(x)$:
$$g'(\theta_x)(x - a) \le g'(a)(x - a).$$

Similarly, if $g(x)$ is convex:
$$g'(\theta_x)(x - a) \ge g'(a)(x - a).$$

For strictly concave or strictly convex functions, the first derivative inequalities are sharp, and so too are the inequalities in the conclusion. ∎

We now turn to an important result related to concave and convex functions known as **Jensen's inequality,** and named after **Johan Jensen** (1859–1925). We assume differentiability of $g(x)$ to be consistent with the above result, but as noted, this is not necessary.

Proposition 4.42 (Jensen's inequality) *Let $g(x)$ be a differentiable function and X a random variable defined on a probability space $(\mathcal{S}, \mathcal{E}, \lambda)$ with range contained in the domain of g, $Rng(X) \subset Dmn(g)$. Then:*

1. *If $g(x)$ is concave:*
$$E[g(X)] \le g(E[X]). \tag{4.92}$$

2. *If $g(x)$ is convex:*
$$E[g(X)] \ge g(E[X]). \tag{4.93}$$

If g is strictly concave or strictly convex, the inequalities are strict.

Proof. *Let $a = E[X]$ in the Proposition 4.41. Then since $E[g'(a)(X - a)] = g'(a)E[(X - a)] = 0$, the result follows.* ∎

4.3.3   Kolmogorov's Inequality

Andrey Kolmogorov (1903–1987) was responsible for introducing a measure-theoretic axiomatic framework for probability theory, and a large number of important results in this field bear his name. Extending Chebyshev's inequality, **Kolmogorov's inequality** addresses a collection of random variables $\{X_i\}_{i=1}^n$, and provides a probability statement regarding the maximum of the associated partial summations. See Remark 4.44.

For notational simplicity, this inequality is conventionally expressed under the assumption that $E[X_j] = 0$ for all j. However, this is not a true restriction since given $\{Y_j\}_{j=1}^n$ with $E[Y_j] = \mu_j$, this result can be applied to $X_j \equiv Y_j - \mu_j$ with no change in variance, since $Var[X_j] = Var[Y_j]$. The result in (4.94) then becomes:

$$\Pr\left\{\max_{1\le k\le n}\left|\sum_{j=1}^k (Y_j - \mu_j)\right| \ge t\right\} \le \sum_{j=1}^n \frac{\sigma_j^2}{t^2}.$$

While this result requires that $\{X_j\}_{j=1}^n$ be independent random variables, it does not require that they be "identically distributed." Consequently this result applies to samples of variates, which are independent and identically distributed (i.i.d.), and to other independent collections of random variables.

Proposition 4.43 (Kolmogorov's inequality) *Let $\{X_i\}_{i=1}^n$ be independent random variables defined on a probability space $(\mathcal{S}, \mathcal{E}, \lambda)$ with all $E[X_j] = 0$ and $Var[X_j] = \sigma_j^2$. Then for $t > 0$:*

$$\Pr\left\{\max_{1\le k\le n}\left|\sum_{j=1}^k X_j\right| \ge t\right\} \le \sum_{j=1}^n \frac{\sigma_j^2}{t^2}. \tag{4.94}$$

Proof. *With $S_k \equiv \sum_{j=1}^k X_j$, define $A_k = \{s \in \mathcal{S} \,|\, |S_k| \ge t \text{ and } |S_j| < t \text{ for } j < k\}$, letting $S_0 \equiv 0$ for A_1. Then $\{A_k\}_{k=1}^n$ are disjoint measurable sets, and $\sum_{k=1}^n \chi_{A_k}(s) \le 1$ since $s \notin \bigcup_{k=1}^n A_k$ if $|S_n(s)| < t$. Recall that $\chi_{A_k}(s)$ is the characteristic function of A_k and defined as 1 for $s \in A_k$ and 0 otherwise.*

Thus:

$$S_n^2 \ge \sum_{k=1}^n \chi_{A_k} S_n^2,$$

and by linearity and monotonicity of E from (4.7), (4.10), and (4.11):

$$E[S_n^2] \ge \sum_{k=1}^n E\left[\chi_{A_k} S_n^2\right].$$

Now $S_n^2 = S_k^2 + 2S_k(S_n - S_k) + (S_n - S_k)^2$, and so by the same steps:

$$\begin{aligned}
E[S_n^2] &\ge \sum_{k=1}^n E\left[\chi_{A_k} S_k^2\right] + 2\sum_{k=1}^n E\left[\chi_{A_k} S_k(S_n - S_k)\right] + E\sum_{k=1}^n \left[\chi_{A_k}(S_n - S_k)^2\right] \\
&\ge \sum_{k=1}^n E\left[\chi_{A_k} S_k^2\right].
\end{aligned} \tag{1}$$

To justify the last step, $E\left[\chi_{A_k}(S_n - S_k)^2\right] \ge 0$ by (4.11), and we claim that $E\left[\chi_{A_k} S_k(S_n - S_k)\right] = 0$. For this, $\chi_{A_k} S_k = \sum_{j=1}^k \chi_{A_k} X_j$ and $S_n - S_k = \sum_{j=k+1}^n X_j$ are independent by Exercise II.3.50 and Proposition II.3.56. Thus by (4.38) and the assumption that all $E[X_j] = 0$:

$$E\left[\chi_{A_k} S_k(S_n - S_k)\right] = E\left[\chi_{A_k} S_k\right] E\left[(S_n - S_k)\right] = 0.$$

Now $\chi_{A_k} S_k^2 \geq t^2 \chi_{A_k}$ *by the definition of* A_k, *and then* $E\left[\chi_{A_k} S_k^2\right] \geq t^2 E\left[\chi_{A_k}\right]$ *by (4.11) and (4.10), and so* $E\left[\chi_{A_k} S_k^2\right] \geq t^2 \lambda\left[A_k\right]$. *Thus by (1) and disjointness of* $\{A_k\}_{k=1}^n$:

$$
\begin{aligned}
E\left[S_n^2\right] &\geq t^2 \sum\nolimits_{k=1}^{n} \lambda\left[A_k\right] \\
&= t^2 \lambda\left[\bigcup\nolimits_{k=1}^{n} A_k\right] \\
&\equiv t^2 \Pr\left[\max_{1 \leq k \leq n} |S_k| \geq t\right].
\end{aligned}
$$

From the independence of $\{X_i\}_{i=1}^n$, $E\left[S_n^2\right] = \sum_{j=1}^n \sigma_j^2$ *by (4.43), and the proof is complete.* ∎

Remark 4.44 (Kolmogorov and Chebyshev) *Kolmogorov's inequality is considerably stronger than Chebyshev's inequality for this probability statement. The Chebyshev inequality would state that for any* k *with* $1 \leq k \leq n$, *since* $Var\left(\sum_{j=1}^k X_j\right) = \sum_{j=1}^k \sigma_j^2$ *as independent random variables:*

$$
\Pr\left\{\left|\sum\nolimits_{j=1}^{k} X_j\right| \geq t\right\} \leq \sum\nolimits_{j=1}^{k} \frac{\sigma_j^2}{t^2}.
$$

As $\sum_{j=1}^k \sigma_j^2/t^2 \leq \sum_{j=1}^n \sigma_j^2/t^2$ *for any* $k \leq n$, *this inequality obtains:*

$$
\max_{1 \leq k \leq n} \Pr\left\{\left|\sum\nolimits_{j=1}^{k} X_j\right| \geq t\right\} \leq \sum\nolimits_{j=1}^{n} \frac{\sigma_j^2}{t^2}.
$$

In other words, Chebyshev's inequality provides a single bound on the probabilities of these n *separate events, but is silent on the question of the simultaneous occurrence of these* n *events.*

Kolmogorov's inequality states that Chebyshev's probability bound is sufficient to bound the probability of the worst case of these n *events. Alternatively, Kolmogorov's inequality states that the Chebyshev probability bound is sufficient to bound the probability that all inequalities are satisfied simultaneously.*

4.3.4   Cauchy-Schwarz Inequality

The **Cauchy-Schwarz inequality** was originally proved by **Augustin-Louis Cauchy** (1759–1857) in 1821, in the context of the n-dimensional real Euclidean space $\mathbb{R}^n$, and generalized 25 years later to all inner product spaces by **Hermann Schwarz** (1843–1921).

To set the stage, assume that X and Y are random variables defined on a probability space $(\mathcal{S}, \mathcal{E}, \lambda)$. Then XY is a random variable by Proposition I.3.30, with expectation defined by (4.7) when (4.8) is satisfied. This latter constraint is that $E[|XY|]$ be finite, where:

$$
E[|XY|] \equiv \int_{\mathcal{S}} |X(s)Y(s)| \, d\lambda(s).
$$

However, the integrability of $|X|$ and $|Y|$ does not in general imply the integrability of $|XY|$.

Example 4.45 (Integrability of XY**)** *With* $(\mathcal{S}, \mathcal{E}, \lambda) = ([0,1], \mathcal{B}([0,1]), m)$, *the random variables* $X(s) = s^{-a}$ *and* $Y(s) = s^{-b}$ *are both integrable for* $0 < a, b < 1$, *but* $XY = s^{-(a+b)}$ *will not be integrable if* $a + b \geq 1$.

But note that if both $E[X^2]$ *and* $E[Y^2]$ *exist, it then follows that* $0 < a, b < 1/2$. *So now* $a + b < 1$ *and* XY *is integrable.*

The Cauchy-Schwarz inequality generalizes this example.

Notation 4.46 (On $|X|^2$) *For random variables on a probability space $(\mathcal{S}, \mathcal{E}, \lambda)$ with range in the real numbers, $X : \mathcal{S} \to \mathbb{R}$, it is redundant to write X^2 as $|X|^2$. However, it is standard convention in the next result to do so, since this result remains true for complex-valued random variables, for which X^2 and $|X|^2$ are in general, not equal. This notational convention is also more consistent with that in **Hölder's inequality** below, for which the absolute value is needed when the exponent 2 is replaced by p for $1 < p < \infty$.*

In Book V, it will be seen that this inequality remains valid for integrable functions defined on various measure spaces, where again, these functions may be real-valued or complex-valued.

Proposition 4.47 (Cauchy-Schwarz inequality) *Assume that $E[|X|^2]$ and $E[|Y|^2]$ exist for random variables X, Y defined on a probability space $(\mathcal{S}, \mathcal{E}, \lambda)$. Then $E[XY]$ exists and:*

$$|E[XY]| \leq \left(E[|X|^2]\right)^{1/2} \left(E[|Y|^2]\right)^{1/2}. \tag{4.95}$$

In addition, there is equality in (4.95) if and only if outside a set of λ-measure 0, $cX + dY \equiv 0$ for real $c, d \in \mathbb{R}$, not both 0.

Proof. *To simplify the notation of the proof, we represent expectations in terms of the integrals on $\mathcal{S}$ as in (4.7).*

For real t, define the function $f(t) = E\left[(tX + Y)^2\right]$:

$$f(t) = \int_{\mathcal{S}} [tX(s) + Y(s)]^2 \, d\lambda(s).$$

It follows from (4.10) and (4.11) that for all t :

$$f(t) = t^2 E[|X|^2] + 2t E[XY] + E[|Y|^2] \geq 0. \tag{1}$$

By the quadratic formula, if $a \neq 0$, then the polynomial $p(t) \equiv at^2 + bt + c \geq 0$ for all t if and only if the discriminant $b^2 - 4ac \leq 0$. This follows because if $b^2 - 4ac > 0$, then $p(t)$ has two distinct real roots at which $p'(t) \neq 0$, and thus by continuity, $p(t) < 0$ in any interval about these roots, a contradiction.

If $E[|X|^2] \neq 0$, then this discriminant inequality obtains:

$$(E[XY])^2 \leq E[|X|^2] E[|Y|^2].$$

If $E[|X|^2] = 0$, then $2t E[XY] + E[|Y|^2] \geq 0$ for all t by (1), and this is possible if and only if $E[XY] = 0$. This proves (4.95).

If $cX + dY \equiv 0$ for $c, d \neq 0$, it follows that $X = \alpha Y$ with $\alpha = -d/c$. Then (4.95) holds with equality by (4.10).

If (4.95) holds with equality, $f(t)$ in (1) can be restated:

$$f(t) = \left[t \left(E[|X|^2]\right)^{1/2} + \left(E[|Y|^2]\right)^{1/2}\right]^2 \geq 0.$$

If $E[|X|^2] \neq 0$, then $f(c) = 0$ for $c = -\left(E[|Y|^2]\right)^{1/2} / \left(E[|X|^2]\right)^{1/2}$, noting that c could be 0. In any case, $E\left[(cX + Y)^2\right] = 0$ by definition. It then follows that $cX + Y = 0$, λ-a.e. by Exercise 4.7. If $E[|X|^2] = 0$ then $X = 0$, λ-a.e., and so $X + dY \equiv 0$, λ-a.e. for $d = 0$. ∎

One may wonder from the above result whether $E[|XY|]$ exists as required by Definition 4.1, and also satisfies (4.95). The **triangle inequality** in (4.12) of Exercise 4.9:

$$|E[XY]| \leq E[|XY|],$$

creates a momentary ambiguity.

Corollary 4.48 (Cauchy-Schwarz inequality) *Under the assumptions above, $E[|XY|]$ exists and:*

$$E[|XY|] \leq \left(E[|X|^2]\right)^{1/2} \left(E[|Y|^2]\right)^{1/2}. \tag{4.96}$$

Proof. *Let $\tilde{X} = |X|$ and $\tilde{Y} = |Y|$, both random variables by Proposition I.3.47. Then $E[|\tilde{X}|^2] = E[|X|^2]$ and $E[|\tilde{Y}|^2] = E[|Y|^2]$ exist, and so by (4.95):*

$$E[\tilde{X}\tilde{Y}] \leq \left(E[|X|^2]\right)^{1/2} \left(E[|Y|^2]\right)^{1/2}.$$

The result now follows by definition. ■

Example 4.49 (Moment inequalities) *The Cauchy-Schwarz inequality provides an upper bound for the odd absolute moments of a random variable in terms of its even moments, and also provides familiar estimates for the covariance and correlation of two random variables.*

1. **Odd absolute moment bounds:**

   Let Z be a random variable defined on a probability space $(\mathcal{S}, \mathcal{E}, \lambda)$, and for integer $k \geq 0$ let $X = Z^k$ and $Y = Z^{k+1}$. Then:

   $$E[|XY|] = E[|Z|^{2k+1}] \equiv \mu'_{|2k+1|},$$

   and so by (4.96):

   $$\mu'_{|2k+1|} \leq \left(\mu'_{2k}\mu'_{2k+2}\right)^{1/2}.$$

   For central moments, let $X = (Z - \mu_Z)^k$ and $Y = (Z - \mu_Z)^{k+1}$, and this obtains:

   $$\mu_{|2k+1|} \leq \left(\mu_{2k}\mu_{2k+2}\right)^{1/2}.$$

   Using (4.12), these inequalities provide upper bounds for the absolute value of the odd moments μ'_{2k+1} and μ_{2k+1}, in terms of the respective even moments:

   $$\begin{aligned} \left|\mu'_{2k+1}\right| &\leq \left(\mu'_{2k}\mu'_{2k+2}\right)^{1/2}, \\ \left|\mu_{2k+1}\right| &\leq \left(\mu_{2k}\mu_{2k+2}\right)^{1/2}. \end{aligned} \tag{4.97}$$

2. **Covariance, correlation bounds**

   Applying the Cauchy-Schwarz inequality in (4.95) to random variables $X - \mu_X$ and $Y - \mu_Y$, we derive:

   $$-\sigma_X\sigma_Y \leq E[(X - \mu_X)(Y - \mu_Y)] \leq \sigma_X\sigma_Y. \tag{4.98}$$

Recalling Definition 4.23, this obtains:

Corollary 4.50 ($|corr(X,Y)| \leq 1$) *Given random variables* X, Y *with finite second moments, the covariance* $cov(X,Y)$ *exists and satisfies:*

$$|cov(X,Y)| \leq \sigma_X \sigma_Y.$$

Hence, the correlation $corr(X,Y)$ *is bounded:*

$$|corr(X,Y)| \leq 1. \tag{4.99}$$

Proof. *Immediate from (4.98) and Definition 4.23.* ∎

4.3.5 Hölder and Lyapunov Inequalities

Hölder's inequality significantly generalizes the **Cauchy-Schwarz inequality,** and was derived by **Otto Hölder** (1859–1937) in 1889, citing the original 1888 derivation by **Leonard James Rogers** (1862–1933). For the current proof of this inequality, we utilize **Young's inequality,** derived by **W. H. Young** (1863–1942) in 1912. W. H. Young was the father of **L. C. Young** (1905–2000), and was noted in Chapter III.4 for a general existence result on Riemann-Stieltjes integration.

From Hölder's inequality, we then derive a useful corollary for moments known as Lyapunov's inequality.

Proposition 4.51 (Young's inequality) *Given* p, q *with* $1 < p, q < \infty$ *and* $\frac{1}{p} + \frac{1}{q} = 1$, *then for all* $a, b > 0$:

$$ab \leq \frac{a^p}{p} + \frac{b^q}{q}. \tag{4.100}$$

Proof. *The function* $g(x) = \ln x$ *is concave on* $(0, \infty)$ *by Proposition 4.40 since* $g'(x) = 1/x$ *is a decreasing function.*

Applying (4.90) with $t = 1/p$:

$$\begin{aligned}
\ln(ab) &\equiv (\ln a^p)/p + (\ln b^q)/q \\
&\leq \ln\left(\frac{a^p}{p} + \frac{b^q}{q}\right),
\end{aligned}$$

and (4.100) follows by exponentiation. ∎

Remark 4.52 (Conjugate indexes) *When* p, q *satisfy* $1 < p, q < \infty$ *and* $\frac{1}{p} + \frac{1}{q} = 1$, *they are called* **conjugate indexes,** *and sometimes* **Hölder conjugate indexes.** *By defining* $\frac{1}{\infty} = 0$, *the pair* $(1, \infty)$ *is also called conjugate. Many results on conjugate indexes can be extended to this limiting pair, including* **Hölder's inequality** *below.*

To set the stage for Hölder's inequality, we generalize Example 4.45.

Example 4.53 (Integrability of XY**)** *With* $(\mathcal{S}, \mathcal{E}, \lambda) = ([0,1], \mathcal{B}([0,1]), m)$, *the random variables* $X(s) = s^{-a}$ *and* $Y(s) = s^{-b}$ *are both integrable for* $0 < a, b < 1$, *but* $XY = s^{-(a+b)}$ *will not be integrable if* $a + b \geq 1$. *However, if both* $E[X^p]$ *and* $E[Y^q]$ *exist with* $1 < p, q < \infty$ *and* $\frac{1}{p} + \frac{1}{q} = 1$, *then it follows that* $0 < a < 1/p$ *and* $0 < b < 1/q$. *In this case,* $a + b < \frac{1}{p} + \frac{1}{q} = 1$, *and* XY *is integrable.*

We now prove Hölder's inequality, and note that this reduces to the **Cauchy-Schwarz inequality** of (4.96) with $p = q = 2$.

Proposition 4.54 (Hölder's inequality) *Given p, q with $1 < p, q < \infty$ and $1/p + 1/q = 1$, assume that $E[|X|^p] < \infty$ and $E[|Y|^q] < \infty$ for random variables X, Y defined on a probability space $(\mathcal{S}, \mathcal{E}, \lambda)$.*

Then $E[|XY|] < \infty$ and:

$$E[|XY|] \le \left(E[|X|^p]\right)^{1/p} \left(E[|Y|^q]\right)^{1/q}. \tag{4.101}$$

For $p = 1$, if $E[|X|] < \infty$ and $\sup[|Y|] < \infty$, then $E[|XY|] < \infty$ and:

$$E[|XY|] \le E[|X|] \sup[|Y|]. \tag{4.102}$$

Proof. *If either or both $E[|X|^p] = 0$ and $E[|Y|^q] = 0$, then either or both $X = 0$ and $Y = 0$ outside a set of λ-measure 0 by Exercise 4.7. Thus $XY = 0$ outside such a set, $E[|XY|] = 0$ by Exercise 4.8, and (4.101) follows in these cases.*

Thus assuming that $E[|X|^p] \ne 0$ and $E[|Y|^q] \ne 0$, apply Young's inequality with $a = |X| / (E[|X|^p])^{1/p}$ and $b = |Y| / (E[|Y|^q])^{1/q}$:

$$\frac{|XY|}{(E[|X|^p])^{1/p} (E[|Y|^q])^{1/q}} \le \frac{1}{p} \frac{|X|^p}{E[|X|^p]} + \frac{1}{q} \frac{|Y|^q}{E[|Y|^q]}.$$

The existence of $E[|X|^p]$ and $E[|Y|^q]$ then assures the existence of $E[|XY|]$ by (4.11). Taking expectations, the right-hand side reduces to 1 for conjugate indexes, and (4.101) again follows.

If $p = 1$ and $q = \infty$, then from $|XY| \le |X| \sup |Y|$ we obtain by (4.11) and (4.10):

$$E[|XY|] \le \sup |Y| \, E[|X|].$$

∎

Remark 4.55 *Two comments on Hölder's inequality:*

1. *When $p = 1$, defining $q = \infty$ in not merely a logical necessity to retain the identity $1/p + 1/q = 1$. It will be seen in Book V in the study of L_p-spaces that if $\sup[|Y|] < \infty$, then $(E[|Y|^q])^{1/q} \to \sup[|Y|]$ as $q \to \infty$. Thus (4.102) generalizes (4.101) as a limiting result.*

2. *Recalling the **triangle inequality** of (4.12), Hölder's inequality also provides an upper bound for $|E[XY]|$ since:*

$$\left| \int_{\mathcal{S}} X(s) d\lambda(s) \right| \le \int_{\mathcal{S}} |X(s)| \, d\lambda(s).$$

This obtains $|E[XY]| \le E[|XY|]$, and Hölder's inequality can be stated:

$$|E[XY]| \le \left(E[|X|^p]\right)^{1/p} \left(E[|Y|^q]\right)^{1/q}. \tag{4.103}$$

There is an important corollary result on moments which is easily proved using Hölder's inequality, called **Lyapunov's inequality**. This result is named for **Aleksandr Lyapunov** (1857–1918) and provides a lower bound on the growth rate of moments.

Corollary 4.56 (Lyapunov's inequality) *Given a random variable X on a probability space $(\mathcal{S}, \mathcal{E}, \lambda)$, then for $0 < \alpha < \beta$ and assuming all moments exist:*

$$\left(E[|X|^{\alpha}]\right)^{1/\alpha} \le \left(E[|X|^{\beta}]\right)^{1/\beta}. \tag{4.104}$$

Proof. Let $p = \beta/\alpha$, then apply Hölder's inequality to $|X|^\alpha$ and $Y \equiv 1$. That $E[Y] = 1$ follows from (4.7) since $\mathcal{S}$ is a probability space. ∎

Example 4.57 (Moment inequalities) *Lyapunov's inequality implies that absolute moments of a random variable X must grow at least geometrically, when such moments exist.*

1. $\left(\mu'_{|m|}\right)^{1/m}$ and $\left(\mu_{|m|}\right)^{1/m}$ *increase with m.*

2. *For $m < n$:*
$$\left(\mu'_{|m|}\right)^{n/m} \leq \mu'_{|n|}.$$

3. *For $m = 1$ in item 2:*
$$\left(\mu'_{|1|}\right)^{n} \leq \mu'_{|n|}.$$

4. *Items 2 and 3 apply to central moments.*

5. *If the random variable $X : \mathcal{S} \to [0, \infty)$ is nonnegative, then items 2 and 3 apply to μ'_m.*

4.4 Uniqueness of Moments

It is often useful and always interesting to know if a distribution function is uniquely determined by its moments, $\{\mu'_n\}_{n=1}^{\infty}$ or $\{\mu_n\}_{n=1}^{\infty}$, assuming all such moments exist. Since (4.49) provides a unique mapping between $\{\mu'_n\}_{n=1}^{\infty}$ and $\{\mu_n\}_{n=1}^{\infty}$, the answer to this question is independent of which set of moments is used.

For example, is the normal distribution the only distribution for which central moments are given by (4.77) as:
$$\mu_{2m} = \frac{\sigma^{2m}(2m)!}{2^m m!}, \qquad \mu_{2m+1} = 0?$$

Is the lognormal distribution the only such distribution for which moments are given by (4.80) as:
$$\mu'_{nL} = e^{n\mu + (n\sigma)^2/2}?$$

It turns out that not all moment collections uniquely identify the underlying distribution functions. As will be seen, the moment collection for the normal distribution is unique to the normal, but there are infinitely many distribution functions with moments equal to those of the lognormal distribution.

The following example was introduced by **C. C. Heyde** (1939–2008) in 1963.

Example 4.58 (Heyde) *The lognormal distribution in (1.67) is defined with $\mu = 0$ and $\sigma = 1$:*
$$f_L(x) = \frac{1}{x\sqrt{2\pi}} \exp\left(-\frac{1}{2}(\ln x)^2\right), \qquad x \geq 0.$$

Now define for $-1 \leq \alpha \leq 1$:
$$f_\alpha(x) = f_L(x)\left[1 + \alpha \sin(2\pi \ln x)\right], \qquad x \geq 0.$$

Then $f_0(x) = f_L(x)$, and $f_\alpha(x) \geq 0$ on $x \geq 0$ because $\sin x \geq -1$.

Further, $f_\alpha(x)$ is a density function for all such α since by a substitution:

$$
\begin{aligned}
\int_0^\infty f_\alpha(x)dx &= (2\pi)^{-1/2} \int_0^\infty \exp\left(-\frac{1}{2}(\ln x)^2\right)[1 + \alpha\sin(2\pi\ln x)]\frac{dx}{x} \\
&= (2\pi)^{-1/2} \int_{-\infty}^\infty \exp\left(-\frac{1}{2}y^2\right)[1 + \alpha\sin(2\pi y)]\,dy \\
&= (2\pi)^{-1/2} \int_{-\infty}^\infty \exp\left(-\frac{1}{2}y^2\right)dy \\
&\quad + (2\pi)^{-1/2}\alpha \int_{-\infty}^\infty \exp\left(-\frac{1}{2}y^2\right)\sin(2\pi y)dy.
\end{aligned}
$$

*The first integral is equal to 1 as this is the normal density of (1.66). The second integral is well defined since $|\sin(2\pi y)| \le 1$, and equals 0 by symmetry since the integrand $g(y) \equiv \exp\left(-y^2/2\right)\sin(2\pi y)$ is an **odd function,** meaning that $g(-y) = -g(y)$.*

To see that $f_\alpha(x)$ has the same moments as $f_L(x)$ by (4.1), we show that for $n = 1, 2, \dots$:

$$
\int_0^\infty x^n dF_L(x) = \int_0^\infty x^n dF_\alpha(x).
$$

Applying item 5 of Proposition III.4.24 and then Proposition III.4.28:

$$
\int_0^\infty x^n dF_\alpha(x) - \int_0^\infty x^n dF_L(x) = \int_0^\infty x^n f_L(x)\sin(2\pi\ln x)dx,
$$

so we prove that for all such n:

$$
I_n \equiv \int_0^\infty x^n f_L(x)\sin(2\pi\ln x)dx = 0.
$$

To this end, making the substitution $x = \exp(y+n)$, and noting that $\sin x = \sin(x+2\pi n)$ for any integer n produces:

$$
\begin{aligned}
I_n &= \int_{-\infty}^\infty \exp(yn + n^2)\exp\left(-(y+n)^2/2\right)\sin(2\pi(y+n))dy \\
&= \exp(n^2/2)\int_{-\infty}^\infty \exp\left(-y^2/2\right)\sin(2\pi y)dy \\
&= 0.
\end{aligned}
$$

*This again follows because the integrand $g(y) = \exp\left(-y^2/2\right)\sin(2\pi y)$ is absolutely integrable by boundedness of $\sin(2\pi y)$, and is an **odd function.***

Hence for any such α, $f_\alpha(x)$ is a density function with the same moments as the lognormal $f_L(x)$.

With the above example in hand, we now begin an investigation toward a positive result on uniqueness of moments. The next proposition states that given the distribution function $F(x)$, if all moments exist and the power series $\sum_{n=0}^\infty \mu'_n t^n/n!$ converges for $t \in (-t_0, t_0)$ with $t_0 > 0$, then $M_F(t)$ exists. Thus by Proposition 4.27, $M_F(t)$ is given by this power series on this interval.

Note that given the next result, we will not be able to conclude that $F(x)$ is the **only** distribution function with these moments. It could well be possible that there exists a second distribution function $G(x)$ with $M_F(t) = M_G(t)$ on $(-t_0, t_0)$ and thus, $F(x)$ and $G(x)$ have the same moments yet $F(x) \ne G(x)$. See Proposition 4.61 and Corollary 4.62.

Proposition 4.59 (Existence of $M_F(t)$) *Given the distribution function $F(x)$, assume that μ'_n exists for all n and that the power series $\sum_{n=0}^{\infty} \mu'_n t^n / n!$ converges absolutely on $(-t_0, t_0)$ with $t_0 > 0$.*

Then the moment generating function $M_F(t)$ of $F(x)$ exists on $(-t_0, t_0)$, and is given by this series.

Proof. *For $t \in (-t_0, t_0)$, the triangle inequality of item 3 of Proposition III.4.24, and then the triangle inequality for sums obtains:*

$$\left| \int_{-\infty}^{\infty} \sum_{j=0}^{n} \frac{(tx)^j}{j!} dF(x) \right| \leq \sum_{j=0}^{n} \int_{-\infty}^{\infty} \left| \frac{(tx)^j}{j!} \right| dF(x) \equiv \sum_{j=0}^{n} \frac{|t|^j}{j!} \mu'_{|j|}, \tag{1}$$

where $\mu'_{|j|}$ denotes the absolute jth moment, $\mu'_{|j|} = E\left[|X|^j \right]$.

We show below that the upper summation in (1) is bounded for all n for any such t. Assuming this, and applying integral properties from Proposition III.4.24:

$$\left| \int_{-\infty}^{\infty} e^{tx} dF(x) - \sum_{j=0}^{n} \frac{\mu'_j t^j}{j!} \right| = \left| \int_{-\infty}^{\infty} e^{tx} dF(x) - \int_{-\infty}^{\infty} \sum_{j=0}^{n} \frac{(tx)^j}{j!} dF(x) \right|$$

$$\leq \int_{-\infty}^{\infty} \left| \sum_{j=n+1}^{\infty} \frac{(tx)^j}{j!} \right| dF(x)$$

$$\leq \sum_{j=n+1}^{\infty} \frac{|t|^j \mu'_{|j|}}{j!}. \tag{2}$$

Thus if the upper summation in (1) bounded for all n, then the upper bound in (2) can be made arbitrarily small for n large. Hence $M_F(t) \equiv \int_{\mathbb{R}} e^{tx} d\mu_F(x)$ exists and is given by this series.

To prove boundedness in (1), by item 1 of Example 4.49:

$$\mu'_{|2n+1|} \leq \left(\mu'_{2n} \mu'_{2n+2} \right)^{1/2} \leq \max[\mu'_{2n}, \mu'_{2n+2}] \equiv \mu'_{2n+}.$$

In other words, the index $2n^+$ is defined:

$$2n^+ \equiv \begin{cases} 2n, & \mu'_{2n} \geq \mu'_{2n+2}, \\ 2n+2, & \mu'_{2n} < \mu'_{2n+2}. \end{cases}$$

Given $t \in (-t_0, t_0)$, let $s = \lambda t$ where $\lambda < 1$. We can assume that $t \neq 0$ since the existence of $M_F(0)$ is always assured. Then:

$$\frac{\mu'_{|2n+1|} |s|^{2n+1}}{(2n+1)!} \leq c_n \frac{\mu'_{2n+} |t|^{2n^+}}{(2n^+)!},$$

where

$$c_n \equiv \begin{cases} |t| \lambda^{2n+1}/(2n+1), & 2n^+ = 2n, \\ (2n+2) \lambda^{2n+1}/|t|, & 2n^+ = 2n+2. \end{cases}$$

In either case, $c_n \to 0$ as $n \to \infty$.

Now for each n, define:

$$d_n^- \equiv \begin{cases} c_n, & 2n^+ = 2n, \\ 0, & 2n^+ = 2n+2. \end{cases} \qquad d_n^+ \equiv \begin{cases} 0, & 2n^+ = 2n, \\ c_n, & 2n^+ = 2n+2. \end{cases}$$

Then since $\mu'_{|2n|} = \mu'_{2n}$ *by definition, splitting the summation in* (1) *into even and odd indexes obtains:*

$$\sum_{j=0}^{2n} \frac{|s|^j \, \mu'_{|j|}}{j!} \leq \sum_{j=0}^{n} \frac{|t|^{2j} \, \mu'_{2j} \left(1 + d^-_{j+1} + d^+_j\right)}{(2j)!}.$$

This partial sum in s *converges as* $n \to \infty$ *for* $|s| < t$ *because* $d^-_{j+1} + d^+_j \to 0$ *and* $\sum_{j=0}^{\infty} t^j \mu'_j / (j)!$ *is absolutely convergent for* $|t| < t_0$ *by assumption.* ∎

The following result provides an alternative test for the existence of $M_F(t)$ given $\{\mu'_n\}_{n=1}^{\infty}$, stated in terms of a growth bound on even moments.

Corollary 4.60 (Existence of $M_F(t)$) *Given the distribution function* $F(x)$, *assume that* μ'_n *exists for all* n *and that:*

$$\limsup \frac{(\mu'_{2n})^{1/2n}}{2n} = r < \infty.$$

Then $M_F(t)$ *exists for* $|t| < 1/r$, *and by Proposition 4.27 is given on this interval by the series in* (4.53).
Proof. *Recall that the limit superior of a sequence* $\{a_n\}_{n=1}^{\infty}$ *is defined:*

$$\limsup a_n \equiv \inf_n \left\{ \sup_{m \geq n} a_m \right\} = \lim_{n \to \infty} \left\{ \sup_{m \geq n} a_m \right\},$$

where the final equality follows since $\{\sup_{m \geq n} a_m\}$ *is decreasing with* n. *See Section I.3.4.2 for more on* $\limsup$.
The assumed bound above then implies that for all but at most finitely many n:

$$\mu'_{2n} \leq (2nr)^{2n}.$$

By (4.97), *this obtains that for all but at most finitely many* n:

$$\begin{aligned} |\mu'_{2n+1}| &\leq \left(\mu'_{2n} \mu'_{2n+2}\right)^{1/2} \\ &\leq \left(2 \, (n+1) \, r\right)^{2n+1}. \end{aligned}$$

Hence in all cases of even or odd m, *with at most finitely many exceptions:*

$$|\mu'_m| \leq ((m+1)\, r)^m,$$

and so:

$$\begin{aligned} \sum_{m=0}^{\infty} \frac{|\mu'_m| \, t^m}{m!} &\leq \sum_{m=0}^{\infty} \frac{(t \, (m+1)\, r)^m}{m!} \\ &= \sum_{m=0}^{\infty} (tr)^m \frac{(m+1)^m}{m!}. \end{aligned}$$

Using the ratio test noted in Example 4.32, it follows that this series is convergent if $|t| < 1/r$, *and the result follows from Proposition 4.59.* ∎

Returning to the investigation of this section, we now know that the existence of all moments, and then either a certain growth bound or the convergence of an associated power series, assures the existence of a moment generating function. So now the question might be reframed:

Does the existence of a moment generating function assure the uniqueness of moments?
In other words, can two different distribution functions have the same moments and moment generating functions? While Example 4.58 showed that different distribution functions can have the same moments, it must be noted that none of those distribution functions have a moment generating function by Example 4.32.

The following proposition settles this question and will be proved in Book VI using properties of the **characteristic function of** $F(x)$. It provides a key test for when a moment collection uniquely defines a distribution function. Corollary 4.62 provides another test that uses this result.

Proposition 4.61 (Uniqueness of Moments) *Given the distribution function $F(x)$, assume that μ'_n exists for all n and that $\sum_{n=0}^{\infty} \mu'_n t^n / n!$ converges absolutely on $(-t_0, t_0)$ with $t_0 > 0$.*

Then $F(x)$ is the only distribution function with these moments.

Proof. *See the section on uniqueness of moments in the Book VI chapter on the characteristic function.* ∎

The following corollary provides a uniqueness result that complements the results of Propositions 4.25 and 4.27. These stated:

1. If $M_X(t)$ exists for $t \in (-t_0, t_0)$ with $t_0 > 0$, then so too do all moments $\{\mu'_n\}_{n=1}^{\infty}$ of X;

2. If $M_X(t)$ exists for $t \in (-t_0, t_0)$ with $t_0 > 0$, then:

$$M_X(t) = \sum_{n=0}^{\infty} \frac{\mu'_n t^n}{n!},$$

and thus $\mu'_n = M_X^{(n)}(0)$.

When $M_X(t)$ exists, it uniquely identifies the underlying distribution function of X, and uniquely identifies the moment collection $\{\mu'_n\}_{n=1}^{\infty}$ associated with this distribution function.

Corollary 4.62 (Uniqueness of $M_F(t)$) *Given the distribution function $F(x)$, assume that $M_F(t) \equiv \int_{-\infty}^{\infty} e^{tx} dF$ exists on $(-t_0, t_0)$ with $t_0 > 0$.*

Then $F(x)$ is the only distribution function with this moment generating function, and the only distribution function with the associated moments..

Proof. *If $F(x)$ has moment generating function $M_F(t)$, which converges on the interval $(-t_0, t_0)$ with $t_0 > 0$, then by Propositions 4.25 and 4.27, $F(x)$ has moments of all orders defined by $\mu'_n = M^{(n)}(0)$, and $M_F(t)$ is given by the power series $\sum_{n=0}^{\infty} \mu'_n t^n / n!$ on $(-t_0, t_0)$. Thus by Proposition 4.61, $F(x)$ is the only distribution function with these moments.*

If $G(x)$ is another distribution function with moment generating function $M_F(t)$ convergent on $(-t'_0, t'_0)$ with $t'_0 > 0$, then by the same argument $G(x)$ has the same moments and convergent power series as $F(x)$. This contradicts Proposition 4.61, and hence $F(x)$ is the only distribution function with this moment generating function. ∎

4.4.1 Applications of Moment Uniqueness

In this section, we apply the above uniqueness of moments results to draw conclusions about various distributional relationships and confirm various statements made in previous sections.

1. **Normal vs. Lognormal:**

   We can now investigate the statements in the introduction to this section concerning the normal and lognormal distributions.

   (a) *The **normal distribution** is uniquely defined by the moments,* $\mu_{2m} = \sigma^{2m}(2m)!/(2^m m!)$ *and* $\mu_{2m+1} = 0$.

   Since the moment generating function $M_N(t)$ exists for all t by (4.78), this distribution function is uniquely determined by these moments by Corollary 4.62.

   Alternatively, it can be checked as an exercise that the even moments satisfy the growth bound in Corollary 4.60 using **Stirling's formula.** Stirling's formula, also known as **Stirling's approximation**, is named for **James Stirling** (1692–1770) and states that as $n \to \infty$:

   $$\frac{n!}{\sqrt{2\pi} n^{n+1/2} e^{-n}} \approx e^{1/12n} \to 1. \tag{4.105}$$

   See for example **Reitano** (2010).

   b. *The **lognormal distribution** is not uniquely defined by the moments* $\mu'_n = e^{n\mu + (n\sigma)^2/2}$.

   As demonstrated in Example 4.32, these moments do not provide a power series $\sum_{n=0}^{\infty} \mu'_n t^n/n!$ that is convergent on an interval $(-t_0, t_0)$ for any $t_0 > 0$. As logically expected, this implies the failure of the even moments test of Corollary 4.60:

   $$\frac{\left(\mu'_{2n}\right)^{1/2n}}{2n} = \frac{e^{\mu + n\sigma^2}}{2n},$$

   which is unbounded in n.

   While this obtains an inconclusive result for confirming if there are other distributions with these moments, the proof of nonuniqueness is obtained with Example 4.58 of **C. C. Heyde**.

2. **Sum of geometric is negative binomial:**

   It was noted in Remark 4.31 that for the geometric distribution with $(1-p)e^t < 1$, or equivalently for $t < -\ln(1-p)$, that $M_G(t) = p/[1-(1-p)e^t]$ by (4.64). Similarly, for the negative binomial and the same range of t, that $M_{NB}(t) = (p/[1-(1-p)e^t])^k$ by (4.66). This latter moment generating function is also recognized to be the moment generating function of a sum of k independent geometric random variables by (4.48).

   By Corollary 4.62, the negative binomial is the only distribution with this moment generating function. Thus a sum of independent geometric random variables is a negative binomial random variable in distribution.

3. **Sum of gamma is gamma:**

   It was demonstrated in Example 2.14 and Exercise 2.15 that a sum of independent exponentials is gamma and then noted in that exercise that the sum of independent gammas with common λ is also gamma.

   To see this, let $\{X_i\}_{i=1}^n$ be independent gammas with parameters λ and $\{\alpha_i\}_{i=1}^n$. By (4.73), $M_i(t) = (1 - t/\lambda)^{-\alpha_i}$ for $t < \lambda$, and thus by (4.48), the moment generating function of $X \equiv \sum_{i=1}^n X_i$ is $M(t) = (1 - t/\lambda)^{-\alpha}$ for $t < \lambda$ with $\alpha \equiv \sum_{i=1}^n \alpha_i$.

   By Corollary 4.62, the gamma is the only distribution with this moment generating function. Thus for n independent gammas $\{X_i\}_{i=1}^n$ with common parameter λ and parameters $\{\alpha_i\}_{i=1}^n$, the sum $\sum_{i=1}^n X_i$ is gamma with parameters λ and $\sum_{i=1}^n \alpha_i$.

4. Sum of independent normals is normal:

As noted in Remark 2.19, a sum of independent normal variates $\{X_i\}_{i=1}^n$ with parameters $\{\mu_i\}_{i=1}^n$ and $\{\sigma_i^2\}_{i=1}^n$ is normal with parameters $\mu = \sum_{i=1}^n \mu_i$ and $\sigma^2 = \sum_{i=1}^n \sigma_i^2$.

By (4.78) and (4.48), if $X = \sum_{i=1}^n X_i$, then for $t \in \mathbb{R}$:

$$
\begin{aligned}
M_X(t) &= \prod_{i=1}^n \exp\left(\mu_i t + \frac{1}{2}\sigma_i^2 t^2\right) \\
&= \exp\left(\sum_{i=1}^n \mu_i t + \frac{1}{2}\sum_{i=1}^n \sigma_i^2 t^2\right).
\end{aligned}
$$

Thus X has the moment generating function of a normal variate with parameters μ and σ^2 as above. By Corollary 4.62, only the normal variate has this moment generating function, and the result is proved.

5. Sum of squared normals is chi-squared:

Recall Remark 1.30, that when a random variable Y has a gamma distribution with $\lambda = 1/2$ and $\alpha = n/2$, it is said to have a **chi-squared distribution with n degrees of freedom,** which is often denoted $\chi_{n\ d.f.}^2$.

Proved in Example 2.4, if X is standard normal, then X^2 is chi-squared with 1 degree of freedom. If $\{X_i\}_{i=1}^n$ are independent standard normals, then $\{X_i^2\}_{i=1}^n$ are independent chi-squared by Proposition II.3.56. Thus as an application of item 3, as a sum of independent gammas with common parameters $\lambda = 1/2$ and $\alpha = 1/2$, $\sum_{i=1}^n X_i^2$ is gamma with $\lambda = 1/2$ and $\alpha = n/2$.

That is, $\sum_{i=1}^n X_i^2$ is $\chi_{n\ d.f.}^2$ as noted in Remark 2.5.

6. Student T construction in Example 2.29:

Recall Example 2.29 on Student's T distribution. Simplifying notation, it was concluded that there are random variables A, B, and C with $A + B = C$, where A and B are independent and both A and C are chi-squared with 1 and n degrees of freedom, respectively.

The claim there was that this assured that B is also Chi-squared, and with $n - 1$ degrees of freedom.

To prove this, the moment generating functions of A and C are given in (4.73) with $\lambda_A = \lambda_C = 1/2$, $\alpha_A = 1/2$, and $\alpha_C = n/2$. Assuming that B has a moment generating function, an application of (4.48) and Corollary 4.62 then obtains that B is also Chi-squared with $n - 1$ degrees of freedom.

This assumption on B is verified by noting that the moments of A and B sum to the moments of C, and thus, the convergence of the moment series for A and C assures the convergence of the moment series for B. Thus by Proposition 4.59, the moment generating function for B exists.

4.5 Weak Convergence and Moment Limits

In this section, we investigate the relationships between weak convergence of distribution functions and the convergence of the associated moment series, or, the convergence of associated moment generating functions. In the Book VI study of weak convergence and expectations, these results will be generalized with the aid of the integration theory of Book V.

Assume that a distribution function $F(x)$ is uniquely determined by its moment collection $\{\mu'_n\}_{n=1}^{\infty}$, and that there exists a sequence of distribution functions $\{F_m\}_{m=1}^{\infty}$ with moment collections $\{\mu'_{m,n}\}_{n=1}^{\infty}$ where $\mu'_{m,n} \to \mu'_n$ for each n. Are we then able to conclude that F_m **converges weakly to the distribution function** F, denoted $F_m \Rightarrow F$? Recalling Definition II.8.2, this means that $F_m(x) \to F(x)$ for every continuity point of F.

As discussed in Section II.8.1, the notion $F_m \Rightarrow F$ can be equivalently stated in other ways:

1. Given random variables $\{X_m\}_{m=1}^{\infty}$ associated with $\{F_m\}_{m=1}^{\infty}$ and X associated with F (recall Proposition II.3.6):

   $F_m \Rightarrow F$ is equivalent to $X_m \Rightarrow X$, meaning that X_m **converges in distribution to the random variable** X.

2. Given the Borel measures $\{\lambda_{F_m}\}_{m=1}^{\infty}$ induced by $\{F_m\}_{m=1}^{\infty}$ and λ_F induced by F (recall Proposition 1.5):

   $F_m \Rightarrow F$ is equivalent to $\lambda_{F_m} \Rightarrow \lambda_F$, meaning that λ_{F_m} **converges weakly to** λ_F.

In probability theory, the **method of moments** is:

- The name often given to the framework within which one can assert that $F_m \Rightarrow F$ by demonstrating that that $\mu'_{m,n} \to \mu'_n$ for each n, or conversely.

- The name given to the process whereby one calculates the moments of a random sample to estimate the parameters of an assumed underlying distribution function. This is accomplished by equating the distribution's parametric moment formulas to the numerical values calculated from the sample and solving.

The focus of this section is on method of moments in the first sense.

Any method of moments in this sense must require that the distribution function F be uniquely determined by its moment collection $\{\mu'_n\}_{n=1}^{\infty}$. For example, assume that $\{F_m\}_{m=1}^{\infty}$ is given and $\mu'_{m,n} \to e^{n\mu+(n\sigma)^2/2}$ for each n. Even though these limits are recognizable as the moments of the lognormal distribution, it was shown in Example 4.58 that these are also the moments of an infinite number of other distribution functions. So certainly there can be no useful statement concerning $F_m \Rightarrow F$.

The main result of this section is in Proposition 4.76 which states that in the case where F is uniquely defined by its moments, if $\mu'_{m,n} \to \mu'_n$ for each n, then $F_m \Rightarrow F$. Proposition 4.77 provides the same conclusion based on convergence of moment generating functions.

For these results, we will first need positive results in the opposite direction. For example, if $F_m \Rightarrow F$ with moment collections $\{\{\mu'_{m,n}\}_{n=1}^{\infty}\}_{m=1}^{\infty}$ and $\{\mu'_n\}_{n=1}^{\infty}$, must it be the case that $\mu'_{m,n} \to \mu'_n$ for each n? Perhaps surprisingly, the answer is in the negative as illustrated in Example 4.64 and subsequent discussion.

Recall Definition II.8.16 on the notion of **tightness** of a family of measures $\{\lambda_m\}_{m=1}^{\infty}$ or distribution functions $\{F_m\}_{m=1}^{\infty}$. The equivalence of definitions in the latter case was addressed in Remark II.8.17 and is here left as an optional exercise.

Definition 4.63 (Tight sequence: $\{\lambda_m\}_{m=1}^{\infty}$ or $\{F_m\}_{m=1}^{\infty}$) *A sequence of probability measures $\{\lambda_m\}_{m=1}^{\infty}$ is said to be **tight** if for any $\epsilon > 0$ there is a finite interval $(a, b]$ so that $\lambda_m((a, b]) > 1 - \epsilon$ for all m.*

*A sequence of distribution functions $\{F_m\}_{m=1}^{\infty}$ is said to be **tight** if for any $\epsilon > 0$ there is a finite interval $(a, b]$ so that $F_m(b) - F_m(a) > 1 - \epsilon$ for all m, or equivalently $F_m(b) > 1 - \epsilon$ and $F_m(a) < \epsilon$ for all m.*

Example 4.64 (Tightness and moment sequences) *We provide an example of a sequence of distribution functions that is not tight, then one that is tight, and see that in neither case does weak convergence imply the convergence of moment sequences. Hence, a property stronger than tightness will be needed for a positive conclusion in the method of moments.*

1. $\{F_m\}_{m=1}^{\infty}$ *is not tight:*

   Define discrete density functions $\{f_m\}_{m=1}^{\infty}$ by $f_m(m) = 1$, $f_m(x) = 0$ for $x \neq m$, and thus $F_m(x) = \chi_{[m,\infty)}(x)$:

   $$F_m(x) = \begin{cases} 0, & x < m, \\ 1, & m \leq x. \end{cases}$$

   Now, $F_m(x) \to F(x) \equiv 0$ for all x, but by definition it does not follow that $F_m \Rightarrow F$ since F is not even a distribution function. That F is not a distribution function is consistent with the result of Exercise II.8.21, since the associated distribution functions $\{F_m\}_{m=1}^{\infty}$ are not tight.

   In this case $\mu'_{m,n} \not\to \mu'_n$ for any n, since $\mu'_{m,n} = m^n$ for all n and $\mu'_n = 0$.

   Conjecture: *It seems natural to hypothesize that if $F_m \Rightarrow F$ with F a distribution function, that we will obtain the positive result that $\mu'_{m,n} \to \mu'_n$ for all n. By Proposition II.8.18, if F is a distribution function and $F_m \Rightarrow F$, then $\{F_m\}_{m=1}^{\infty}$ must be tight, so we next look at such an example.*

2. $\{F_m\}_{m=1}^{\infty}$ *is tight:*

   Define discrete density functions $\{f_m\}_{m=1}^{\infty}$ by $f_m(0) = 1 - 1/m$, $f_m(m) = 1/m$, and $f_m(x) = 0$ for $x \neq 0, m$. Then:

   $$F_m(x) = \begin{cases} 0, & x < 0, \\ 1 - 1/m, & 0 \leq x < m, \\ 1, & m \leq x. \end{cases}$$

   This collection of distribution functions $\{F_m\}_{m=1}^{\infty}$ is tight. Given $\epsilon > 0$, choose $1/b < \epsilon$ and $a < 0$, and then $F_m(b) - F_m(a) > 1 - \epsilon$ for all m.

   Further, $F_m \Rightarrow F$ with distribution function $F(x) = \chi_{[0,\infty)}(x)$. However, $\mu'_{m,n} = m^{n-1}$ and $\mu'_n = 0$, and so it is again the case that $\mu'_{m,n} \not\to \mu'_n$ for any n.

   Conclusion: *If $F_m \Rightarrow F$, an additional restriction is needed on the sequence $\{F_m\}_{m=1}^{\infty}$ beyond tightness to assure convergence of moments.*

   Before proceeding to a solution, we investigate the second example further, and demonstrate why the desired conclusion was not achieved even with tightness of the sequence $\{F_m\}_{m=1}^{\infty}$.

Example 4.65 (First analysis of Example 4.64: $\{F_m\}_{m=1}^{\infty}$ is tight) *Assume that $\{F_m\}_{m=1}^{\infty}$ is tight sequence of distribution functions and $F_m \Rightarrow F$ for a distribution function F. By **Skorokhod's representation theorem** of Proposition II.8.30, we can define random variables $\{X_m\}_{m=1}^{\infty}$ and X on the Lebesgue measure space $((0,1), \mathcal{B}(0,1), m)$ with these distribution functions, and for which $X_m \to X$ for all $t \in (0,1)$. As in (4.7), but now as **Lebesgue integrals**:*

$$\mu'_{m,n} = \int_0^1 X_m^n(t)\,dm_L, \qquad \mu'_n = \int_0^1 X^n(t)\,dm_L, \tag{1}$$

denoting $dm = dm_L$ for notational clarity.

In the second example above, an application of the Skorokhod construction yields that $X_m(t) \equiv 0$ for $t \in (0, 1 - 1/m]$ and $X_m(t) \equiv m$ for $t \in (1 - 1/m, 1)$, while $X(t) \equiv 0$. As assured by the Skorokhod result, $X_m \to X$ for all $t \in (0, 1)$. Not surprisingly, $\mu'_{m,n} = m^{n-1}$ and $\mu'_n = 0$ as before, but now calculated as in (1).

The assumption that $\{F_m\}_{m=1}^{\infty}$ is tight implies that for any $\epsilon > 0$ there is an interval, say $(-N_\epsilon, N_\epsilon]$, such that for all m :

$$F_m(N_\epsilon) - F_m(-N_\epsilon) \geq 1 - \epsilon.$$

By definition of the distribution function, that $F_m(x) \equiv m_L \left[X_m^{-1}(-\infty, x] \right]$, this obtains:

$$m_L \left[X_m^{-1}(-N_\epsilon, N_\epsilon] \right] \geq 1 - \epsilon.$$

Hence:

$$m_L \left[X_m^{-1} \left[(-\infty, -N_\epsilon] \bigcup (N_\epsilon, \infty) \right] \right] \leq \epsilon. \tag{2}$$

We can replace $(-N_\epsilon, N_\epsilon]$ by a slightly larger open interval for notational simplicity, and then have using the same notation:

$$\mu'_{m,n} = \int_{|X_m| < N_\epsilon} X_m^n(t) dm_L + \int_{|X_m| \geq N_\epsilon} X_m^n(t) dm_L. \tag{3}$$

As will be seen in the proof of Proposition 4.72, the first integral will convergence as $m \to \infty$ by the bounded convergence theorem of Proposition III.2.22. This result applies because for any N_ϵ, $m_L \left[|X_m| < N_\epsilon \right] \leq m_L \left[(0, 1) \right] = 1$, while $|X_m(t)| \leq N_\epsilon$ on this set for all m by definition, and $X_m \to X$ for all $t \in (0, 1)$. Thus the bounded convergence theorem applies and obtains that the first integral converges:

$$\int_{|X_m| < N_\epsilon} X_m^n(t) dm_L \to \int_{|X| < N_\epsilon} X^n(t) dm_L. \tag{4}$$

Hence given tightness, the challenge of achieving the desired result that $\mu'_{m,n} \to \mu'_n$ for each n, is the convergence of the second integral in (3) with unbounded integrands. In other words, the outstanding question is as follows.

Given that $X_m(t) \to X(t)$ for all t and $m_L\{t \mid |X_m(t)| \geq N_\epsilon\} \leq \epsilon$ by (2), what assumption will assure that:

$$\int_{|X_m| \geq N_\epsilon} X_m^n(t) dm_L \to \int_{|X_m| \geq N_\epsilon} X^n(t) dm_L = 0? \tag{5}$$

Certainly, tightness is inadequate for such an assurance, as the second example above illustrates. That $m_L\{t \mid |X_m(t)| \geq N_\epsilon\} \leq \epsilon$ for all m does not even assure that the integrals of $[X_m(t)]^n$ over these sets remain bounded. In detail, given any such N_ϵ and any $m > N_\epsilon$:

$$\int_{|X_m| \geq N_\epsilon} X_m^n(t) dm_L = \int_{1-1/m}^{1} m^n dm_L = m^{n-1}. \tag{6}$$

For this example, and as will be the case in general, the failure of $\mu'_{m,n} \to \mu'_n$ for each n will be linked to the second integral in (3) not converging as needed in (5). The first integral in (3) will always be assured to converge as in (4) by the bounded convergence theorem.

Returning to Book II and keeping the above notation, there were 3 "integration to the limit" results whereby pointwise convergence of $X_m^n \to X^n$ implied convergence of the associated Lebesgue integrals:

1. **Bounded convergence theorem (Proposition III.2.22):** This is not applicable to the second integral in (3) since $X_m^n(t)$ need not be bounded in m.

2. **Lebesgue's monotone convergence theorem (Proposition III.2.37):** Applicable only if $\{X_m^n(t)\}_{m=1}^{\infty}$ are nonnegative and the convergence $X_m^n(t) \to X^n(t)$ is monotonically increasing for all t.

3. **Lebesgue's dominated convergence theorem (Proposition III.2.52):** Applicable only if $|X_m^n(t)| \leq Y(t)$ for Lebesgue integrable $Y(t)$.

Though items 2 or 3 may apply in special situations, none of these results is generally applicable to the current investigation because the criteria needed are very strong.

As it turns out, there is another general criterion, called **uniform integrability,** which will provide another integration to the limit result which will serve our purpose. We introduce this definition next in the context of a general probability space, even though we currently only require this notion for Lebesgue integrals. The general integrals in this definition were introduced in Section 4.1.2 and will be developed in Book V.

Definition 4.66 (Uniform integrability) *A sequence of random variables $\{X_m\}_{m=1}^{\infty}$ defined on a probability space $(\mathcal{S}, \mathcal{E}, \lambda)$ is said to be **uniformly integrable (U.I.)** if:*

$$\lim_{N \to \infty} \sup_m \int_{|X_m| \geq N} |X_m(s)| \, d\lambda(s) = 0. \tag{4.106}$$

Remark 4.67 (On uniform integrability) *A few comments on this definition are warranted.*

1. *If $N' \leq N$, then for all m, then by (4.11):*

$$\int_{|X_m| \geq N} |X_m(s)| \, d\lambda(s) \leq \int_{|X_m| \geq N'} |X_m(s)| \, d\lambda(s),$$

and thus the same is true of the associated suprema.

Consequently,

$$A(N) \equiv \sup_m \int_{|X_m| \geq N} |X_m(s)| \, d\lambda(s),$$

is a nonnegative, monotonically decreasing function, so the limit in (4.106) is well-defined.

Further, the limit $N \to \infty$ in this definition can be interpreted quite generally, with N denoting all real numbers, or integers, or any increasing collection of reals.

2. *Given uniformly integrable $\{X_m\}_{m=1}^{\infty}$, this terminology reflects the fact that if N is large enough to ensure that the above supremum is less that 1, then:*

$$
\begin{aligned}
\int_{\mathcal{S}} |X_m(s)| \, d\lambda(s) &= \int_{|X_m| < N} |X_m(s)| \, d\lambda(s) + \int_{|X_m| \geq N} |X_m(s)| \, d\lambda(s) \\
&\leq N + 1.
\end{aligned}
$$

*Thus $\{X_m\}_{m=1}^{\infty}$ are all **integrable**, and the associated integrals are **uniformly bounded**.*

3. *Uniform integrability is assured if $\{X_m\}_{m=1}^{\infty}$ are dominated by integrable Y, meaning that $|X_m(s)| \leq |Y(s)|$ for all m and $\int_{\mathcal{S}} |Y(s)| \, d\lambda(s) < \infty$. In Book V, this result will be seen to be true in a general measure space, and not just in a probability space. But uniform integrability is weaker than the assumption that $|X_m(s)| \leq |Y(s)|$ for such Y. See Exercises 4.68 and 4.69.*

4. If $\{X_m\}_{m=1}^{\infty}$ are integrable and uniformly bounded, so $|X_m(s)| \leq c$ for all m, this implies uniform integrability in a probability space. But this conclusion is not valid in a general measure space as the example of $X_m \equiv \chi_{[m.m+1]}(x)$ defined on $(\mathbb{R}, \mathcal{B}(\mathbb{R}), m)$ illustrates.

The following exercises are stated for the probability space $((0,1), \mathcal{B}(0,1), m_L)$ so the more familiar Lebesgue integration theory can be used. But note that these results generalize to arbitrary probability spaces using the integration theory introduced in Section 4.1.2.

Exercise 4.68 (All $|X_m| \leq |Y|$ for integrable $Y \Rightarrow \{X_m\}_{m=1}^{\infty}$ U.I.) *Prove that if $\{X_m\}_{m=1}^{\infty}$ are defined on the Lebesgue probability space $((0,1), \mathcal{B}(0,1), m_L)$ and $|X_m(t)| \leq |Y(t)|$ for all m where $\int_0^1 |Y(t)| \, dm_L(t) < \infty$, then $\{X_m\}_{m=1}^{\infty}$ are uniformly integrable. Hint: Item 4 of Proposition III.2.49.*

Exercise 4.69 ($\{X_m\}_{m=1}^{\infty}$ U.I. $\not\Rightarrow$ All $|X_m| \leq |Y|$ for integrable Y) *On the Lebesgue probability space $((0,1), \mathcal{B}(0,1), m_L)$, develop an example of uniformly integrable random variables $\{X_m\}_{m=1}^{\infty}$ for which there is no integrable Y with $|X_n| \leq |Y|$. Hint: $\{X_m\}_{m=1}^{\infty}$ must be unbounded and uniformly integrable. Identify variates that are tightly bounded by $Y(t) = 1/t$, which is not integrable.*

Exercise 4.70 ($\sup_m \int_{(0,1)} |X_m|^{1+\epsilon} \, dm_L < \infty \Rightarrow \{X_m\}_{m=1}^{\infty}$ U.I.) *Let $\{X_m(t)\}_{m=1}^{\infty}$ be a random variable sequence on the Lebesgue probability space $((0,1), \mathcal{B}(0,1), m_L)$. Prove that if for some $\epsilon > 0$:*

$$\sup_m \int_{(0,1)} |X_m(t)|^{1+\epsilon} \, dm_L(t) = K < \infty,$$

then $\{X_m(t)\}_{m=1}^{\infty}$ are uniformly integrable. Hint: For any $N > 0$:

$$\sup_m \int_{|X_m(t)| \geq N} |X_m(t)| \, dm_L(t) \leq c(N) \sup_m \int_{|X_m(t)| \geq N} |X_m(t)|^{1+\epsilon} \, dm_L(t).$$

Determine $c(N)$.

Example 4.71 Second analysis of Example 4.64: On uniform integrability *Before we turn to general proofs, we return to Example 4.64 for a second analysis.*

Certainly the variates $\{X_m^n\}_{m=1}^{\infty}$ there were not uniformly integrable for any $n > 1$ as noted in (6) of that proof.

On the other hand, if this example was modified so that $\{X_m^n\}_{m=1}^{\infty}$ were uniformly integrable for all n, then we would have obtained the desired result that $\mu'_{m,n} \to \mu'_n$ for each n. We do not provide this detail here, because it will be developed starting with (3) of the proof of the next proposition.

Turning to the next result, it will perhaps be surprising that it is stated without a mention of the uniform integrability of the variates $\{X_m^n\}_{m=1}^{\infty}$ for all n. Indeed, this result is completely silent on these variates, and simply assumes that $F_m \Rightarrow F$, and that all distribution functions have all moments.

To obtain the necessary uniform integrability of the unmentioned variates $\{X_m^n\}_{m=1}^{\infty}$, it makes a more easily verifiable assumption on the boundedness of the even moments $\{\mu'_{m,2n}\}_{m=1}^{\infty}$ for each n. As will be seen in the proof, this then assures the uniform integrability assumption, and this is enough to obtain the desired conclusion that $\mu'_{m,n} \to \mu'_n$ for all n, as noted in Example 4.71.

Proposition 4.72 (When $F_m \Rightarrow F$ implies that $\mu'_{m,n} \to \mu'_n$ for all n) *Let $\{F_m\}_{m=1}^{\infty}$ and F be distribution functions with respective moment collections $\{\{\mu'_{m,n}\}_{n=1}^{\infty}\}_{m=1}^{\infty}$ and $\{\mu'_n\}_{n=1}^{\infty}$.*

If $F_m \Rightarrow F$ and $\{\mu'_{m,2n}\}_{m=1}^{\infty}$ is bounded for every n, then $\mu'_{m,n} \to \mu'_n$ for all n.

Proof. *By Skorokhod's representation theorem of Proposition II.8.30, define random variables $\{X_m\}_{m=1}^{\infty}$ and X on the Lebesgue measure space $((0,1), \mathcal{B}(0,1), m_L)$ with respective distribution functions $\{F_m\}_{m=1}^{\infty}$ and F, and for which $X_m \to X$ for all $t \in (0,1)$. For given N define:*

$$X_m^{(N)} = \begin{cases} X_m, & |X_m| < N, \\ 0, & |X_m| \geq N, \end{cases} \qquad X^{(N)} = \begin{cases} X, & |X| < N, \\ 0, & |X| \geq N. \end{cases}$$

For each n and N, $\left[X_m^{(N)}\right]^n \to \left[X^{(N)}\right]^n$ for all $t \in (0,1)$, and so by the bounded convergence theorem of Proposition III.2.22:

$$\int_{(0,1)} \left[X_m^{(N)}(t)\right]^n dm_L \to \int_{(0,1)} \left[X^{(N)}\right]^n dm_L. \tag{1}$$

Also, using properties of the Lebesgue integral from Proposition III.2.49:

$$\begin{aligned} \int_{(0,1)} [X_m(t)]^n dm_L - \int_{(0,1)} \left[X_m^{(N)}(t)\right]^n dm_L &= \int_{|X_m(t)| \geq N} [X_m(t)]^n dm_L, \\ \int_{(0,1)} [X(t)]^n dm_L - \int_{(0,1)} \left[X^{(N)}(t)\right]^n dm_L &= \int_{|X(t)| \geq N} [X(t)]^n dm_L. \end{aligned} \tag{2}$$

From (1) and (2) it follows that:

$$\begin{aligned} \limsup_m &\left| \int_{(0,1)} [X_m(t)]^n dm_L - \int_{(0,1)} [X(t)]^n dm_L \right| \\ &\leq \sup_m \int_{|X_m(t)| \geq N} |X_m(t)|^n dm_L + \int_{|X(t)| \geq N} |X(t)|^n dm_L. \end{aligned} \tag{3}$$

Existence of μ'_n assures that the last integral $\int_{|X(t)| \geq N} |X(t)|^n dm_L$ in (3) can be made arbitrarily small as $N \to \infty$. To complete the proof, we show that the assumption on boundedness of even moments assures that $\{[X_m(t)]^n\}$ are uniformly integrable for any n, and thus the first term on the right in (3) converges to zero as $N \to \infty$ by Definition 4.66.

To this end, the assumed boundedness of $\{\mu'_{m,2n}\}_{m=1}^{\infty}$ implies that

$$\sup_m \int_{(0,1)} |X_m(t)|^{2n} dm_L = K_n < \infty.$$

Hence:,

$$\begin{aligned} \sup_m \int_{|X_m(t)| \geq N} |X_m(t)|^n dm_L &\leq \sup_m \int_{|X_m(t)| \geq N} |X_m(t)|^{2n} dm_L / N^n \\ &\leq K_n / N^n, \end{aligned}$$

and $\{[X_m(t)]^n\}_{m=1}^{\infty}$ are uniformly integrable.

It then follows from (3) that for all n:

$$\limsup_m \left| \int_{(0,1)} [X_m(t)]^n dm_L - \int_{(0,1)} [X(t)]^n dm_L \right| = 0,$$

and since these terms are nonnegative, this assures that the corresponding limit is 0. Thus $\mu'_{m,n} \to \mu'_n$ for all n. ∎

The existence of all moments for F was assumed above, but could have been part of the conclusion.

Corollary 4.73 (When $F_m \Rightarrow F$ implies that $\mu'_{m,n} \to \mu'_n$ for all n) Let $\{F_m\}_{m=1}^\infty$ be distribution functions with moment collections $\{\{\mu'_{m,n}\}_{n=1}^\infty\}_{m=1}^\infty$. If $F_m \Rightarrow F$ and $\{\mu'_{m,2n}\}_{m=1}^\infty$ is bounded for every n, then F has moments of all orders $\{\mu'_n\}_{n=1}^\infty$, and $\mu'_{m,n} \to \mu'_n$ for all n.

Proof. Once the existence of moments for F is demonstrated, the above proof applies.

To this end, let $\{X_m\}_{m=1}^\infty$ and X be defined as above. Then since $X_m \to X$ for all $t \in (0,1)$, Fatou's lemma of Proposition III.2.34 obtains:

$$E\left[|X|^{2n}\right] \le \liminf_m E\left[|X_m|^{2n}\right] = \liminf_m \left[\mu'_{m,2n}\right] < \infty.$$

For odd moments, recalling Example 4.49:

$$E\left[|X_m|^{2n+1}\right] \le \left(E\left[|X_m|^{2n}\right] E\left[|X_m|^{2n+2}\right]\right)^{1/2} < \infty.$$

∎

In addition to convergence of moments, the next result seeks to conclude convergence of moment generating functions from $F_m \Rightarrow F$. Here a condition is needed on $\{M_m(t)\}_{m=1}^\infty$ that will provide uniform integrability of these functions.

Proposition 4.74 (When $F_m \Rightarrow F$ implies that $M_m(t) \to M(t)$) Let $\{F_m\}_{m=1}^\infty$ and F be distribution functions with with respective moment generating functions $\{M_m(t)\}_{m=1}^\infty$ and $M(t)$, all defined on a common interval $(-t_0, t_0)$ with $t_0 > 0$.

If $F_m \Rightarrow F$ and $\{M_m(t)\}_{m=1}^\infty$ is bounded for each $t \in (-t_0, t_0)$, then $M_m(t) \to M(t)$ for $t \in (-t_0, t_0)$.

Proof. Using the notation of the proof of Proposition 4.72:

$$\int_{(0,1)} \exp\left[tX_m^{(N)}(s)\right] dm_L(s) \to \int_{(0,1)} \exp\left[tX^{(N)}(s)\right] dm_L(s),$$

by the bounded convergence theorem of Proposition III.2.22, and hence:

$$\limsup_m \left| \int_{(0,1)} \exp\left[tX_m(s)\right] dm_L(s) - \int_{(0,1)} \exp\left[tX(s)\right] dm_L(s) \right|$$

$$\le \sup_m \int_{|X_m(s)| \ge N} \exp\left[tX_m(s)\right] dm_L(s) + \int_{|X(s)| \ge N} \exp\left[tX(s)\right] dm_L(s).$$

The last integral converges to 0 as $N \to \infty$ for all $t \in (-t_0, t_0)$ by the existence of $M(t)$. To complete the proof we show that the first term in this upper bound will also have limit 0 as $N \to \infty$ by proving that $\{\exp[tX_m(s)]\}_{m=1}^\infty$ are uniformly integrable for $t \in (-t_0, t_0)$. Exercise 4.70 proves that uniform integrability follows if for some $\epsilon > 0$,

$$\sup_m \int_{(0,1)} (\exp[tX_m(s)])^{1+\epsilon} dm_L(s) = K < \infty.$$

By definition:

$$\int_{(0,1)} (\exp[tX_m(s)])^{1+\epsilon} dm_L(t) = M_m(t[1+\epsilon]),$$

and hence is bounded by assumption when $t(1+\epsilon) \in (-t_0, t_0)$. For $t \in (-t_0, t_0)$, such $\epsilon > 0$ always exists.

The proof is complete by the same final step as for Proposition 4.72. ∎

As was the case for Proposition 4.72 as confirmed in Corollary 4.73, the existence of $M(t)$ could have been part of the conclusion of the prior result.

Corollary 4.75 (When $F_m \Rightarrow F$ implies that $M_m(t) \to M(t)$) *Let $\{F_m\}_{m=1}^{\infty}$ be distribution functions with moment generating functions $\{M_m(t)\}_{m=1}^{\infty}$ defined on a common interval $(-t_0, t_0)$ with $t_0 > 0$. If $F_m \Rightarrow F$ and $\{M_m(t)\}_{m=1}^{\infty}$ is bounded for each $t \in (-t_0, t_0)$, then F has a moment generating function $M(t)$, and $M_m(t) \to M(t)$ for $t \in (-t_0, t_0)$.*
Proof. *As for Corollary 4.73, the existence of $M(t)$ for each $t \in (-t_0, t_0)$ follows from Fatou's lemma of Proposition III.2.34.* ∎

We are now ready for the main results on **method of moments**. These address the reverse implications to those above. Namely, when can we infer that $F_m \Rightarrow F$ from convergence of moments, or convergence of moment generating functions?

As noted in the introduction, it will always be assumed that the distribution function F is uniquely determined by its moment collection $\{\mu'_n\}_{n=1}^{\infty}$. Proposition 4.61 and Corollary 4.62 provide criteria which assure this.

While Proposition 4.76 on convergence of moments is addressed first, it is Proposition 4.77 on convergence of the associated moment generating functions that will provide the result in the form most commonly used in applications. Many important results will be derived based on this proposition in Section 6.2.

Proposition 4.76 (Method of moments: When $\mu'_{m,n} \to \mu'_n$ for each n implies $F_m \Rightarrow F$) *Assume that a distribution function F is uniquely determined by its moment collection $\{\mu'_n\}_{n=1}^{\infty}$.*

If $\{F_m\}_{m=1}^{\infty}$ is a sequence of distribution functions with moment collections $\{\{\mu'_{m,n}\}_{n=1}^{\infty}\}_{m=1}^{\infty}$, and $\mu'_{m,n} \to \mu'_n$ for each n, then $F_m \Rightarrow F$.
Proof. *Since $\mu'_{m,2} \to \mu'_2$, the collection $\{\mu'_{m,2}\}_{m=1}^{\infty}$ is bounded, say by K. Let $\{X_m\}_{m=1}^{\infty}$ be random variables with distribution functions $\{F_m\}_{m=1}^{\infty}$, constructed for example using Skorokhod's representation theorem of Proposition II.8.30. Then by Chebyshev's inequality in (4.83), for all m:*

$$\Pr[|X_m| \geq t] \leq K/t.$$

This implies by Definition 4.63 that the collection of distribution functions $\{F_m\}_{m=1}^{\infty}$ is tight.

By Helly's selection theorem of Proposition II.8.14, there exists a subsequence $\{F_{m_k}\}_{k=1}^{\infty}$, and a right continuous, increasing function $\widetilde{F}$, so that $F_{m_k}(x) \to \widetilde{F}(x)$ at all continuity points of $\widetilde{F}$. By Proposition II.8.20, $\widetilde{F}$ is a distribution function because $\{F_m\}_{m=1}^{\infty}$ is tight, and thus $F_{m_k} \Rightarrow \widetilde{F}$.

Now since $\mu'_{m,n} \to \mu'_n$ for each n, it follows that $\mu'_{m_k,n} \to \mu'_n$ for each n for this subsequence. If it can be proved that μ'_n is the nth moment of $\widetilde{F}$, it would follow that $\widetilde{F} = F$ by the assumption that F is uniquely determined by its moment collection. Then by Corollary II.8.22 to Helly's selection theorem, since every such Helly subsequence satisfies $F_{m_k} \Rightarrow F$, we can conclude that $F_m \Rightarrow F$.

To prove that μ'_n is the nth moment of $\widetilde{F}$, note that the convergence assumption $\mu'_{m,n} \to \mu'_n$ for each n assures that $\{\mu'_{m_k,2n}\}_{m=1}^{\infty}$ is bounded for all n. Thus from Corollary 4.73, $F_{m_k} \Rightarrow \widetilde{F}$ assures that $\widetilde{F}$ has moments $\{\widetilde{\mu}'_n\}_{n=1}^{\infty}$ and $\mu'_{m_k,n} \to \widetilde{\mu}'_n$ for each n. But since $\mu'_{m,n} \to \mu'_n$ for each n, we conclude that $\widetilde{\mu}'_n = \mu'_n$ for all n. ∎

Proposition 4.77 (Method of moments: When $M_m(t) \to M(t)$ implies $F_m \Rightarrow F$)
Assume that a distribution function F has moment generating function $M(t)$ which exists on $(-t_0, t_0)$ with $t_0 > 0$, and that $\{F_m\}_{m=1}^{\infty}$ is a sequence of distribution functions with respective moment generating functions $\{M_m(t)\}_{m=1}^{\infty}$ that exist on the same interval.

If $M_m(t) \to M(t)$ for $t \in (-t_0, t_0)$, then $F_m \Rightarrow F$.

Proof. *The existence of $M(t)$ on $(-t_0, t_0)$ assures that F is uniquely determined by its moments by Corollary 4.62, and that its moments are given by $\mu_n' = M^{(n)}(0)$ by Proposition 4.26. Let $\{X_m\}_{m=1}^\infty$ be random variables with distribution functions $\{F_m\}_{m=1}^\infty$.*

By (4.85), for any $t' \in (-t_0, t_0)$:

$$\Pr[|X_m| \geq t] \leq e^{-tt'} \left[M_m(t') + M_m(-t')\right] \leq c(t')e^{-tt'}.$$

This upper bound exists since $M_m(\pm t') \to M(\pm t')$ for all t'. Letting $t \to \infty$ obtains that the collection of distribution functions $\{F_m\}_{m=1}^\infty$ is tight.

As in the above proof, there is a subsequence $\{F_{m_k}\}_{k=1}^\infty$ and a distribution function $\widetilde{F}$ so that $F_{m_k} \Rightarrow \widetilde{F}$. The proof that $\widetilde{M}(t)$ exists and $M_{m_k}(t) \to \widetilde{M}(t)$ for $t \in (-t_0, t_0)$, and thus that $\widetilde{M}(t) = M(t)$, follows from Corollary 4.75 as for the proof above. Corollary 4.62 on the uniqueness of the moment generating functions now assures that $\widetilde{F} = F$ and thus $F_{m_k} \Rightarrow F$. As this conclusion is true for any Helly subsequence, it follows as above that $F_m \Rightarrow F$. ∎

5

Simulating Samples of RVs – Examples

We begin by recalling Proposition II.4.9. For this result, a continuous uniformly distributed random variable U is defined on some probability space:

$$U : (\mathcal{S}', \mathcal{E}', \lambda') \to ((0,1), \mathcal{B}(0,1), m), \tag{5.1}$$

with m Lebesgue measure. The distribution function $F_U(y) = y$ for $y \in (0,1)$, and this is an example of the uniform distribution of item 1 of Example 1.28.

Proposition II.4.9: *Let $(\mathcal{S}, \mathcal{E}, \lambda)$ be given, and $X : (\mathcal{S}, \mathcal{E}, \lambda) \to (\mathbb{R}, \mathcal{B}(\mathbb{R}), m)$ a random variable with distribution function $F(x)$ and left-continuous inverse $F^*(y)$ as given in (1.52) of Definition 1.29:*

$$F^*(y) = \inf\{x | F(x) \geq y\}.$$

1. *If $\{U_j\}_{j=1}^M$ are independent, continuous uniformly distributed random variables, then:*

$$\{X_j\}_{j=1}^M \equiv \{F^*(U_j)\}_{j=1}^M,$$

   are independent random variables with distribution function $F(x)$.

2. *If F is continuous and $\{X_j\}_{j=1}^M$ are independent random variables with distribution function $F(x)$, then:*

$$\{U_j\}_{j=1}^M \equiv \{F(X_j)\}_{j=1}^M,$$

   are independent, continuous uniformly distributed random variables.

In the next section we apply this result to develop random samples for a number of the distributions introduced in Section 1.4. Following that, we investigate the generation of **random ordered samples.**

We will assume throughout this chapter that one has access to computer software for generating the collections $\{U_j\}_{j=1}^M$ of independent, continuous uniformly distributed random variables. Generating samples of such U-variates is now very easy since most mathematical software has a built-in function which does exactly this. For example, in Microsoft Excel this function is called **RAND()**, while in MathWorks MATLAB it is **rand**, and so forth.

See Chapter II.4 for the theoretical framework for random samples, or the introduction to Chapter 6 on Limit Theorems for a discussion of the Book II results.

5.1 Random Samples

5.1.1 Discrete Distributions

Let X be a discrete random variable defined on a probability space $(\mathcal{S}, \mathcal{E}, \lambda)$, by which is meant that the range of X, denoted $Rng(X)$, is a finite or countable set $\{a_i\} \subset \mathbb{R}$. In other words, $X(\mathcal{S}) = \{a_i\}$. Since X is measurable by definition, $X^{-1}(a_i) \in \mathcal{E}$ for all i, and the **probability density function** $f(x)$ is defined on $\mathbb{R}$ by:

$$f(x) \equiv \lambda \left[X^{-1}(x) \right].$$

Thus $f(x) = 0$ for $x \notin \{a_i\}$, and otherwise $f(a_i)$ is often denoted p_i.

In theory, a discrete random variable can have $a \in \{a_i\}$ with $\lambda \left[X^{-1}(a) \right] = 0$, but this is never seen in practice. One can simple define a new random variable X' to equal X everywhere except on $X^{-1}(a)$, and on this set define X' to be any other a_i-value. Then all probability statements about X and X' will be identical.

Thus it is always assumed that $p_i = f(a_i) > 0$ for all i, and it is commonly said that $f(a_i) \equiv \lambda \left[X^{-1}(a_i) \right]$ **is the probability that** $X = a_i$:

$$f(a_i) \equiv \Pr[X = a_i].$$

The associated **distribution function** $F(x)$ is defined:

$$F(x) \equiv \lambda \left[X^{-1}(-\infty, x] \right],$$

and it is commonly said that $F(x)$ **is the probability that** $X \leq x$:

$$F(x) \equiv \Pr[X \leq x].$$

Consequently:

$$F(x) = \sum_{a_i \leq x} f(a_i) = \sum_{a_i \leq x} p_i, \tag{5.2}$$

and it follows that for any i :

$$\sum_{j \leq i} p_j \in [0, 1].$$

In almost all applications, such discrete distribution functions are defined relative to a collection $\{a_i\} \subset \mathbb{R}$, which if not finite, will have **no accumulation points**, and can thus be assumed to be strictly increasing. These points induce a collection of right semi-closed intervals $\{[a_i, a_{i+1})\}$, and there are then three or four cases possible, depending on how one counts case 2:

1. Finite collection: $\{[a_i, a_{i+1})\}_{i=0}^{N-1}$;

2. One-sided infinite collection: $\{[a_i, a_{i+1})\}_{i=0}^{\infty}$ or $\{[a_i, a_{i+1})\}_{i=-\infty}^{-1}$;

3. Two-sided infinite collection: $\{[a_i, a_{i+1})\}_{i=-\infty}^{\infty}$.

It follows from (5.2) that the **distribution function** $F(x)$ is defined as the bounded step function:

$$F(x) = \sum_{j \leq i} p_j, \quad x \in [a_i, a_{i+1}). \tag{5.3}$$

Note that $F(x)$ is right continuous, as is every distribution function by Notation II.3.5 and Proposition I.3.60.

Remark 5.1 (On $F(x)$ and $\{[a_i, a_{i+1})\}$) *When $\{a_i\}$ has no accumulation points, it then follows that this collection is unbounded if an only if it is infinite. Thus it will only be true in case 3 that:*

$$\bigcup_i [a_i, a_{i+1}) = \mathbb{R}.$$

However, by extension $F(x)$ is nonetheless well defined on $\mathbb{R}$ by (5.3) in the bounded or semi-bounded cases. In essence, we interpret $x \in [a_i, a_{i+1})$ to be well-defined as long as at least one of a_i and a_{i+1} are in the collection $\{a_i\}$. For example in case 1, $[a_N, a_{N+1}) \equiv [a_N, \infty)$, while $[a_{-1}, a_0) \equiv (-\infty, a_0)$. Thus:

1. *Finite collection:* $F(x) = 0$ *on* $(-\infty, a_0)$ *and* $F(x) = 1$ *on* $[a_N, \infty)$.

2. *One-sided infinite collection:*

    (a) *In the first case:* $F(x) = 0$ *on* $(-\infty, a_0)$, *while* $a_i \to \infty$ *and* $F(a_i) \to 1$ *as* $i \to \infty$;

    (b) *In the second case:* $F(x) = 1$ *on* $[a_0, \infty)$, *while* $a_i \to -\infty$ *and* $F(a_i) \to 0$ *as* $i \to -\infty$.

3. *Two-sided infinite collection:* $F(a_i) \to 0$ *as* $i \to -\infty$ *and* $F(a_i) \to 1$ *as* $i \to \infty$.

In theory, though rarely encountered in practice, it is possible for a discrete distribution function to have no such parametrization because the collection $\{a_i\}$ is dense in $\mathbb{R}$, or on an interval of $\mathbb{R}$, or has accumulation points.

For example, with $\{r_n\}_{n=1}^{\infty} \subset \mathbb{R}$ an arbitrary ordering of the rationals, define the probability density function $f(x)$ on this set by $f(r_n) = 2^{-n}$, and thus:

$$F(x) = \sum_{r_n \leq x} 2^{-n}.$$

As this type of distribution function is rarely encountered in applications, we will not dwell on it, but note that the approach of this section can be adapted to these cases.

Given $F(x)$ as in (5.3) and defined by finite or infinite $\{a_i\}$, with no accumulation points in the latter case, the **left-continuous inverse** $F^*(y)$ as defined in *(1.52)* can be expressed as follows. With $F(a_i) \equiv \sum_{j \leq i} p_j$ by (5.2), for $y \in (0, 1)$:

$$F^*(y) = \sum_i a_i \chi_{(F(a_{i-1}), F(a_i)]}(y), \tag{5.4}$$

where $\chi_{(F(a_{i-1}), F(a_i)]}(y)$ is the **characteristic function** of the interval $(F(a_{i-1}), F(a_i)]$, and defined to be 1 if $y \in (F(a_{i-1}), F(a_i)]$, and 0 otherwise.

Remark 5.2 (On $F^*(y)$ and $\{(F(a_{i-1}), F(a_i)]\}$) *As was noted in Remark 5.1 for (5.3), we interpret $(F(a_{i-1}), F(a_i)]$ to be well-defined as long as one variable is in the collection $\{a_i\}$. For example in case 1,*

$$(F(a_{-1}), F(a_0)] \equiv (F(-\infty), F(a_0)] \equiv (0, F(a_0)].$$

Thus $F^*(y) = a_0$ for $y \in (0, F(a_0)]$.
    Note also that:

- *While $F^*(y)$ is formally represented as a finite or infinite sum, there is only one nonzero term for any y due to the disjointedness of the interval collection $\{(F(a_{i-1}), F(a_i)]\}$.*

- *$F^*(y)$ is seen to be increasing and left continuous as is true in general by Proposition II.3.16.*

- *By Proposition II.4.9 noted in the introduction to this chapter, if U is a continuous uniformly distributed random variable on $(0, 1)$, then $\tilde{X} \equiv F^*(U)$ has the same distribution function as the random variable X that we started with. This follows because $\tilde{X} = a_i$ if $U \in (F(a_{i-1}), F(a_i)]$, and this occurs with probability:*

$$F_U[F(a_i)] - F_U[F(a_{i-1})] = p_i.$$

Here $F_U(x)$ is the distribution function of U and so $F_U(x) = x$ by definition of this variate.

With this general set-up, we next turn to examples.

Remark 5.3 (Independence of simulated variates) *Below we derive various examples of simulating random variables starting with independent, continuous uniform $\{U_j\}_{j=1}^M$. That the resulting simulated variates are independent follows from Proposition II.3.56 since F^* is Borel measurable by Proposition II.3.16.*

1. **Discrete Rectangular Distribution on** $[0,1]$, parameter n, as defined in (1.38):

   This is an example of case 1 of Remark 5.1 with a finite collection, here $\{a_i\} \equiv \{k/n\}_{k=1}^n$, which defines the collection $\{[a_i, a_{i+1})\}$, supplemented as noted in Remark 5.2 to include $(-\infty, 0)$ and $[1, \infty)$.

   Given independent, continuous uniform $\{U_j\}_{j=1}^M$, the variates $X_j^R \equiv F^*(U_j)$ are defined for $k = 1, ..., n$ by (5.4):

   $$X_j^R = \frac{k}{n}, \quad \frac{k-1}{n} < U_j \le \frac{k}{n}.$$

   These values of $F^*(U_j)$ can be formulaically obtained with a built-in function that is contained in most mathematical software:

   $$F^*(U_j) = \frac{\lceil nU_j \rceil}{n}.$$

   Here $\lceil y \rceil$ denotes the **ceiling function:**

   $$\lceil y \rceil = \min\{m \in \mathbb{N} | m \ge y\}. \tag{5.5}$$

   This follows because $\lceil nU_j \rceil = k$ if $\frac{k-1}{n} < U_j \le \frac{k}{n}$.

2. **Standard Binomial Distribution**, parameter p, as defined in (1.39):

   Here $\{a_i\} \equiv \{0, 1\}$, which defines the the single interval $[a_i, a_{i+1}) \equiv [0, 1)$, supplemented to include $(-\infty, 0)$ and $[1, \infty)$.

   Given independent, continuous uniform $\{U_j\}_{j=1}^M$, the variates $X_j^B \equiv F^*(U_j)$ are defined by (5.4):

   $$X_j^B = \begin{cases} 0, & 0 < U_j \le 1 - p, \\ 1, & 1 - p < U_j < 1. \end{cases}$$

   By symmetry, $F^*(U_j)$ can also be defined to be 1 for $0 < U_j \le p$ and 0 otherwise.

3. **General Binomial Distribution**, parameters p, n, defined in (1.40):

   Here $\{a_i\} \equiv \{k\}_{k=0}^n$, which defines the collection $\{[a_i, a_{i+1})\}$, supplemented to include $(-\infty, 0)$ and $[n, \infty)$.

   Given independent, continuous uniform $\{U_j\}_{j=1}^M$, the variates $X_{n,j}^B \equiv F^*(U_j)$ are defined by (5.4):

   $$X_{n,j}^B = k, \quad b_{k-1} < U_j \le b_k,$$

   where:

   $$b_k \equiv F_B(k) = \sum_{j=0}^k \binom{n}{j} p^j (1-p)^{n-j}, \qquad k = 0, 1, ..., n.$$

   Alternatively, since $X_{n,j}^B$ is the sum of n independent standard binomials, it can also be simulated by sums of variates simulated as in item 2. However, this is not a computationally efficient approach unless one is interested in both the component variates as well as the sum.

4. **Geometric Distribution**, parameter p, defined in (1.43):

   An example of case 2 of Remark 5.1, here $\{a_i\} \equiv \{k\}_{k=0}^\infty$, which defines the collection $\{[a_i, a_{i+1})\}$, supplemented to include $(-\infty, 0)$.

Given independent, continuous uniform $\{U_j\}_{j=1}^M$, the variates $X_j^G \equiv F^*(U_j)$ are defined as in (5.4), but this calculation is simplified by the fact that $F_G^*(y)$ can be explicitly calculated. With $F_G(j)$ defined in (1.44) by $F_G(j) = 1 - (1-p)^{j+1}$, the definition in (1.52) produces:

$$F_G^*(y) = \inf\left\{j \,\middle|\, \frac{\ln(1-y)}{\ln(1-p)} \le j+1\right\}.$$

Using the **ceiling function** $\lceil y \rceil$ of (5.5), it follows that:

$$F_G^*(y) = \left\lceil \frac{\ln(1-y)}{\ln(1-p)} \right\rceil - 1.$$

Hence given $\{U_j\}_{j=1}^M$:

$$X_j^G = \left\lceil \frac{\ln(1-U_j)}{\ln(1-p)} \right\rceil - 1.$$

Since U_j has the same distribution as $1 - U_j$, this can be expressed:

$$X_j^G = \left\lceil \frac{\ln U_j}{\ln(1-p)} \right\rceil - 1.$$

5. **Negative Binomial Distribution,** parameters p, k, defined in (1.46):

Another example of case 2 of Remark 5.1, here again $\{a_i\} \equiv \{k\}_{k=0}^\infty$, which defines the collection $\{[a_i, a_{i+1})\}$, supplemented to include $(-\infty, 0)$.

Given $\{U_j\}_{j=1}^M$, the variates $X_j^{NB} \equiv F^*(U_j)$ can be simulated directly in terms of characteristic functions of the intervals defined by $\{F_{NB}(j)\}_{j=0}^\infty$ as in (5.4).

Also, item 2 of Section 4.4.1 proved that the negative binomial is the sum of k independent geometric random variables. Hence given independent continuous uniform $\{U_j\}_{j=1}^{kM}$, X_j^{NB} can be defined in terms of k-sums of geometric variables as defined in item 4:

$$X_j^{NB} = \sum_{k=(j-1)k+1}^{jk} \left\lceil \frac{\ln U_j}{\ln(1-p)} \right\rceil - k.$$

6. **Poisson Distribution,** parameter λ, defined in (1.47):

Again $\{a_i\} \equiv \{k\}_{k=0}^\infty$, with $\{[a_i, a_{i+1})\}$ supplemented to include $(-\infty, 0)$.

Given $\{U_j\}_{j=1}^M$, the variates $X_j^P \equiv F^*(U_j)$ can be simulated directly in terms of characteristic functions of the intervals defined by $\{F_P(j)\}_{j=0}^\infty$ by (5.4).

Alternatively, we can use the development in item 3 of Example 2.14 which proved that with X_j^E denoting independent exponential random variables with parameter λ, then:

$$N \equiv \max\left\{n \,\middle|\, \sum_{j=1}^n X_j^E \le 1\right\},$$

is a Poisson random variable with parameter λ. As derived in the next section, from independent, continuous uniform $\{U_j\}_{j=1}^M$, such exponentials are generated by:

$$X_j^E = -\frac{\ln U_j}{\lambda}.$$

Further, these exponentials are independent by Proposition II.3.56.

Thus given independent continuous uniform $\{U_j\}_{j=1}$, independent Poisson random variables can be produced by:

$$X^P = \max\left\{n \,\middle|\, \prod_{j=1}^n U_j \ge e^{-\lambda}\right\}.$$

5.1.2 Simpler Continuous Distributions

For many continuous distributions, $F(x)$ is strictly increasing and one can in theory by Proposition II.3.22 determine F^* by inverting $F(x) = y$. In other words, $F^* = F^{-1}$ in such cases, and for several common distribution functions this inversion is algebraically tractable. Thus Proposition II.4.9 in the introduction to this chapter can be applied directly, and in this section we address the simpler continuous distributions this way. The normal, lognormal and Student T distributions are deferred to the following sections, as these require special methods.

Recall Remark 5.3 on independence of the simulated variates below.

1. **Continuous Uniform Distribution on** $[a, b]$, defined in (1.51):

   Since $F(x) = \frac{x-a}{b-a}$ for $a \leq x \leq b$, given independent, continuous uniform $\{U_j\}_{j=1}^{M}$, the variates $X_j^U \equiv F^{-1}(U_j)$ are given by:

   $$X_j^U = \{a + (b-a)U_j\}_{j=1}^{M}.$$

2. **Exponential Distribution,** parameter λ, defined in (1.53):

   Since $F_E(x) = 1 - e^{-\lambda x}$ for $x \geq 0$, it follows that $F_E^{-1}(y) = -\frac{1}{\lambda}\ln(1-y)$. Thus given independent, continuous uniform $\{U_j\}_{j=1}^{M}$:

   $$X_j^E = -\frac{1}{\lambda}\ln(1 - U_j),$$

   or equivalently:

   $$X_j^E = -\frac{1}{\lambda}\ln U_j,$$

   since U_j has the same distribution as $1 - U_j$.

3. **Gamma Distribution,** parameters λ, α, defined in (1.55):

   The gamma distribution function is not explicitly invertible for general α, but for $\alpha = k \in \mathbb{N}$ a positive integer, Exercise 2.15 proves that this random variable is the sum of k independent exponential random variables with parameter λ. Hence, given independent, continuous uniform $\{U_j\}_{j=1}^{kM}$, we can define $\{X_j^{\Gamma}\}_{j=1}^{M}$ in terms of k-sums or k-products of exponential variables from item 2:

   $$X_j^{\Gamma} = -\frac{1}{\lambda}\sum_{k=(j-1)k+1}^{jk}\ln U_j = -\frac{1}{\lambda}\ln\left(\prod_{k=(j-1)k+1}^{jk}U_j\right).$$

4. **Beta Distribution,** parameters, v, w, defined in (1.61):

   The beta distribution is also not explicitly invertible, but an important connection between the beta and gamma distributions makes generating beta samples feasible when v, w are positive integers.

   Recalling Proposition 2.30, if X^{Γ_1} and X^{Γ_2} are independent gamma random variables with parameters v, λ and w, λ, then the random variable $\frac{X^{\Gamma_1}}{X^{\Gamma_1}+X^{\Gamma_2}}$ has a beta distribution with parameters v, w, and this is independent of λ.

   Hence when $v, w \in \mathbb{N}$ are positive integers, two such gammas can be generated as in item 3 with a total of $v + w$ independent, continuous uniform random variables. Taking $\lambda = 1$ for simplicity, it follows that $\{X_j^{\beta}\}_{j=1}^{M}$ can then be generated from $\{U_j\}_{j=1}^{(v+w)M}$.

5. **Cauchy Distribution,** parameters x_0, γ, defined in (1.68):

The substitution $(y - x_0)/\gamma = \tan z$ produces:

$$
\begin{aligned}
F_C(x) &= \frac{1}{\pi \gamma} \int_{-\infty}^{x} \frac{dy}{1 + ((y - x_0)/\gamma)^2} \\
&= \frac{1}{\pi} \int_{-\pi/2}^{\arctan((y-x_0)/\gamma)} dz \\
&= \frac{1}{\pi} \left[\arctan\left(\frac{y - x_0}{\gamma} \right) + \pi/2 \right].
\end{aligned}
$$

Thus $F_C(x)$ can be inverted to obtain $F_C^*(x)$.

Given independent continuous uniform $\{U_j\}_{j=1}^{M}$, define $X_j^C \equiv F_C^{-1}(U_j)$ by:

$$
X_j^C = x_0 + \gamma \tan\left[\pi \left(U_j - 1/2 \right) \right].
$$

6. **Extreme Value Distribution,** parameter γ, defined in (7.14) and (7.15):

When $\gamma \neq 0$, $G_\gamma^{-1}(y)$ is given by:

$$
G_\gamma^{-1}(y) = \frac{[- \ln y]^{-\gamma} - 1}{\gamma},
$$

while for $\gamma = 0$:

$$
G_0^{-1}(y) = - \ln[- \ln y].
$$

Note that the result for G_0^{-1} equals the result for $\lim_{\gamma \to 0} G_\gamma^{-1}$ since this limit is seen to be $f'(0)$ with $f(x) = [- \ln y]^{-x}$.

Given independent continuous uniform $\{U_j\}_{j=1}^{M}$, define:

$$
X_j^{EV} = G_\gamma^{-1}(U_j),
$$

using the appropriate formula above.

5.1.3 Normal and Lognormal Distributions

If Z is a normal variate with parameters μ and σ^2 in (1.65), and Y is a lognormal variate with the same parameters in (1.67), then $Y =_d \exp Z$. In other words, these variates are "equal in distribution," meaning they have the same distribution function. Thus simulating an independent collection, say $\{Z_j\}_{j=1}^{M}$, obtains an independent collection $\{Y_j\}_{j=1}^{M}$ by Proposition II.3.56, with Y defined as above. Also, it is enough to simulate either variate with $\mu = 0$ and $\sigma^2 = 1$. For example, if Z were such a standard normal variate, then $Z' \equiv \sigma Z + \mu$ is normal with these general parameters.

Unfortunately, the associated distribution functions are not analytically tractable and the inversion required by Proposition II.4.9 is only approximately implementable. Two early approaches to the simulation of standard normal Z used numerical approximations are:

- Generate numerical estimates of $\Phi(x)$ for discrete x values, then approximate $\Phi^{-1}(U)$ with interpolation.

- Applying the central limit theorem of Proposition 6.13 obtains that for n large, if $\{X_j\}_{j=1}^n$ are independent and identically distributed random variables with finite mean and variance, then distribution function of the random variable:

$$Z_n \equiv \frac{\sum_{j=1}^n X_j - nE[X]}{\sqrt{nVar[X]}},$$

is approximately standard normal. Further, the distribution of Z_n converges to the distribution of this normal Z as $n \to \infty$. In the notation of Definition II.5.19, $Z_n \to_d Z$, or $Z_n \Rightarrow Z$, as $n \to \infty$.

For example, if X_j is continuous uniform on $[0,1]$, then $E[X] = 1/2$ and $Var[X] = 1/12$ by (4.70). Choosing $n \geq 30$ say, we conclude that given independent, continuous uniform $\{U_j\}_{j=1}^{Mn}$, we can approximate independent standard normal variates $\{Z_j^\Phi\}_{j=1}^M$ by:

$$Z_j^\Phi = \frac{\sum_{k=(j-1)n+1}^{jn} U_j - n/2}{\sqrt{n/12}}.$$

A very powerful approach to generating exact pairs of independent normal variates was introduced in 1958 by **G. E. P. Box** (1919–2013) and **Mervin E. Muller** (1928–2018). This approach is often referred to as the **Box–Muller transform,** since it transforms independent pairs (U_1, U_2) of continuous, uniform variates into independent pairs (Z_1, Z_2) of standard normal variates.

The following proof requires a couple of results. The first has been noted before, while the second generalizes the result in (2.4) of Proposition 2.1. Both will be formally derived in Book VI using the integration theory of Book V:

1. **Densities of independent random variables:** In Proposition 1.14 it was noted that if $X \equiv (X_1, ..., X_n)$ is a random vector defined on a probability space $(\mathcal{S}, \mathcal{E}, \lambda)$ with distribution function $F(x_1, ..., x_n)$, then the random variables $\{X_i\}_{i=1}^n$ are independent if and only if:

$$F(x_1, ..., x_n) = \prod_{i=1}^n F_i(x_i),$$

where $F_i(x_i)$ is the distribution function of X_i. If all distribution functions are continuously differentiable with associated continuous density functions, then this implies that as Riemann integrals:

$$\int_{R_x} f(y)dy = \prod_{i=1}^n \int_{-\infty}^{x_i} f_i(y_i)dy_i,$$

where $y = (y_1, ..., y_n)$, dy reflects the Riemann integral in $\mathbb{R}^n$ of Section III.1.3, and with $x = (x_1, ..., x_n)$, $R_x \equiv \prod_{i=1}^n (-\infty, x_i]$.

An application of Corollary III.1.77 for the integral on the left, justifying the lower limit extension to $-\infty$, obtains that for all $(x_1, ..., x_n)$:

$$\int_{-\infty}^{x_n} \cdots \int_{-\infty}^{x_1} f(y_1, ..., y_n)dy_1...dy_n = \int_{-\infty}^{x_n} \cdots \int_{-\infty}^{x_1} \prod_{i=1}^n f_i(y_i)dy_1...dy_n.$$

Then n applications of the fundamental theorem of calculus of Proposition III.1.33 yields that $\{X_i\}_{i=1}^n$ are independent if and only if:

$$f(x_1, ..., x_n) = \prod_{i=1}^n f_i(x_i). \tag{5.6}$$

It will be seen in Book VI that this conclusion does not require continuity of densities, but then requires the Lebesgue generalization of the Riemann Corollary III.1.77, to be derived in Fubini's or Tonelli's theorems. Further, with these results, it is only necessary to assume that $\{X_i\}_{i=1}^n$ have density functions, and then the density function for X is assured to exist. This last observation is needed below.

2. **Densities of transformed random vectors:** Let $X \equiv (X_1, ..., X_n)$ be a random vector defined on a probability space $(\mathcal{S}, \mathcal{E}, \lambda)$ with distribution function $F(x_1, ..., x_n)$ and density $f(x_1, ..., x_n)$. Define the random variable $Y = g(X)$ where $g : \mathbb{R}^n \to \mathbb{R}^n$ is continuously differentiable and one-to-one. What can be said about the distribution and density functions of Y?

Generalizing (2.4) of Proposition 2.1, it will be proved in Book VI that:

$$f_Y(y) = f\left(g^{-1}(y)\right)\left|\det\left(\frac{\partial g^{-1}}{\partial y}\right)\right|, \tag{5.7}$$

where $\det\left(\frac{\partial g^{-1}}{\partial y}\right)$ is the determinant of the Jacobian matrix of the transformation g^{-1}. This transformation is well-defined by the assumption that g is one-to-one.

Given a transformation $g^{-1}(y)$, which we denote by $T : \mathbb{R}^n \to \mathbb{R}^n$ for notational simplicity, denote:

$$T(y) = (t_1(y), t_2(y), \cdots, t_n(y)),$$

where $t_i : \mathbb{R}^n \to \mathbb{R}$ for all i. When this transformation has differentiable component functions, the **Jacobian matrix** associated with T, denoted $\left(\frac{\partial T}{\partial y}\right)$ or $T'(y)$, is defined by:

$$\left(\frac{\partial T}{\partial y}\right) \equiv \begin{pmatrix} \frac{\partial t_1}{\partial y_1} & \frac{\partial t_1}{\partial y_2} & \cdots & \frac{\partial t_1}{\partial y_n} \\ \frac{\partial t_2}{\partial y_1} & \frac{\partial t_2}{\partial y_2} & \cdots & \frac{\partial t_2}{\partial y_n} \\ \vdots & \vdots & \vdots & \vdots \\ \frac{\partial t_n}{\partial y_1} & \frac{\partial t_n}{\partial y_2} & \cdots & \frac{\partial t_n}{\partial y_n} \end{pmatrix}. \tag{5.8}$$

This matrix and its determinant are named for **Carl Gustav Jacob Jacobi** (1804–1851), an early developer of determinants and their applications in analysis.

We are now ready for the main result.

Proposition 5.4 (Box–Muller transform) *Let* (U_1, U_2) *be independent, continuous uniform random variables, and define:*

$$\begin{aligned} Z_1 &= \sqrt{-2\ln U_1}\cos(2\pi U_2), \\ Z_2 &= \sqrt{-2\ln U_1}\sin(2\pi U_2). \end{aligned} \tag{5.9}$$

Then Z_1 *and* Z_2 *are independent, standard normal variates.*
Proof. *Define a random variable* $R \equiv \sqrt{-2\ln U_1}$. *Recall from (5.1) that* $U : (\mathcal{S}', \mathcal{E}', \lambda') \to ((0,1), \mathcal{B}(0,1), m)$, *and thus* R *is well defined. A calculation obtains the distribution* $F_R(r)$:

$$F_R(r) = 1 - F_U\left(\left[\exp(-r^2/2)\right]^-\right),$$

where $F(x^-)$ *denotes the left limit at* x. *Since* $F_U(x) = x$ *is continuous, it follows that:*

$$F_R(r) = 1 - \exp(-r^2/2),$$

a continuously differentiable distribution. The associated density function is thus given by Proposition III.1.33:

$$f_R(r) = r \exp(-r^2/2).$$

Define the random variable $S = 2\pi U_2$, and so $f_S(s) = 1/2\pi$ on $(0, 2\pi)$. Independence of U_1 and U_2 assures independence of R and S by Proposition II.3.56. Thus as noted above, the joint density $f_{R,S}(r, s)$ exists on $(0, \infty) \times (0, 2\pi)$ and is given by:

$$f_{R,S}(r, s) = \frac{1}{2\pi} r \exp(-r^2/2).$$

Define a transformation $g : (0, \infty) \times (0, 2\pi) \to \mathbb{R}^2$ by:

$$g(r, s) = (r \cos s, r \sin s).$$

The transformation g uniquely identifies the Cartesian coordinates (x, y) associated with the polar coordinates (r, s). Thus:

$$g^{-1}(x, y) = \left(\sqrt{x^2 + y^2}, \arctan\left(\frac{y}{x}\right) \right).$$

Defined as in (5.9):

$$(Z_1, Z_2) = g(R, S),$$

and the joint density function $f_{Z_1, Z_2}(x, y)$ can be obtained from (5.7):

$$\begin{aligned}
f_{Z_1, Z_2}(x, y) &= f_{R,S}\left(\sqrt{x^2 + y^2}, \arctan\left(\frac{y}{x}\right) \right) \left| \det\left(\frac{\partial g^{-1}}{\partial(x, y)} \right) \right| \\
&= \frac{1}{2\pi} \sqrt{x^2 + y^2} \exp(-(x^2 + y^2)/2) \left| \det\left(\frac{\partial g^{-1}}{\partial(x, y)} \right) \right|.
\end{aligned}$$

The Jacobian matrix is calculated:

$$\left(\frac{\partial g^{-1}}{\partial(x, y)} \right) = \begin{pmatrix} \frac{x}{\sqrt{x^2+y^2}} & \frac{y}{\sqrt{x^2+y^2}} \\ \frac{-y}{x^2+y^2} & \frac{x}{x^2+y^2} \end{pmatrix},$$

and this matrix has determinant $1/\sqrt{x^2 + y^2}$.

It follows that:

$$\begin{aligned}
f_{Z_1, Z_2}(x, y) &= \frac{1}{2\pi} \sqrt{x^2 + y^2} \exp(-(x^2 + y^2)/2) \\
&= f_{Z_1}(x) f_{Z_2}(y),
\end{aligned}$$

a product of standard normal density functions. Thus Z_1 and Z_2 are independent, standard normal variates as noted above. ∎

Exercise 5.5 (Independent $\{Z_j\}_{j=1}^{2M}$) *Given independent, continuous uniform $\{U_j\}_{j=1}^{2M}$, assume that (5.9) is applied in pairs, so that Z_{2k} and Z_{2k+1} are generated from U_{2k} and U_{2k+1}. Prove that $\{Z_j\}_{j=1}^{2M}$ are independent, standard normals. Hint: Proposition II.3.56.*

5.1.4   Student T Distribution

Like the normal and lognormal, the distribution function of Student's T with parameter $\nu > 0$ in (2.25) is not analytically tractable, and the inversion provided by Proposition II.4.9 is not feasible. However, a simulation approach reminiscent of the Box-Muller transform was developed by **Ralph W. Bailey** in 1994, which we call the Bailey transform.

Again this method generates pairs of T-variates from pairs of independent, continuous uniform variates, but in this case these T-variates are **not independent**. See Exercise 5.8.

We utilize the same technical tools identified in the prior section.

Proposition 5.6 (Bailey transform) *Let (U_1, U_2) be independent, continuous uniform random variables, and define for $\nu > 0$:*

$$T_1 = \sqrt{\nu \left(U_1^{-2/\nu} - 1 \right)} \cos(2\pi U_2),$$
$$T_2 = \sqrt{\nu \left(U_1^{-2/\nu} - 1 \right)} \sin(2\pi U_2). \tag{5.10}$$

Then T_1 and T_2 are Student T variates with ν degrees of freedom, but are not independent random variables.

Proof. *Define a random variable $R \equiv \sqrt{\nu \left(U_1^{-2/\nu} - 1 \right)}$. Since $U : (\mathcal{S}', \mathcal{E}', \lambda') \to ((0,1), \mathcal{B}(0,1), m)$ from (5.1), R is well defined. A calculation obtains the distribution $F_R(r)$:*

$$F_R(r) = 1 - F_U \left(\left(\exp\left[-\frac{\nu}{2} \ln\left(\frac{r^2 + \nu}{\nu} \right) \right] \right)^- \right),$$

where $F(x^-)$ denotes the left limit at x. By continuity of $F_U(x) = x$ it follows that:

$$F_R(r) = 1 - \left(\frac{r^2 + \nu}{\nu} \right)^{-\nu/2},$$

and the associated density function is given by Proposition III.1.33:

$$f_R(r) = r \left(\frac{r^2 + \nu}{\nu} \right)^{-\nu/2 - 1}.$$

Define the random variable $S = 2\pi U_2$, and thus $f_S(s) = 1/2\pi$ on $(0, 2\pi)$. Independence of U_1 and U_2 assures independence of R and S by Proposition II.3.56. Thus as noted above, the joint density $f_{R,S}(r, s)$ exists on $(0, \infty) \times (0, 2\pi)$ and is given by:

$$f_{R,S}(r, s) = \frac{1}{2\pi} r \left(\frac{r^2 + \nu}{\nu} \right)^{-\nu/2 - 1}.$$

Recall the transformation $g : (0, \infty) \times (0, 2\pi) \to \mathbb{R}^2$ in the proof of Proposition 5.4:

$$g(r, s) = (r \cos s, r \sin s),$$

with inverse:

$$g^{-1}(x, y) = \left(\sqrt{x^2 + y^2}, \arctan(y/x) \right).$$

Defined as in (5.10):

$$(T_1, T_2) = g(R, S),$$

and the joint density function $f_{T_1, T_2}(x, y)$ can be obtained from (5.7):

$$\begin{aligned}
f_{T_1, T_2}(x, y) &= f_{R,S}\left(\sqrt{x^2 + y^2}, \arctan(y/x) \right) \left| \det\left(\frac{\partial g^{-1}}{\partial (x, y)} \right) \right| \\
&= \frac{1}{2\pi} \sqrt{x^2 + y^2} \left(\frac{x^2 + y^2}{\nu} + 1 \right)^{-\nu/2 - 1} \left| \det\left(\frac{\partial g^{-1}}{\partial (x, y)} \right) \right|.
\end{aligned}$$

The Jacobian matrix has determinant $1/\sqrt{x^2 + y^2}$ as above, and thus:

$$f_{T_1, T_2}(x, y) = \frac{1}{2\pi} \left(1 + \frac{x^2 + y^2}{\nu} \right)^{-\nu/2 - 1}. \tag{1}$$

It follows from this that T_1 and T_2 are not independent.

The density function for T_1 is the density of the marginal distribution function, and this follows from Definition 1.11. In detail, the distribution function $F_{T_1,T_2}(x,y)$ is defined:

$$F_{T_1,T_2}(x,y) = \int_{-\infty}^{x} \int_{-\infty}^{y} f_{T_1,T_2}(u,w)\,dudw.$$

The marginal distribution $F_{T_1}(x)$ is then defined:

$$F_{T_1}(x) = \int_{-\infty}^{x} \int_{-\infty}^{\infty} f_{T_1,T_2}(u,w)\,dudw,$$

and thus by Proposition III.1.33, the associated density function is seen to be:

$$f_{T_1}(x) = \int_{-\infty}^{\infty} f_{T_1,T_2}(x,w)\,dw.$$

By symmetry, an analogous expression defines $f_{T_2}(y)$.

To prove that T_1 is a Student T variate, we integrate the joint density in (1):

$$
\begin{aligned}
f_{T_1}(x) &= \frac{1}{2\pi} \int_{-\infty}^{\infty} \left(1 + \frac{x^2 + w^2}{\nu}\right)^{-\nu/2-1} dw \\
&= \frac{1}{\pi} \int_{0}^{\infty} \left(1 + \frac{x^2 + w^2}{\nu}\right)^{-\nu/2-1} dw.
\end{aligned}
\tag{2}
$$

The second step reflects that the integrand is an even function of w, and thus the integral over $(-\infty, 0]$ and $[0, \infty)$ agree.

Substituting $u = \frac{w^2}{\nu}\left(1 + \frac{x^2}{\nu}\right)^{-1}$ obtains:

$$f_{T_1}(x) = \frac{\sqrt{\nu}}{2\pi} \left(1 + \frac{x^2}{\nu}\right)^{-(\nu+1)/2} \int_{0}^{\infty} (1+u)^{-\nu/2-1} u^{-1/2} du.$$

A substitution of $w = \frac{1}{1+u}$ converts this integral into $B\left(\frac{\nu+1}{2}, \frac{1}{2}\right)$. Then using (1.63), (1.57) and (1.59), it can be verified that $f_{T_1}(x)$ is the density function of Student's T with ν degrees of freedom of (2.25). ∎

Remark 5.7 ($\nu \to \infty$) *Note that for $x > 0$, that as $\nu \to \infty$:*

$$\nu\left(x^{-2/\nu} - 1\right) \to -2\ln x.$$

This is equivalent to the observation that as $t \to 0$:

$$\frac{x^{-t} - 1}{t} \to -\ln x,$$

which is seen to follow from the derivative of $f(t) = x^{-t}$ at $t = 0$. Thus for ν large, the Bailey transform is very similar to the Box-Muller transform.

This is consistent with the result of Proposition 6.4 that as $\nu \to \infty$, Student's T converges in distribution to the standard normal. In the notation of Definition II.5.19, $T_\nu \to_d Z$, or $T_\nu \Rightarrow Z$, as $\nu \to \infty$.

Exercise 5.8 (Independent $\{T_j\}_{j=1}^{M}$) *Given independent, continuous uniform $\{U_j\}_{j=1}^{2M}$, assume that (5.10) is applied in pairs, so that T_{2k} and T_{2k+1} are generated from U_{2k} and U_{2k+1}. Prove that $\{T(k)\}_{k=1}^{M}$ are independent, Student T variates with ν degrees of freedom, where $T(k) \in \{T_{2k}, T_{2k+1}\}$ for each k. Hint: Proposition II.3.56.*

5.2 Ordered Random Samples

Because any simulated random sample $\{X_j\}_{j=1}^{M}$ can be reordered into $\{X_{(k)}\}_{k=1}^{M}$ where $X_{(k)} \le X_{(k+1)}$, an ordered random sample can always be created by first generating a random M-sample as in the prior section, then reordering. However, this is inefficient for applications in which we are primarily interested in generating a collection such as $\{X_{(k)_j}\}_{j=1}^{N}$, a random sample of kth **order statistics**. That is, these random variates would each be the kth largest of a random M-sample.

An example from finance is the estimation of the so-called **value at risk (VaR)** at a given percentile or quantile. Here we might be interested in financial losses in what would be the **worst** one year event say, with 95% or 99% confidence. In this case, we are interested in simulating M-samples of one-year losses using large M, and then selecting $X_{(k)}$ from each with $k \simeq 0.95M$ or $k \simeq 0.99M$, respectively. While simply generating N such M-samples, reordering, and choosing the appropriate variates is a feasible approach, it is apparent that this is not an efficient approach as it results in discarding most of the simulated variates.

There are also situations in which we are interested in properties of the conditional distribution of variates, conditional on $X \ge X_{(k)}$. For example, the average value of variates greater than or equal to the identified $X_{(k)}$ variate if often of interest. In the above finance application, the average of such losses is called the **expected shortfall (ES)**, or **conditional value at risk (CVaR)**. In this case, we are interested in generating collections: $\{X_{(k)_j}, X_{(k+1)_j}, ..., X_{(M)_j}\}_{j=1}^{N}$. Again, one can in theory generate full M-samples and select those of interest, though in practice this is often not the most efficient option.

5.2.1 Direct Approaches

Given a random variable X with distribution function F, assume that our goal is to simulate $\{X_{(k)_j}\}_{j=1}^{N}$, an N-random sample of kth **order statistics**. Thus each random variate represents the kth largest of a random M-sample. As noted above, such k may be a function of M such as $k \simeq 0.95M$ or $k \simeq 0.99M$, and then $X_{(k)_j}$ represents an estimate of the 95th or 99th quantile of X.

There are several direct approaches to achieving this objective:

1. **Complete/Partial Ordered Samples:**

   Beginning with MN continuous, uniformly distributed random variables $\{U_i\}_{i=1}^{MN}$, generate N sets of M-samples $\{X_i\}_{i=1}^{M}$ of X variates as in the prior sections:

   $$\left\{ \{F^*(U_i)\}_{i=(j-1)M+1}^{jM} \right\}_{j=1}^{N}.$$

   Here we assume that $X_i = F^*(U_i)$ for notational simplicity, but this is not essential. In the case of normal and Student T variates, for example, these variates would be generated from pairs of uniform variates and the formulas in (5.9) and (5.10).

   Each M-sample $\{F^*(U_i)\}_{i=(j-1)M+1}^{jM}$ of X variates is then reordered to determine $\{X_{(k)_j}\}_{k=1}^{M}$:

   $$\{F^*(U_i)\}_{i=(j-1)M+1}^{jM} \to \left\{X_{(k)_j}\right\}_{k=1}^{M}, \qquad j = 1, ..., N,$$

resulting in N complete samples of order statistics: $\{\{X_{(k)_j}\}_{k=1}^M\}_{j=1}^N$, from which specific variates, or ranges of variates can be selected.

Alternatively, since $F^*(y)$ is increasing by Proposition II.3.16, one can instead reorder each collection $\{U_i\}_{i=(j-1)M}^{jM}$ to $\{U_{(k)_j}\}_{k=1}^M$, and then note that:

$$\left\{\left\{X_{(k)_j}\right\}_{k=1}^M\right\}_{j=1}^N = \left\{\left\{F^*\left(U_{(k)_j}\right)\right\}_{k=1}^M\right\}_{j=1}^N.$$

Thus it is not necessary to evaluate $F^*(U_i)$ for every U_i in the various M-samples unless all order statistics are required. For example, if only a sample of $X_{(k)}$ is required for a fixed k, then only $F^*\left(U_{(k)_j}\right)$ need be evaluated for such $U_{(k)_j}$.

When seeking ordered samples for normal, lognormal and Student T, this alternative approach is not an option since the ordering of these variates is not predictable based on the orderings of the pairs of continuous, uniform variates.

Summary: When $X_i = F^*(U_i)$, this approach requires MN continuous, uniformly distributed random variables $\{U_i\}_{i=1}^{MN}$, and KN evaluations of $F^*\left(U_{(k)_j}\right)$ for $1 \le K \le M$, to produce an N-sample of the order statistics $\{\{X_{(k_i)_j}\}_{i=1}^K\}_{j=1}^N$.

When X_i is generated from pairs of uniform variates, this approach requires MN (normal, lognormal) or $2MN$ (Student T) continuous, uniformly distributed random variables, and an equal number of evaluations of the formulas in (5.9) and (5.10).

2. **Direct kth Order Statistics 1; when $X = F^*(U)$ is calculable**

Recall Example 3.9, that the kth order statistic $U_{(k)}$ from an M-sample of continuous, uniform variates has a beta distribution with $v = k$ and $w = M-k+1$. This would appear promising for the generation of N-samples $\{U_{(k)_j}\}_{j=1}^N$ of kth order uniform variates, and then N-samples $\{X_{(k)_j}\}_{j=1}^N$ in the case where $X_i = F^*(U_i)$.

But based on item 4 of the prior section, to generate one beta random variable with $v = k$ and $w = M - k + 1$ requires $v + w = M + 1$ continuous, uniformly distributed random variables. Thus to generate $\{U_{(k)_j}\}_{j=1}^N$ requires $\{U_i\}_{i=1}^{(M+1)N}$, and this is more than that needed for the complete sample method in item 1 despite seeking only a sample of one order variate.

Summary: When $X_i = F^*(U_i)$, this approach requires $(M + 1)N$ continuous, uniformly distributed random variables $\{U_i\}_{i=1}^{(M+1)N}$, with each $(M + 1)$-sequence used to produce one beta $U_{(k)_j}$ as above, and then N evaluations of $F^*\left(U_{(k)_j}\right)$ to produce an N-sample of the kth ordered statistics $\{X_{(k)_j}\}_{j=1}^N$.

3. **Direct kth Order Statistics 2; when $X_{(k)} = F_{(k)}^*(U)$ is calculable**

Since the distribution function $F_{(k)}(x)$ for $X_{(k)}$ in (3.5) is known, if it is possible to determine its left-continuous inverse $F_{(k)}^*(y)$, then by definition only one N-sample $\{U_j\}_{j=1}^N$ would be needed to obtain a sample of $X_{(k)}$:

$$\left\{X_{(k)_j}\right\}_{j=1}^N = \left\{F_{(k)}^*(U_j)\right\}_{j=1}^N.$$

However, since even the mathematically simple uniform distribution produced $F_{(k)}(x)$ with the complexity of the beta distribution, it is unlikely that many examples will be encountered for which one can calculate $F_{(k)}^*(y)$ directly. But in general, to determine $F_{(k)}^*(U)$ using (3.5) requires two steps, the first of which is almost surely numerical:

(a) Solve for V:

$$U = \sum_{j=k}^{M} \binom{M}{j} V^j (1-V)^{M-j}.$$

(b) Evaluate $X_{(k)}$:

$$X_{(k)} = F^*(V).$$

Summary: Requires N uniformly distributed continuous random variables $\{U_j\}_{j=1}^N$, and N evaluations or estimations of $\{V_j\}_{j=1}^N$ from step a, then N estimates of $\{X_{(k)_j}\}_{j=1}^N$ from step b, to produce an N-sample of the kth order statistics $\{X_{(k)_j}\}_{j=1}^N$.

5.2.2 The Rényi Representation

Given a random variable X with distribution function F, the goal of this section is to again generate $\{X_{(k)_j}\}_{j=1}^N$, a random sample of kth **order statistics** from random M-samples. While several approaches to achieving this objective were discussed above, in this section, we provide a fourth approach based on the **Rényi representation theorem** of Proposition 3.30. While this representation theorem generates ordered **exponential** random variables, it is then a simple matter to convert these variates to ordered continuous, uniform random variables.

This will involve the following result, which is an application of results from Book II.

Proposition 5.9 (Exponenial $\Leftrightarrow$ Uniform) *For any n, if the collection $\{U_j\}_{j=1}^n$ are independent, continuous uniform random variables, then:*

$$\left\{X_j^E\right\}_{j=1}^n \equiv \left\{-\ln\left(1-U_j\right)\right\}_{j=1}^n,$$

are independent, standard exponential random variables, meaning with parameter $\lambda = 1$. Conversely, given independent standard exponentials $\left\{X_j^E\right\}_{j=1}^n$, then:

$$\{U_j\}_{j=1}^n \equiv \left\{1-\exp\left(-X_j^E\right)\right\}_{j=1}^n,$$

are independent, continuous uniform random variables.
Proof. *By Proposition II.3.22, if a distribution function F is continuous and strictly increasing on $D = \{x|0 < F(x) < 1\}$, then $F^* = F^{-1}$. By (1.54), $F_E(x) = 1 - e^{-x}$ has these properties on $D = (0, \infty)$, and so $F_E^{-1}(y) = -\ln(1-y)$. The result above can thus be stated:*

1. *If $\{U_j\}_{j=1}^n$ are independent, continuous uniform random variables, then $\{F_E^*(U_j)\}_{j=1}^n$ are independent with distribution function F_E.*

2. *If $\{X_j\}_{j=1}^n$ are independent random variables with distribution function F_E, then $\{F_E(X_j)\}_{j=1}^n$ are independent continuous uniform random variables.*

Both the distributional results and the independence results are Proposition II.4.9 in the introduction to this chapter, since F_E is continuous. ∎

Remark 5.10 $(U \Leftrightarrow 1 - U)$ *Because U and $1 - U$ have the same distribution, given $\{U_j\}_{j=1}^{n}$, we can in Proposition 5.9 optionally generate independent standard exponentials using:*

$$\{X_j^E\}_{j=1}^{n} \equiv \{-\ln(U_j)\}_{j=1}^{n}.$$

By the same argument, given exponential $\{X_j^E\}_{j=1}^{n}$, we can optionally generate independent uniform variables using:

$$\{U_j\}_{j=1}^{n} \equiv \{\exp(-Y_j^E)\}_{j=1}^{n}.$$

*But in both cases, this **reverses the order statistics**.*

For example, if exponential kth order statistics $\left\{X_{(k)}^E\right\}_{k=1}^{M}$ are simulated using the Rényi representation theorem, then $\{U_{(k)}\}_{k=1}^{M} \equiv \left\{1 - \exp\left(-X_{(k)}^E\right)\right\}_{k=1}^{M}$ will be kth order statistics of the uniform distribution. However:

$$\left\{\widehat{U}_{(k)}\right\}_{k=1}^{M} \equiv \left\{\exp\left(-X_{(k)}^E\right)\right\}_{k=1}^{M} = \left\{U_{(M-k+1)}\right\}_{k=1}^{M},$$

will be uniformly distributed kth order statistics in the reverse order.

Returning to the goal of generating $\{X_{(k)_j}^E\}_{j=1}^{N}$, a random sample of N, kth order exponential statistics, we add a fourth approach to those in the above section.

4. Direct kth Order Statistics 3: Rényi Representation

Assume that we want to simulate kth order statistics of a random variable X with distribution function $F(x)$ for which we can use Proposition II.4.9 from the introduction to simulate independent $\{X_j\}_{j=1}^{n}$ by $X_j = F^*(U_j)$, where $\{U_j\}_{j=1}^{n}$ are independent, continuous uniform random variates. As above, these kth order statistics are defined relative to a sample of M variates.

Here we use the Rényi representation to simulate $\{X_{(k)_j}^E\}_{j=1}^{N}$, N independent kth order exponential statistics, then Proposition 5.9 to obtain $\{U_{(k)_j}\}_{j=1}^{N}$, independent kth order, continuous uniform statistics, and finally applying Proposition II.4.9, obtain $\{X_{(k)_j}\}_{j=1}^{N}$, independent kth order statistics for the random variable X. Remark 5.10 can then be applied to minimize simulations depending on the relative size of k versus M.

Recall that by Corollary 3.31 to the Rényi representation theorem, each exponential kth order statistic $X_{(k)}^E$ requires the generation of k standard exponentials and an application of (3.27).

(a) When k is small relative to M, we implement the following N times:

    i. Given independent, continuous uniform variates $\{U_j\}_{j=1}^{k}$, generate independent standard exponential random variables $\{X_j^E\}_{j=1}^{k}$ by defining $X_j^E = -\ln(1 - U_j)$.

    ii. Calculate $X_{(k)}^E$ as in (3.27) with $\lambda = 1$, defining $E_j = X_j^E$.

    iii. Define $U_{(k)} = 1 - \exp\left(-X_{(k)}^E\right)$, producing a kth order statistic of the continuous, uniform distribution.

    iv. Evaluate $X_{(k)} = F^*\left(U_{(k)}\right)$.

(b) When k is large relative to M, implement the following N times:

    i. Given independent, continuous uniform variates $\{U_j\}_{j=1}^{M-k+1}$, generate independent standard exponential random variables $\{X_j^E\}_{j=1}^{M-k+1}$ by defining $X_j^E = -\ln(1 - U_j)$.

    ii. Calculate $X_{(M-k+1)}^E$ as in (3.27) with $\lambda = 1$, defining $E_j = X_j^E$.

    iii. Define $U_{(k)} = \exp\left(-X_{(M-k+1)}^E\right)$, producing a kth order statistic of the continuous, uniform distribution.

    iv. Evaluate $X_{(k)} = F^*\left(U_{(k)}\right)$.

Summary: For an N-sample of kth order statistics of X, this requires $N' \equiv N \times \min\{k, M - k + 1\}$ uniformly distributed continuous random variables $\{U_i\}_{i=1}^{N'}$, and N evaluations of $F^*\left(U_{(k)}\right)$

6

Limit Theorems

Continuing the investigations of Chapters II.5 and II.8, in this chapter we study limit theorems grouped into three categories:

- Weak Convergence of Distributions

- Laws of Large Numbers

- Convergence of Empirical Distribution Functions

This study will be supplemented in Book VI with more general results related to weak convergence of measures and the central limit theorem.

6.1 Introduction

Throughout this chapter we will repeatedly encounter the notion of a collection of n independent random variables, often identically distributed but sometimes not, and will typically want to address measure-theoretic questions related to the summation of such variables, and the limits of such sums as $n \to \infty$. This appears intuitively plausible and yet requires some justification to avoid building the house of emerging theories on a foundation of sand.

Among the questions that need to be addressed are:

Q1. Given a random variable $X : (\mathcal{S}, \mathcal{E}, \lambda) \to (\mathbb{R}, \mathcal{B}(\mathbb{R}), m)$ with distribution function $F(x)$, we will want to study properties of a sample $\{X_j\}_{j=1}^{n}$ of X. By a sample or n-sample is meant (Definition 3.2) that these are independent random variables defined on some probability space, that are identically distributed as X.

The fundamental questions are:

How is this sample constructed, and on what probability space is this sample defined?

If $\{X_j\}_{j=1}^{n}$ is then such a sample of X defined on the space identified, then an expression like $\sum_{j=1}^{n} X_j$ is again a random variable on this same space by Proposition I.3.30. Thus the measure on this space provides meaning to probability statements on such summations, and underlies the definition of the associated distribution function.

Q2. If $\{X_j\}_{j=1}^{n}$ is a sample of X defined on the probability space identified in Q1, what does it mean to let $n \to \infty$? Is this increasing collection still defined on the same space so that probability statements regarding this sum, or properties of its distribution function, are all defined relative to a given probability measure?

Q3. More generally, if $X_j : (\mathcal{S}_j, \mathcal{E}_j, \lambda_j) \to (\mathbb{R}, \mathcal{B}(\mathbb{R}), m)$ are random variables with distribution functions $F_j(x)$, is there a common probability space on which all can be defined, and on which these will be independent random variables? If so, then an expression like

DOI: 10.1201/9781003264583-6

$\sum_{j=1}^{n} X_j$ is well defined. In this more general context, Q2 again applies: what does it mean to let $n \to \infty$?

In Chapter II.4, the construction sought in Q1 was developed using the infinite dimensional probability space theory of Chapter I.9. While this construction targeted the application to Q1, the Book I theory applied equally well to the more general construction needed for Q3. We summarize below the Book II construction in this more general context, noting that the Book II result can be viewed as a change of notation. While the Book II construction allowed for probability spaces for either finite of infinite samples, we summarize the infinite dimensional models to simultaneously address Q2.

To this end, assume that we are given random variables $\{X_j\}_{j=1}^{\infty}$ defined on probability spaces $\{(\mathcal{S}_j, \mathcal{E}_j, \lambda_j)\}_{j=1}^{\infty}$ with distribution functions $\{F_j\}_{j=1}^{\infty}$. In the Q1 application of Book II, we started with an infinite collection of copies of X and $(\mathcal{S}, \mathcal{E}, \lambda)$, and the indexing just allowed a basis for referencing members of these collections. As summarized in Propositions 1.3 and 1.5, each F_j is increasing and right continuous, and gives rise to a Borel measure μ_{F_j} defined on $\mathcal{B}(\mathbb{R})$ that is defined on right semi-closed intervals by:

$$\mu_{F_j} ((a, b]) \equiv F_j(b) - F_j(a) \equiv \lambda_j \left(X_j^{-1} ((a, b]) \right).$$

It then follows by extension that for all $A \in \mathcal{B}(\mathbb{R})$:

$$\mu_{F_j} (A) = \lambda_j \left(X_j^{-1}(A) \right). \tag{6.1}$$

Thus $\mu_{F_j} (\mathbb{R}) = 1$ and $(\mathbb{R}, \mathcal{B}(\mathbb{R}), \mu_{F_j})$ is a probability space for all j.

We can now apply the construction of Chapter I.9 to $\{(\mathbb{R}_j, \mathcal{B}(\mathbb{R}_j), \mu_{F_j})\}_{j=1}^{\infty}$, where $\mathbb{R}_j$ and $\mathcal{B}(\mathbb{R}_j)$ are indexed only for notational purposes. In addition, all μ_{F_j} will equal μ_F in the Q1 context of a single random variable, but are also allowed to be different to accommodate the Q3 model.

The infinite dimensional probability space $(\mathbb{R}^{\mathbb{N}}, \sigma(\mathcal{A}^+), \mu_{\mathbb{N}})$, and a complete counterpart $(\mathbb{R}^{\mathbb{N}}, \sigma(\mathbb{R}^{\mathbb{N}}), \mu_{\mathbb{N}})$, can now be constructed as in Chapter I.9. For this probability space:

$$\mathbb{R}^{\mathbb{N}} \equiv \{(x_1, x_2, ...) | x_j \in \mathbb{R}_j\},$$

and $\mu_{\mathbb{N}}$ is uniquely defined on $\sigma(\mathcal{A}^+)$, the smallest sigma algebra containing the algebra $\mathcal{A}^+$ of **general finite dimensional measurable rectangles** or **general cylinder sets in $\mathbb{R}^{\mathbb{N}}$**. Thus:

$$\mathcal{A}^+ \subset \sigma(\mathcal{A}^+) \subset \sigma(\mathbb{R}^{\mathbb{N}}). \tag{6.2}$$

In more detail, $H \in \mathcal{A}^+$ if for some positive integer n and n-tuple of positive integers $J = (j(1), j(2), ..., j(n))$:

$$H = \{x \in \mathbb{R}^{\mathbb{N}} | (x_{j(1)}, x_{j(2)}, ... x_{j(n)}) \in A\},$$

where $A \in \mathcal{B}(\mathbb{R}^n)$, the Borel sigma algebra on $\mathbb{R}^n$. Further, $\mu_{\mathbb{N}}$ is defined on $\mathcal{A}^+$ by:

$$\mu_{\mathbb{N}}(H) = \mu_F^{(n)}(A),$$

where $\mu_F^{(n)}$ is the product measure on $\mathbb{R}^n$ induced by $\{\mu_{F_{j(k)}}\}_{k=1}^{n}$ and constructed in Chapter I.7.

In the special case where A in the definition of $H \in \mathcal{A}^+$ is given by a rectangle, $A = \prod_{k=1}^{n} A_{j(k)}$ for $A_{j(k)} \in \mathcal{B}(\mathbb{R}_{j(k)})$:

$$\mu_{\mathbb{N}}(H) = \prod_{k=1}^{n} \mu_{F_{j(k)}}(A_{j(k)}). \tag{6.3}$$

Now define $X_j' : \mathbb{R}^{\mathbb{N}} \to \mathbb{R}$ by:

$$X_j' : (x_1, x_2, ...) = x_j, \tag{6.4}$$

which is the **projection mapping defined** on $\mathbb{R}^{\mathbb{N}}$ to the *jth* coordinate, and often denoted π_j. We verify below that such X_j' is measurable, and thus a random variable on this space.

The following result is a modest generalization of Proposition II.4.4. It uses the same Book I theory, and answers the above questions.

For the application to Q1, all $(\mathcal{S}_j, \mathcal{E}_j, \lambda_j) = (\mathcal{S}, \mathcal{E}, \lambda)$, all $X_j = X$, meaning each has the distribution of X, and thus all $F_j = F$. Then for the conclusion, each X_j' has the distribution of X. In other words, $\{X_j'\}_{j=1}^{\infty}$ are independent and identically distributed in this case. This is essentially the context of Proposition II.4.4.

Proposition 6.1 (A probability space for independent $\{X_j\}_{j=1}^{\infty}$) *Let probability spaces $\{(\mathcal{S}_j, \mathcal{E}_j, \lambda_j)\}_{j=1}^{\infty}$ and random variables $X_j : \mathcal{S}_j \longrightarrow \mathbb{R}$ with distribution functions $\{F_j\}_{j=1}^{\infty}$ be given. With the notation above, let $(\mathcal{S}', \mathcal{E}', \lambda')$ denote $(\mathbb{R}^{\mathbb{N}}, \sigma(\mathbb{R}^{\mathbb{N}}), \mu_{\mathbb{N}})$ or $(\mathbb{R}^{\mathbb{N}}, \sigma(\mathcal{A}^{+}), \mu_{\mathbb{N}})$ of Proposition I.9.20.*

Then $\{X_j'\}_{j=1}^{\infty}$ as defined in (6.4) is a sample of $\{X_j\}_{j=1}^{\infty}$, meaning these are independent random variables defined on $(\mathcal{S}', \mathcal{E}', \lambda')$ with respective distribution functions $\{F_j\}_{j=1}^{\infty}$.

Proof. *First, X_j' is measurable and thus a random variable on $\mathbb{R}^{\mathbb{N}}$ since for $A \in \mathcal{B}(\mathbb{R})$, (6.4) yields:*

$$\left(X_j'\right)^{-1}(A) = \{x \in \mathbb{R}^{\mathbb{N}} | x_j \in A\} \in \mathcal{A}^{+}.$$

Thus $\left(X_j'\right)^{-1}(A)$ is a general cylinder set and an element of either sigma algebra by (6.2).

Further, if $A \in \mathcal{B}(\mathbb{R})$ and $H \equiv \left(X_j'\right)^{-1}(A)$, then from (6.3) and (6.1):

$$\mu'\left[\left(X_j'\right)^{-1}(A)\right] = \mu_{F_j}(A) \equiv \lambda_j(X_j^{-1}(A)),$$

and thus X_j' and X_j have the same distribution.

Next, let $J = (j(1), j(2), ..., j(n))$ and $A_{j(k)} \in \mathcal{B}(\mathbb{R}_{j(k)})$ be given. By (6.4):

$$\bigcap_{k=1}^{n} \left(X_{j(k)}'\right)^{-1}(A_{j(k)}) = \left\{x \in \mathbb{R}^{\mathbb{N}} | (x_{j(1)}, x_{j(2)}, ... x_{j(n)}) \in \prod_{k=1}^{n} A_{j(k)}\right\}.$$

Thus by (6.3) and (6.1):

$$\mu'\left(\bigcap_{k=1}^{n} \left(X_{j(k)}'\right)^{-1}(A_{j(k)})\right) = \prod_{k=1}^{n} \mu_{F_{j(k)}}(A_{j(k)}) \equiv \prod_{k=1}^{n} \mu'\left(\left(X_{j(k)}'\right)^{-1}(A_{j(k)})\right),$$

and thus $\{X_{j(k)}'\}_{k=1}^{n}$ are independent random variables. ∎

Notation 6.2 *It is somewhat of a notational burden to distinguish between $\{X_j\}_{j=1}^{\infty}$ defined on $\{(\mathcal{S}_j, \mathcal{E}_j, \lambda_j)\}_{j=1}^{\infty}$, and independent $\{X_j'\}_{j=1}^{\infty}$ defined on $(\mathcal{S}', \mathcal{E}', \lambda')$. So this formality is often suppressed and statements are made about independent, and possibly identically distributed random variables $\{X_j\}_{j=1}^{\infty}$ defined on a probability space $(\mathcal{S}, \mathcal{E}, \mu)$.*

With the construction of Proposition 6.1, the above questions can be answered:

If $\{X_j\}_{j=1}^{n}$ are random variables defined on the new space $(\mathcal{S}, \mathcal{E}, \mu)$, then $\sum_{j=1}^{n} X_j$ is a measurable function on $(\mathcal{S}, \mathcal{E}, \mu)$ by Proposition I.3.30 and thus a random variable. Since probability statements on such $\sum_{j=1}^{n} X_j$ are made in this probability space for any n, these statements remain well defined as $n \to \infty$.

Remark 6.3 *The above framework will not be formally mentioned again in the next two sections on weak convergence of distributions and laws of large numbers.*

In the final section on convergence of empirical distributions functions, some additional comments will be required.

6.2   Weak Convergence of Distributions

In this section, we develop several important results relating to the **weak convergence** (Definition II.8.2) of certain sequences of distribution functions. Recall that given a sequence $\{F_n\}_{n=1}^{\infty}$ of distribution functions, we say that F_n **converges weakly** to a distribution function F, denoted:

$$F_n \Rightarrow F,$$

if $F_n(x) \to F(x)$ for every continuity point x of F. This same terminology is often used when these functions are increasing, and not necessarily distribution functions.

If X_n and X are random variables defined on a probability space $(\mathcal{S}, \mathcal{E}, \lambda)$ with such distributions, which exist by Proposition 1.4, this notion of weak convergence can also be expressed by:

$$X_n \Rightarrow X, \text{ or, } X_n \to_d X,$$

as noted in Remark II.8.3.

The first section provides a proof that **Student's T distribution** converges weakly to the normal distribution as the number of degrees of freedom $n \to \infty$. This result belongs in this section, but it is also somewhat out of place because this proof utilizes a law of large numbers from the next section.

The next result is the **Poisson limit theorem**, which was stated and proved in Proposition II.1.11 using explicit calculations and the given probability density functions. In this section, we prove this result using Proposition 4.77 by showing convergence of the associated moment generating functions.

This same approach will then be applied to a generalization of the Poisson result often called the **weak law of small numbers,** and then to the **De Moivre-Laplace theorem,** and to a special case of the **central limit theorem.** Using other tools we then turn to results on quantiles and order statistics.

6.2.1   Student T to Normal

Recall the density function of Student's T distribution with $n \in \mathbb{N}$ degrees of freedom in (2.24):

$$f_T(t) = \frac{\Gamma\left((n+1)/2\right)}{\sqrt{\pi n}\,\Gamma\left(n/2\right)} \left(1 + \frac{t^2}{n}\right)^{-(n+1)/2},$$

where the gamma function $\Gamma\left(\alpha\right)$ is defined in (1.56), and several properties identified there. This density is well defined with n replaced by real $\nu > 0$ as seen in (2.25).

Below we prove that if F_{T_n} is the associated distribution function of this variate, and Φ is the distribution function of the standard normal, then as $n \to \infty$:

$$F_{T_n} \Rightarrow \Phi,$$

meaning $F_{T_n}(x) \to \Phi(x)$ **for all** x since $\Phi(x)$ is continuous. It is intriguingly easy to get close to a proof of this, but most approaches run into technical problems without more advanced tools.

For example, a proof using moment generating functions and Proposition 4.77 as for other results below might seem appealing until one realizes that F_{T_n} has no moment generating function. Indeed, for no n does F_{T_n} have all moments as required by Proposition 4.25. Using (4.15), the *mth* moment of this distribution is given by:

$$E[T_n^m] = \frac{\Gamma\left((n+1)/2\right)}{\sqrt{\pi n}\,\Gamma\left(n/2\right)} \int t^m \left(1 + \frac{t^2}{n}\right)^{-(n+1)/2} dt.$$

At $\pm\infty$, this integrand has order of magnitude $O(t^{m-(n+1)})$, and thus this integral will converge only when $m-(n+1) < -1$, or for $m \le n-1$. However, using the more powerful tools of **characteristic functions** in Book VI, a proof of this result in the flavor of Proposition 4.77 will be possible.

More directly, for all $s \in \mathbb{R}$:

$$n \ln (1 + s/n) \longrightarrow s \text{ as } n \to \infty.$$

This follows from the definition of $f'(0)$ for $f(x) = \ln(1 + sx)$, noting that this function is well-defined in an open interval about $x = 0$ for any s. This then obtains:

$$(1 + s/n)^n \longrightarrow e^s \text{ as } n \to \infty. \tag{6.5}$$

Applying this to $f_{T_n}(t)$:

$$\left(1 + \frac{t^2}{n}\right)^{-(n+1)/2} = \left[\left(1 + \frac{t^2}{n}\right)^n\right]^{-(n+1)/2n} \to e^{-t^2/2},$$

which is the functional part of the standard normal density function $\varphi(t)$ in (1.66). The coefficient derivation, that:

$$\frac{\Gamma\left((n+1)/2\right)}{\sqrt{\pi n}\,\Gamma\left(n/2\right)} \to \frac{1}{\sqrt{2\pi}},$$

will certainly be messier and is not pursued here. But even once done it will still take an integration to the limit result (recall Section III.2.6) to prove that $f_{T_n}(t) \to \varphi(t)$ **for all** t, implies that for all x :

$$F_{T_n}(x) = \int_{-\infty}^x f_{T_n}(t)dt \to \int_{-\infty}^x \varphi(t)dt = \Phi(x).$$

These details can be settled successfully, but there is a proof that avoids such technical details. It uses a weak law of large numbers result (6.30) of Proposition 6.35, and then Slutsky's theorem from Book II.

Proposition 6.4 ($F_{T_n} \Rightarrow \Phi$) *Let $F_{T_n}(x)$ denote the distribution function of Student's T with $n \in \mathbb{N}$ degrees of freedom, and $\Phi(x)$ the distribution function of the standard normal. Then as $n \to \infty$:*

$$F_{T_n} \Rightarrow \Phi. \tag{6.6}$$

Proof. *Recall from item 3 of Example 2.28 that $f_{T_n}(t)$ is the density function of the random variable:*

$$T_n \equiv \frac{X}{\sqrt{Y/n}},$$

where X is standard normal, and Y is chi-squared with n degrees of freedom.

If $\{X_i\}_{i=1}^n$ *are independent standard normals, then $\{X_i^2\}_{i=1}^n$ are independent by Proposition II.3.56, and then $\sum_{i=1}^n X_i^2$ is chi-squared with n degrees of freedom by item 5 of Section 4.4.1. Thus by (6.30):*

$$\frac{Y}{n} = \frac{1}{n}\sum_{i=1}^n X_i^2 \to_P E\left[X_1^2\right] = 1.$$

Convergence in probability was introduced in Definition II.5.11, and the above results means that for every $\epsilon > 0$:

$$\lim_{n\to\infty} \Pr\left(\left|\frac{Y}{n} - 1\right| \ge \epsilon\right) = 0.$$

It then follows that $\sqrt{Y/n} \to_P 1$ since:

$$\left\{ \left| \sqrt{\frac{Y}{n}} - 1 \right| \geq \epsilon \right\} \subset \left\{ \left| \frac{Y}{n} - 1 \right| \geq 2\epsilon + \epsilon^2 \right\}.$$

Further, $X \to_d X$ by definition, since by defining $X'_n = X$ for all n obtains that $X'_n \to_d X$.

Summarizing, $\sqrt{Y/n} \to_P 1$ and $X \to_d X$. By item 3 of Slutsky's theorem of Proposition II.5.29, it then follows that:

$$\frac{X}{\sqrt{Y/n}} \to_d X,$$

and the proof is complete. ■

6.2.2 Poisson Limit Theorem

The following result, named for **Siméon-Denis Poisson** (1781–1840), was proved in Proposition II.1.11 using explicit calculations with density functions and limiting arguments. Here we implement the method of moments to prove that properly parametrized, the binomial distribution converges weakly to the Poisson distribution.

Note that the binomial variate $B_{n,\lambda/n}$ in (6.7) of this result is by definition a summation:

$$B_{n,\lambda/n} \equiv \sum_{m=1}^{n} X_{1,m}^{B},$$

where for each n, $\{X_{1,m}^{B}\}_{m=1}^{n}$ are independent, standard binomial variables with $p = \lambda/n$. Then by (4.39), (4.43), and (4.62):

$$\mu_{B_{n,\lambda/n}} = \lambda, \quad \sigma_{B_{n,\lambda/n}}^2 = \lambda(1 - \lambda/n).$$

Thus:

$$\mu_{B_{n,\lambda/n}} = \mu_{P_\lambda}, \quad \sigma_{B_{n,\lambda/n}}^2 \to \sigma_{P_\lambda}^2 \text{ as } n \to \infty,$$

where μ_{P_λ} and $\sigma_{P_\lambda}^2$ are the mean and variance of the Poisson with parameter λ by (4.69). Remarkably, this next result proves that not only do a couple of binomial moments converge to the Poisson moments, but the entire binomial distribution function converges to the Poisson distribution.

We will see in the next section that the Poisson can also be the weak limit of sums of independent but nonhomogeneous binomials, meaning with different values of p.

Proposition 6.5 (Poisson limit theorem) *Let $F_{B_{n,p}}$ denote the distribution function of the general binomial with parameters n, p associated with the density function in (1.40), and $F_{P_\lambda}(j)$ the distribution function of the Poisson with parameter λ associated with the density function in (1.47).*

If $\lambda = np$ is fixed, then as $n \to \infty$:

$$F_{B_{n,\lambda/n}} \Rightarrow F_{P_\lambda}. \tag{6.7}$$

Proof. *Let $M_{B_{n,\lambda/n}}(t)$ and $M_{P_\lambda}(t)$ denote the respective moment generating functions of (4.63) and (4.68):*

$$M_{B_{n,\lambda/n}}(t) = \left(1 + \frac{\lambda}{n}(e^t - 1)\right)^n,$$

and:

$$M_{P_\lambda}(t) = \exp[\lambda(e^t - 1)].$$

By (6.5), $M_{B_{n,\lambda/n}}(t) \to M_{P_\lambda}(t)$ for all t, and thus $F_{B_{n,\lambda/n}} \Rightarrow F_{P_\lambda}$ by Proposition 4.77. ■

Example 6.6 (Modeling bond defaults; insurance losses) *The Poisson limit theorem has immediate applications in finance in any situation in which one is modeling the binomial outcomes of a large group with each member of the group having the same or similar probabilities of the event being observed.*

For example, in a relatively large portfolio of bonds with similar credit ratings, or similarly rated bank loans of various types, the event of default over a given period is fundamentally binomial with p equalling the probability of default over this period. Assume we have n = 200 loans, each with default probability p = 0.02 in one year, and define the random variable N equal to the number of defaults. One can then model N exactly as the sum of 200 binomials, or approximately as a Poisson random variable with $\lambda = np = 4$. It is then true that in either model, the assumption of independence from loan to loan is often reasonable, though the assumed value of p is highly dependent on the economic cycle.

One can similarly model a variety of insurance loss events this way. For example, death, disability, hospitalization, etc., can be modeled as sums of binomials or approximately Poisson random variables, as long as the group being modeled is reasonably homogeneous and individuals have similar values for p. Similarly, various automobile and homeowner insurance loss events can be modeled with binomials or Poisson random variables when the groups modeled are reasonably homogeneous in terms of claim probabilities. Independence is again often a reasonable assumption.

6.2.3 "Weak Law of Small Numbers"

The weak law of small numbers is the tongue-in-cheek name given to various generalizations of Poisson's limit theorem which weaken the assumption underlying (6.7), that $B_{n,\lambda/n} = \sum_{m=1}^{n} X_m^B$ where $\{X_m^B\}_{m=1}^n$ are independent and **identically distributed** standard binomial variables with $p = \lambda/n$. Two versions of this small law assume that for each n, $\{X_{n_m}^B\}_{m=1}^n$ are independent, standard binomial variables with parameters $\{p_{n_m}\}_{m=1}^n$, where:

- $\sum_{m=1}^{n} p_{n_m} = \lambda$ for each n;

- $\sum_{m=1}^{n} p_{n_m} \to \lambda$ as $n \to \infty$.

This result can even be generalized further to more general "binomials" which assume that:

$$\Pr\left[X_{n_m}^B = 1\right] = p_{n_m}$$
$$\Pr\left[X_{n_m}^B = 0\right] = 1 - p_{n_m} - \epsilon_{n_m},$$
$$\Pr\left[X_{n_m}^B \geq 2\right] = \epsilon_{n_m},$$

where $\max_m\{\epsilon_{n_m}\} \to 0$ as $n \to \infty$.

In order to ensure that all of the probabilities p_{n_m} become small as $n \to \infty$, and thus this remains a law of "small numbers," it is necessary to also require that $\max_m\{p_{n_m}\} \to 0$ as $n \to \infty$.

Below we state and prove a version of intermediate generality, but first a definition.

Definition 6.7 (Triangular array) *A collection of random variables $\{\{X_{n,m}\}_{m=1}^n\}_{n=1}^\infty$ is called a **triangular array** if for each n, the random variables $\{X_{n,m}\}_{m=1}^n$ are independent. The same definition applies if $1 \leq m \leq m_n$ where $m_n \to \infty$ as $n \to \infty$, though we will have no need of the more general notion.*

Proposition 6.8 (Weak law of small numbers) *Let $\{\{X_{n,m}\}_{m=1}^{n}\}_{n=1}^{\infty}$ be a triangular array where for each n, $\{X_{n_m}^{B}\}_{m=1}^{n}$ are independent, standard binomial variables with parameters $\{p_{n_m}\}_{m=1}^{n}$, where as $n \to \infty$:*

$$\sum_{m=1}^{n} p_{n_m} \to \lambda > 0 \quad and \quad \max_{m}\{p_{n_m}\} \to 0.$$

If $S_n = \sum_{m=1}^{n} X_{n_m}^{B}$, then as $n \to \infty$:

$$F_{S_n} \Rightarrow F_{P_\lambda}, \tag{6.8}$$

where F_{P_λ} denotes the distribution function of the Poisson with parameter λ.
Proof. *By (4.48):*

$$M_{S_n}(t) = \prod_{m=1}^{n} \left(1 + p_{n_m}(e^t - 1)\right).$$

Similarly,

$$M_{B_{n,\lambda/n}}(t) = \left(1 + \frac{\lambda}{n}(e^t - 1)\right)^n,$$

where $M_{B_{n,\lambda/n}}(t)$ is the moment generating function of the sum of n independent, identically distributed binomials with $p = \lambda/n$.

By the Poisson limit theorem, $M_{B_{n,\lambda/n}}(t) \to M_{P_\lambda}(t)$ for all t, and hence (6.8) will be proven if it can be shown that $M_{S_n}(t)/M_{B_{n,\lambda/n}}(t) \to 1$ for all t. For this we prove that $\ln\left[M_{S_n}(t)/M_{B_{n,\lambda/n}}(t)\right] \to 0$ for all t. This logarithm is well-defined since $M(t) > 0$, and this limit obtains the needed result because the exponential function is continuous.

To this end, fix t and for arbitrary $\epsilon < \max[1, e^t - 1]$ define N so that for all $n \geq N$ and all $m \leq n$:

$$\left|p_{n_m}(e^t - 1)\right| < \epsilon \quad and \quad \left|\frac{\lambda}{n}(e^t - 1)\right| < \epsilon.$$

The first bound is possible since $\max_m\{p_{n_m}\} \to 0$ as $n \to \infty$, and the second is apparent.
Recall the Taylor series for $\ln(1 + x)$, that for $|x| < 1$:

$$\ln(1 + x) = \sum_{j=1}^{\infty}(-1)^{j+1}\frac{x^j}{j}. \tag{6.9}$$

By the above bounds, both $\ln\left(1 + p_{n_m}(e^t - 1)\right)$ and $\ln\left(1 + \frac{\lambda}{n}(e^t - 1)\right)$ can be expanded as absolutely convergent Taylor series.
Simplifying notation with $a \equiv e^t - 1$ and $A_n \equiv M_{S_n}(t)/M_{B_{n,\lambda/n}}(t)$:

$$
\begin{aligned}
\ln A_n &= \sum_{m=1}^{n}\sum_{j=1}^{\infty}(-1)^{j+1}a^j\left(p_{n_m}^j - \left(\frac{\lambda}{n}\right)^j\right)\bigg/ j \\
&= \sum_{j=1}^{\infty}(-1)^{j+1}a^j\left[\sum_{m=1}^{n}\left(p_{n_m}^j - \left(\frac{\lambda}{n}\right)^j\right)\right]\bigg/ j \\
&= a\sum_{m=1}^{n}\left(p_{n_m} - \frac{\lambda}{n}\right) + \sum_{j=2}^{\infty}(-1)^{j+1}a^j\left[\sum_{m=1}^{n}\left(p_{n_m}^j - \left(\frac{\lambda}{n}\right)^j\right)\right]\bigg/ j.
\end{aligned}
$$

Here the interchange in summations is justified by the absolute convergence of these series for $n \geq N$.

Now for $j \geq 2$, since $p_{n_m} < \epsilon/|a|$ and $\lambda/n < \epsilon/|a|$:

$$p_{n_m}^j - \left(\frac{\lambda}{n}\right)^j = \left[p_{n_m} - \frac{\lambda}{n}\right]\sum_{k=0}^{j-1} p_{n_m}^{j-k-1}\left(\frac{\lambda}{n}\right)^k$$

$$< \left[p_{n_m} - \frac{\lambda}{n}\right]\sum_{k=0}^{j-1}\left(\frac{\epsilon}{|a|}\right)^{j-1}$$

$$\leq j\left|p_{n_m} - \frac{\lambda}{n}\right|\left(\frac{\epsilon}{|a|}\right)^{j-1}.$$

Finally by the triangle inequality:

$$\left|\ln A_n - a\sum_{m=1}^n\left(p_{n_m} - \frac{\lambda}{n}\right)\right| \leq \sum_{j=2}^\infty \frac{|a|^j}{j}\sum_{m=1}^n j\left|p_{n_m} - \frac{\lambda}{n}\right|\left(\frac{\epsilon}{|a|}\right)^{j-1}$$

$$= |a|\sum_{m=1}^n\left|p_{n_m} - \frac{\lambda}{n}\right|\sum_{j=2}^\infty \epsilon^{j-1}$$

$$= |a|\sum_{m=1}^n\left|p_{n_m} - \frac{\lambda}{n}\right|\frac{\epsilon}{1-\epsilon}$$

$$\leq |a|\left(\sum_{m=1}^n p_{n_m} + \lambda\right)\frac{\epsilon}{1-\epsilon}.$$

Since ϵ was arbitrary and $\sum_{m=1}^n p_{n_m} \to \lambda$ by assumption, it follows that $|\ln A_n| \to 0$. Hence, $M_{S_n}(t) \to M_{P_\lambda}(t)$ for all t, and $F_{S_n} \Rightarrow F_{P_\lambda}$ by Proposition 4.77. ∎

Example 6.9 (Modeling bond defaults) *This theorem allows the examples in the prior section to be more generally applied. Specifically, a portfolio of bonds with various credit ratings can also be modeled with a Poisson random variable. Similarly for various insurance applications. The requirement that $\max_m\{p_{n_m}\} \to 0$ as $n \to \infty$ provides a constraint, however, that the default or insurance claim probabilities should individually be small.*

For example, given $n = 400$ bonds, 100 with $p_1 = 0.002$, 200 with $p_2 = 0.005$ and 100 with $p_3 = 0.01$, one could reasonably model the number of defaults in this portfolio as a Poisson random variable with $\lambda = 100p_1 + 200p_2 + 100p_3 = 2.2$.

6.2.4 De Moivre-Laplace Theorem

The **De Moivre-Laplace theorem** is a special case of a very general result discussed below that is known as the **central limit theorem**, and addresses another property of the limiting distribution of sums of binomials as $n \to \infty$. The property identified is useful for making various probability estimates about such sums. See Remark 6.12.

Let $X_n^B \equiv \sum_{j=1}^n X_1^B$ be a binomially distributed random variable with parameters n and p, where X_1^B are independent, standard binomial variables. Then from (1.40), for any integers a and b :

$$\Pr[a \leq X_n^B \leq b] = \sum_{j=a'}^{b'}\binom{n}{j}p^j(1-p)^{n-j},$$

where $a' = \max(a,0)$ and $b' = \min(b,n)$. It makes little sense to investigate the limit of such probabilities as $n \to \infty$, because the range of the random variable X_n^B is $[0,n]$ which grows with n. Thus this probability will certainly converge to 0 for any fixed a and b.

Exercise 6.10 ($\Pr[a \leq X_n^B \leq b] \to 0$) *Prove that for fixed a and b that $\Pr[a \leq X_n^B \leq b] \to 0$ as $n \to \infty$. Hint: Stirlings formula of (4.105).*

Put another way, we have from (4.62) that:

$$E[X_n^B] = np, \qquad Var[X_n^B] = np(1-p).$$

In contrast to the Poisson limit theorem, here p is fixed and hence $np \to \infty$. Thus both the mean and variance of X_n^B grow without bound as $n \to \infty$.

To investigate quantitatively the limiting probabilities under this distribution as $n \to \infty$, some form of "scaling" is necessary to stabilize this distribution. The approach used by **Abraham de Moivre** (1667–1754) in the special case of $p = \frac{1}{2}$, and many years later generalized by **Pierre-Simon Laplace** (1749–1827) to all p with $0 < p < 1$, was to consider what is often called the **normalized random variable** Y_n^B. This variable is defined:

$$Y_n^B \equiv \frac{X_n^B - E[X_n^B]}{\sqrt{Var[X_n^B]}} = \frac{X_n^B - \mu_B}{\sigma_B}. \tag{6.10}$$

Since $\mu_B = np$ and $\sigma_B = \sqrt{np(1-p)}$ are constants for each n, the random variable Y_n^B has the same binomial probabilities as does X_n^B in the sense that:

$$\Pr\left[Y_n^B = j' \equiv \frac{j - np}{\sqrt{np(1-p)}}\right] = \Pr\left[X_n^B = j\right].$$

The range of Y_n^B is:

$$Rng\left[Y_n^B\right] = \left\{ \frac{j - np}{\sqrt{np(1-p)}} \,\middle|\, 0 \le j \le n \right\},$$

and is contained in $[-a\sqrt{n}, \sqrt{n}/a]$ with $a = \sqrt{p/q}$.

This may not seem to be much of an improvement over that for X_n^B with range $[0, n]$. However, a calculation using (4.3) yields that:

$$E[Y_n^B] = 0, \qquad Var[Y_n^B] = 1.$$

In other words, in terms of mean and variance, the distribution of Y_n^B "stays put," in contrast the the distribution of X_n^B that wanders off with n.

Consequently, with mean and variance both constant and independent of n, the question of investigating and potentially identifying the limiting distribution of Y_n^B as $n \to \infty$ is better defined and its pursuit more compelling. The following proposition identifies this limiting result.

Note that the variate Y_n^B in (6.11) can also be expressed as a summation:

$$Y_n^B \equiv \sum_{m=1}^{n} Y_{1,m}^B,$$

where for each n and independent standard binomials $\{X_{1,m}^B\}_{m=1}^n$:

$$Y_{1,m}^B \equiv \frac{X_{1,m}^B - p}{\sqrt{np(1-p)}}.$$

Hence $\{Y_{1,m}^B\}_{m=1}^n$ are independent, binomial variables with fixed p, and with $\mu_{Y_1} = 0$ and $\sigma_{Y_1} = 1/\sqrt{n}$.

Proposition 6.11 (De Moivre-Laplace theorem) *Let $F_{Y_n^B}(j')$ denote the distribution function of the normalized general binomial Y_n^B with parameters n, p, where $0 < p < 1$. In other words,*

$$F_{Y_n^B}(j') \equiv \sum_{k' \le j'} f_{Y_n^B}(k'),$$

where $f_{Y_n^B}(j') \equiv f_{B_n}(j)$ with $j \equiv j'\sqrt{np(1-p)} + np$ for $0 \le j \le n$.
  Then as $n \to \infty$:

$$F_{Y_n^B} \Rightarrow \Phi, \qquad (6.11)$$

where $\Phi(x)$ denotes the distribution function of the standard normal associated with the density function in (1.66).

Proof. Let $M_{B_{n,p}}(t)$ denote the moment generating function of the general binomial, which from (4.63) is given for all t by:

$$M_{B_{n,p}}(t) = \left(1 + p(e^t - 1)\right)^n.$$

From (4.30) and (4.63), the moment generating function for Y_n^B is defined for all t as follows, denoting $q \equiv 1 - p$:

$$
\begin{aligned}
M_{Y_n^B}(t) &= \exp\left[-\frac{npt}{\sqrt{npq}}\right] M_{B_{n,p}}\left(\frac{t}{\sqrt{npq}}\right) \\
&= \exp\left[-\frac{npt}{\sqrt{npq}}\right] \left[1 + p\left(\exp\left[\frac{t}{\sqrt{npq}}\right] - 1\right)\right]^n \\
&= \left[q\exp\left(-\frac{pt}{\sqrt{npq}}\right) + p\exp\left(\frac{qt}{\sqrt{npq}}\right)\right]^n \\
&= \left[q\exp\left(-t\sqrt{p/q}\Big/\sqrt{n}\right) + p\exp\left(t\sqrt{q/p}\Big/\sqrt{n}\right)\right]^n.
\end{aligned}
$$

  The moment generating function of the standard normal $M_\Phi(t)$ is defined for all t and given in (4.79) by:

$$M_\Phi(t) = \exp\left(t^2/2\right).$$

To prove that $M_{Y_n^B}(t) \to M_\Phi(t)$ for all t, it is sufficient to demonstrate that:

$$\ln M_{Y_n^B}(t) \to t^2/2,$$

for all t, since $M_{Y_n^B}(t)$ is nonnegative and the exponential function is continuous.
  To this end:

$$\ln M_{Y_n^B}(t) = n \ln\left[q\exp\left[-t\sqrt{p/q}\Big/\sqrt{n}\right] + p\exp\left[t\sqrt{q/p}\Big/\sqrt{n}\right]\right].$$

Define the function:

$$f(x) = \ln\left[q\exp\left[-t\sqrt{p/q}\sqrt{x}\right] + p\exp\left[t\sqrt{q/p}\sqrt{x}\right]\right].$$

With $\Delta x = 1/n$, note that by definition:

$$\lim_{n\to\infty} \ln M_{Y_n^B}(t) = f'(0),$$

assuming that this derivative exists.
  To investigate existence, $f(x) = \ln g(x)$, so if $g(x) > 0$ for all x, then $f(x)$ is differentiable everywhere $g(x)$ is differentiable. But as $\sqrt{x}$ is not differentiable at $x = 0$, some analysis is needed to establish differentiability of $g(x)$.
  Expanding the exponentials in a Taylor series:

$$g(x) \equiv q\left[1 - t\sqrt{p/q}\sqrt{x} + O_1(x)\right] + p\left[1 + t\sqrt{q/p}\sqrt{x} + O_2(x)\right],$$

where $O_j(x)$ denotes the absolutely convergent remainder series in powers of $t\sqrt{x}$ with leading terms of the form ct^2x. A calculation verifies that the $\sqrt{x}$-terms cancel and that $g(x) = 1 + ex + O(x^{3/2})$. Thus $g(x)$ is differentiable at $x = 0$, and then so too is $f(x)$. A calculation of this derivative produces $f'(0) = t^2/2$.
  Hence, $M_{Y_n^B}(t) \to M_\Phi(t)$ for all t, and $F_{Y_n^B} \Rightarrow \Phi$ by Proposition 4.77. ∎

Remark 6.12 (On binomial approximations with the normal) *Because $\Phi(x)$ is continuous everywhere, the De Moivre-Laplace theorem conclusion of $F_{Y_n^B} \Rightarrow \Phi$ implies that **for all** y :*

$$F_{Y_n^B}(y) \to \Phi(y).$$

With $j' \equiv \frac{j - \mu_B}{\sigma_B}$, where $\mu_B \equiv np$, $\sigma_B \equiv \sqrt{npq}$ and $0 \leq j \leq n$, it follows that for any real number w :

$$
\begin{aligned}
F_{B_{n,p}}(w) &= \sum_{j \leq w} \frac{n!}{j!(n-j)!} p^j q^{n-j} \\
&= \sum_{j' \leq (w - \mu_B)/\sigma_B} \frac{n!}{j!(n-j)!} p^j q^{n-j} \\
&= F_{Y_n^B}\left(\frac{w - \mu_B}{\sigma_B}\right).
\end{aligned}
$$

In applications where by definition $n << \infty$, the above limiting result for the binomial provides an approximation that for n "large," meaning that $n \geq 30$ or so for most applications:

$$F_{B_{n,p}}(w) \approx \Phi\left(\frac{w - \mu_B}{\sigma_B}\right).$$

The corresponding approximation for $F_{B_{n,p}}(w) - F_{B_{n,p}}(v)$ when $w > v$:

$$F_{B_{n,p}}(w) - F_{B_{n,p}}(v) \approx \Phi\left(\frac{w - \mu_B}{\sigma_B}\right) - \Phi\left(\frac{v - \mu_B}{\sigma_B}\right). \tag{6.12}$$

In applications one is typically interested in integer values of w and/or v, and this approximation can not therefore be uniformly useful. For example, if $w = j$ an integer and real $v < j$, we would conclude that as $v \to j$ that:

$$F_{B_{n,p}}(j) - F_{B_{n,p}}(v) \to F_{B_{n,p}}(j) - F_{B_{n,p}}(j^-) = f_{B_{n,p}}(j),$$

while

$$\Phi\left(\frac{j - \mu_B}{\sigma_B}\right) - \Phi\left(\frac{v - \mu_B}{\sigma_B}\right) \to 0.$$

*While not apparent from the above approach to the proof, the problem here can be investigated with a more direct analysis of the binomial probabilities using Stirling's formula in (4.105), and some careful calculations as in Proposition 8.24 of **Reitano** (2010). This reveals that if $j'_n \to y$ as $n \to \infty$, where each j'_n is in the range of the random variable Y_n^B, then:*

$$\lim_{n \to \infty} \sqrt{npq}\, \Pr\{Y_n^B = j'_n\} \to \frac{1}{\sqrt{2\pi}} e^{-y^2/2}. \tag{6.13}$$

This produces the approximation:

$$f_{Y_n^B}(j'_n) \approx \frac{1}{\sqrt{2\pi npq}} \exp\left[-(j'_n)^2/2\right],$$

which for $0 \leq j \leq n$ is equivalent to:

$$f_{B_{n,p}}(j) \approx \frac{1}{\sqrt{2\pi npq}} \exp\left[-\frac{1}{2}\left(\frac{j - \mu_B}{\sigma_B}\right)^2\right]. \tag{6.14}$$

This approximation for $f_{Y_n^B}(j)$ can be interpreted as a single term in the Riemann summation which approximates the integral of the standard normal density function $e^{-\frac{1}{2}y^2}/\sqrt{2\pi}$

using $\Delta y = 1/\sigma_B$ where $\sigma_B = \sqrt{npq}$. This value of Δy is seen to equal $j'_n(k+1) - j'_n(k)$ where $\{j'_n(k)\}_{k=0}^n$ denote the $n+1$ values of this variate.

Hence, the discrete probability $f_{B_{n,p}}(j)$ in (6.14) is approximated by the standard normal probability of an interval of length $1/\sigma_B$ which contains $(j - \mu_B)/\sigma_B$. Alternatively, for some λ with $0 < \lambda < 1$, one wants to approximate $f_{B_{n,p}}(j)$ by the integral of the standard normal density over $[a, b]$ with $a = [j - \mu_B - (1 - \lambda)]/\sigma_B$ and $b = [j - \mu_B + \lambda]/\sigma_B$, an interval of length $1/\sigma_B$.

The conventional solution is $\lambda = 1/2$, which makes a **half interval adjustment**, or **half integer adjustment** for the above approximation. This produces:

$$f_{B_{n,p}}(j) \approx \Phi\left(\frac{j + 1/2 - \mu_B}{\sigma_B}\right) - \Phi\left(\frac{j - 1/2 - \mu_B}{\sigma_B}\right), \tag{6.15}$$

from which one obtains:

$$F_{B_{n,p}}(w) - F_{B_{n,p}}(v) \approx \Phi\left(\frac{w + 1/2 - \mu_B}{\sigma_B}\right) - \Phi\left(\frac{v - 1/2 - \mu_B}{\sigma_B}\right) \tag{6.16}$$

6.2.5 The Central Limit Theorem 1

There are many versions of "the" central limit theorem, which is almost universally denoted CLT. All versions generalize the De Moivre-Laplace theorem in one remarkable way or another. In essence, what every version states, and what makes any such version indeed a "central" limit theorem, is this:

Under a wide variety of assumptions, the distribution function of the sum of n independent random variables, normalized as in (6.10), converges to the standard normal distribution as $n \to \infty$.

Remarkably, these random variables need not be identically distributed, just independent, although the need for normalization demands that these random variables have at least two moments: mean and variance. When not identically distributed there is a requirement that:

1. The sequence of variances does not grow too fast, to preclude latter terms in the series from increasingly dominating the summation, and,

2. A requirement that the sequence of variances does not converge to 0 so quickly that the average variance converges to 0.

These theorems can be equivalently stated in terms of the **sum** of independent random variables or their **average**. This is because by (4.39) and (4.3):

$$E\left[\frac{1}{n}\sum_{j=1}^n X_j\right] = \frac{1}{n}E\left[\sum_{j=1}^n X_j\right],$$

and from (4.43):

$$Var\left[\frac{1}{n}\sum_{j=1}^n X_j\right] = \frac{1}{n^2}Var\left[\sum_{j=1}^n X_j\right].$$

Thus normalized sums equal normalized averages:

$$\frac{\sum_{j=1}^n X_j - E\left[\sum_{j=1}^n X_j\right]}{\sqrt{Var\left[\sum_{j=1}^n X_j\right]}} = \frac{\frac{1}{n}\sum_{j=1}^n X_j - E\left[\frac{1}{n}\sum_{j=1}^n X_j\right]}{\sqrt{Var\left[\frac{1}{n}\sum_{j=1}^n X_j\right]}}. \tag{6.17}$$

So while the ranges of the sum and average of independent random variables are quite different, the associated normalized random variables are identical.

Consequently, central limit theorems in general, and the De Moivre-Laplace theorem in particular, apply to the sums of random variables if and only if they apply to the averages of random variables. More generally, if a given version of a central limit theorem applies to $\sum_{j=1}^{n} X_j$ for independent $\{X_j\}$, then it applies to $\sum_{j=1}^{n} Y_j$ where $Y_j = aX_j + b$ for arbitrary constants $a \neq 0$ and b.

In this section, we provide a proof of a simplified version of the central limit theorem in the case of independent, identically distributed random variables which have moments of all orders and a well-defined moment generating function. Mechanically, the proof will be quite similar to that of the De Moivre-Laplace theorem, except that we will have to accommodate a more general form of $M_X(t)$.

In Book VI, using more powerful tools than the moment generating function, we will present more general versions of this result, with far weaker assumptions.

Proposition 6.13 (Central limit theorem 1) *Let F_X denote the distribution function of a random variable X with mean and variance denoted μ and σ^2, and moment generating function $M_X(t)$ which exists for $t \in (-t_0, t_0)$ with $t_0 > 0$. Let Y_n denote the normalized random variable associated with the sum or average of n independent values of X, respectively defined as in (6.10):*

$$Y_n = \frac{\sum_{j=1}^{n} X_j - n\mu}{\sqrt{n}\sigma} = \frac{\frac{1}{n}\sum_{j=1}^{n} X_j - \mu}{\sigma/\sqrt{n}}.$$

Then as $n \to \infty$,

$$F_{Y_n} \Rightarrow \Phi, \tag{6.18}$$

where $\Phi(x)$ denotes the distribution function of the standard normal.
Proof. *By (4.48) and (4.30):*

$$M_{Y_n}(t) = \left[\exp\left(-\frac{\mu t}{\sqrt{n}\sigma}\right) M_X\left(\frac{t}{\sqrt{n}\sigma}\right)\right]^n, \tag{1}$$

and our goal is to prove that:

$$M_{Y_n}(t) \to \exp\left(\frac{t^2}{2}\right) = M_{\Phi}(t).$$

Equivalently as above, we show that:

$$\ln M_{Y_n}(t) \to \frac{t^2}{2}.$$

By (4.53):

$$M_X\left(\frac{t}{\sqrt{n}\sigma}\right) = \sum_{j=0}^{\infty} \frac{1}{j!} \left(\frac{t}{\sqrt{n}\sigma}\right)^j \mu_j'.$$

Recalling that $\mu_0' = 1$, $\mu_1' = \mu$ and $\mu_2' = \sigma^2 + \mu^2$:

$$M_X\left(\frac{t}{\sqrt{n}\sigma}\right) = 1 + \frac{\mu}{\sqrt{n}\sigma}t + \left(\frac{\sigma^2 + \mu^2}{2n\sigma^2}\right)t^2 + n^{-\frac{3}{2}}E_1(n), \tag{2}$$

where:

$$E_1(n) = \sum_{j=3}^{\infty} \frac{1}{j!}\mu_j'\left(\frac{t}{\sigma}\right)^j n^{\frac{3-j}{2}}.$$

Since $M_X(t)$ is by assumption absolutely convergent for $|t| < t_0$, $M_X(t/[\sqrt{n}\sigma])$ and hence $E_1(n)$ are absolutely convergent for $|t| < \sqrt{n}\sigma t_0$, and so for any t, $E_1(n) \to \mu'_3 (t/\sigma)^3 /6$ as $n \to \infty$.

Similarly, using the Taylor series in (4.54) obtains:

$$\exp\left[-\frac{\mu t}{\sqrt{n}\sigma}\right] = \sum_{j=0}^{\infty} \frac{1}{j!}\left[-\frac{\mu t}{\sqrt{n}\sigma}\right]^j$$

$$= 1 - \frac{\mu}{\sqrt{n}\sigma}t + \frac{\mu^2}{2n\sigma^2}t^2 + n^{-\frac{3}{2}}E_2(n), \tag{3}$$

where:

$$E_2(n) = \sum_{j=3}^{\infty} \frac{1}{j!}\left[-\frac{\mu t}{\sigma}\right]^j n^{\frac{3-j}{2}}.$$

Again this series is absolutely convergent for all t, and $E_2(n) \to -\mu^3 (t/\sigma)^3 /6$ for any t as $n \to \infty$.

With a bit of algebra on (2) and (3):

$$\exp\left[-\frac{\mu t}{\sqrt{n}\sigma}\right] M_X\left(\frac{t}{\sqrt{n}\sigma}\right) = 1 + \frac{t^2}{2n} + n^{-\frac{3}{2}}E_3(n),$$

where the error term $E_3(n)$ is is absolutely convergent for $|t| < \sqrt{n}\sigma t_0$, and $E_3(n) \to -\mu^3\mu'_3 (t/\sigma)^6 /36$ as $n \to \infty$. To obtain $M_{Y_n}(t)$ in (1), this expression can now be raised to the nth power, a logarithm taken, and the function $\ln(1+x)$ expanded in a Taylor series as in (6.9).

To simplify, note that we only need to keep track of the powers of n that are needed for the final limit, meaning only those terms that will not converge to zero as $n \to \infty$. This produces:

$$\ln M_{Y_n}(t) = n\ln\left[1 + \frac{t^2}{2n} + n^{-\frac{3}{2}}E_3(n)\right]$$

$$= n\left[\left(\frac{t^2}{2n} + n^{-\frac{3}{2}}E_3(n)\right) - \frac{1}{2}\left(\frac{t^2}{2n} + n^{-\frac{3}{2}}E_3(n)\right)^2 + O\left[n^{-3}\right]\right]$$

$$= t^2/2 + O\left[n^{-1/2}\right].$$

To justify the second step in which the power series expansion for $\ln(1+x)$ is invoked with $x = t^2/(2n) + n^{-\frac{3}{2}}E_3(n)$, it must be verified that $|x| < 1$ for any t for n large enough. But since:

$$|x| \le \frac{t^2}{2n} + n^{-\frac{3}{2}}|E_3(n)|,$$

where $E_3(n)$ is continuous and bounded, the conclusion follows.

Hence for all t :

$$\ln M_{Y_n}(t) \to \frac{t^2}{2},$$

as $n \to \infty$. Equivalently $M_{Y_n}(t) \to M_\Phi(t)$ for all t, so $F_{Y_n} \Rightarrow \Phi$ by Proposition 4.77. ∎

The assumption in this version of the central limit theorem, that $M_X(t)$ exists and is convergent for all $t \in (-t_0, , t_0)$ with $t_0 > 0$, is quite strong. As was proved in Proposition

4.25, this implies that the associated distribution function of X has finite moments of all orders.

For independent and identically distributed random variables, the conclusion in (6.18) will be seen to be valid under the assumption that X has only two finite moments, a mean and variance. It is also valid more generally for sums of independent random variables which are not identically distributed. But in neither case will the tools of this book suffice for the demonstration.

The problem with the current approach is that the moment generating function is a blunt instrument. If it exists on an open interval $(-t_0, , t_0)$ with $t_0 > 0$, then all moments exist by Proposition 4.25. There is no way to adapt this argument in the case of random variables with only finitely many moments.

In Book V, following the development of a general integration theory, the **Fourier transform** of a measurable function will be introduced. In the same way that the moment generating function of a random variable is defined relative to the Laplace transform of this measurable function as noted in Remark 4.15, in Book VI, the **characteristic function** of a random variable will be defined in terms of the Fourier transform. It will there be seen that unlike the moment generating function, which may or may not exist, characteristic functions always exist.

Characteristic functions will also be seen to uniquely identify distribution functions, so they are useful in proofs in the same way that moment generating functions are useful. But more generally, the associated proofs can also be implemented for random variables with only finitely many moments. The moments of a distribution, to the extent they exist, will be seen to appear in the series expansion of the characteristic function in a familiar way, reminiscent of (4.53).

6.2.6   Smirnov's Theorem on Uniform Order Statistics

In this section, we investigate the limiting distribution of the order statistics of **independent, continuous uniform** random variables on $[0, 1]$. This result is named for its discoverer, **N. V. (Nikolaĭ Vasil'evich) Smirnov** (1900–1966), and will be generalized in Book VI using a generalized version of the central limit theorem introduced above. **Smirnov's limit theorem** addresses the limiting distribution as $n \to \infty$ of the kth order statistic from sample $\{U_j\}_{j=1}^n$. It will require that both $k \to \infty$ and $n - k \to \infty$.

For example, this result applies if $k/n \to q \in (0, 1)$, the qth quantile of this distribution, which implies that both $k \to \infty$ and $n - k \to \infty$ as $n \to \infty$. On the other hand, the requirement that both $k \to \infty$ and $n - k \to \infty$ as $n \to \infty$ also allows that $k/n \to \{0, 1\}$ by choosing $k \approx \sqrt{n}$, or $k = n - \sqrt{n}$, respectively. Hence, this result allows k to converge in relative terms to the tails of the distribution of order statistics.

More generally, this theorem allows k/n to not converge to any value as $n \to \infty$, and remains valid as long as both $k \to \infty$ and $n - k \to \infty$.

Remark 6.14 (Normalized U) *Recall from Example 3.9 that the distribution function for $U_{(k)}$ is beta with $v = k$ and $w = n - k + 1$. A calculation using (4.75) will show that as $n \to \infty$, using the notation of Proposition 6.15:*

$$\frac{b_n}{E\left[U_{(k)}\right]} \to 1, \quad \frac{a_n^2}{Var\left[U_{(k)}\right]} \to 1.$$

*This follows since both $k \to \infty$ and $n - k \to \infty$. So $U_{(k)}'$ below is effectively a **normalized random variable** as defined in (6.10).*

In fact by (9.21) of Proposition II.9.16, (6.19) assures that:

$$\widetilde{F}_{(k)} \Rightarrow \Phi,$$

where $\widetilde{F}_{(k)}$ denotes the distribution function of the truly normalized variate $U''_{(k)}$, defined as in (6.10) in terms of the moments of $U_{(k)}$:

$$U''_{(k)} \equiv \frac{U_{(k)} - E\left[U_{(k)}\right]}{\sqrt{Var\left[U_{(k)}\right]}}.$$

In Corollary 6.17 we will see another application of this Book II result.

Proposition 6.15 (Smirnov's Limit Theorem) *Let $\{U_{(k)}\}_{k=1}^{n}$ be the order statistics from a continuous uniform distribution on $[0,1]$. Define:*

$$U'_{(k)} \equiv \frac{U_{(k)} - b_n}{a_n},$$

where:

$$b_n = \frac{k-1}{n-1}, \qquad a_n = \sqrt{\frac{b_n(1 - b_n)}{n-1}},$$

and let $F_{(k)}$ denote the distribution function of $U'_{(k)}$.
Then if $k \to \infty$ and $n - k \to \infty$ as $n \to \infty$:

$$F_{(k)} \Rightarrow \Phi, \tag{6.19}$$

where $\Phi(x)$ is the distribution function of the standard normal.
Proof. *We begin by proving pointwise convergence of the associated density functions, and then justify an application of Lebesgue's dominated convergence theorem.*
The density function $g_{(k)}(y)$ of $U_{(k)}$ is defined on $[0,1]$ by (3.7) and (3.8):

$$g_{(k)}(y) = \frac{n!}{(k-1)!\,(n-k)!} y^{k-1} (1-y)^{n-k}.$$

The density $f_{(k)}(y)$ of $U'_{(k)}$ is thus defined on $[-b_n/a_n, (1-b_n)/a_n]$ by:

$$
\begin{aligned}
f_{(k)}(y) &= \frac{n!}{(k-1)!\,(n-k)!} (a_n y + b_n)^{k-1} (1 - a_n y - b_n)^{n-k} \\
&= \frac{n!}{(k-1)!\,(n-k)!} a_n b_n^{k-1} (1 - b_n)^{n-k} \left(1 + \frac{a_n}{b_n} y\right)^{k-1} \left(1 - \frac{a_n}{1 - b_n} y\right)^{n-k}. \tag{1}
\end{aligned}
$$

The coefficient in (1) simplifies with Stirling's formula of (4.105). Here we use the notation "$\sim$" to denote that as $n \to \infty$, the expressions have the same limit.

$$
\begin{aligned}
&\frac{n!}{(k-1)!\,(n-k)!} a_n b_n^{k-1} (1 - b_n)^{n-k} \\
&\sim \frac{n^{n+1/2} e^{-1}}{\sqrt{2\pi}\,(k-1)^{k-1/2} (n-k)^{n-k+1/2}} \left(\frac{k-1}{n-1}\right)^{k-1/2} \left(\frac{n-k}{n-1}\right)^{n-k+1/2} \left(\frac{1}{n-1}\right)^{1/2} \\
&= \frac{1}{\sqrt{2\pi}} \left(1 + \frac{1}{n-1}\right)^{n+1/2} e^{-1}.
\end{aligned}
$$

This last expression converges to $\frac{1}{\sqrt{2\pi}}$ as $n \to \infty$ by an application of (6.5).
The next step is to prove that the polynomial expression in (1) for y converges to $e^{-y^2/2}$.

Taking logs, and applying the Taylor series expansion in (6.9), which is justified noting that both $\frac{a_n y}{b_n}$ and $\frac{a_n y}{1-b_n}$ have absolute value less than 1 :

$$\ln\left[\left(1+\frac{a_n}{b_n}y\right)^{k-1}\left(1-\frac{a_n}{1-b_n}y\right)^{n-k}\right]$$

$$= (k-1)\left[\sum_{j=1}^{\infty}(-1)^{j+1}\left(\frac{a_n}{b_n}\right)^{j}\frac{y^j}{j}\right]+(n-k)\left[-\sum_{j=1}^{\infty}\left(\frac{a_n}{1-b_n}\right)^{j}\frac{y^j}{j}\right].$$

The coefficient c_j of $\frac{y^j}{j}$ for $j \geq 1$ is:

$$c_j \equiv (-1)^{j+1}(k-1)\left(\frac{a_n}{b_n}\right)^{j}-(n-k)\left(\frac{a_n}{1-b_n}\right)^{j}$$

$$= (-1)^{j+1}(k-1)\left(\frac{n-k}{(n-1)(k-1)}\right)^{j/2}-(n-k)\left(\frac{k-1}{(n-1)(n-k)}\right)^{j/2}$$

$$= (-1)^{j+1}(k-1)\left(\frac{-1}{n-1}+\frac{1}{k-1}\right)^{j/2}-(n-k)\left(\frac{-1}{n-1}+\frac{1}{n-k}\right)^{j/2}.$$

From this expression it follows that $c_1 = 0$, $c_2 = -1$, and since $k \to \infty$ and $n-k \to \infty$, the coefficient of $\frac{y^j}{j}$ converges to zero as $n \to \infty$ for $j \geq 3$.
   In summary, for all y :

$$f_{(k)}(y) \to \varphi(y) \equiv \frac{1}{\sqrt{2\pi}}e^{-y^2/2},$$

as $n \to \infty$.
   The final step is to prove convergence of the associated distribution functions, $F_{(k)}(y) \to \Phi(y)$ for all y. To this end, let $h_{(k)}(y) \equiv \max\left[\varphi(y)-f_{(k)}(y),0\right]$, and note that:

$$\left|f_{(k)}(y)-\varphi(y)\right| = f_{(k)}(y)-\varphi(y)+2h_{(k)}(y).$$

Since $f_{(k)}(y)$ and $\varphi(y)$ are densities that integrate to 1 :

$$\int\left|f_{(k)}(y)-\varphi(y)\right|dy = 2\int h_{(k)}(y)dy. \tag{2}$$

Now $0 \leq h_{(k)}(y) \leq \varphi(y)$ and $h_{(k)}(y) \to 0$ pointwise, so $\int h_{(k)}(y)dy \to 0$ by Lebesgue's dominated convergence theorem of Proposition III.2.52, and then by (2):

$$\int\left|f_{(k)}(y)-\varphi(y)\right|dy \to 0.$$

Definition II.2.9 and item 4 of Proposition II.2.49 obtain that for all measurable A:

$$\int_A\left|f_{(k)}(x)-\varphi(x)\right|dx \to 0.$$

Taking $A = (-\infty, y]$ and using the the triangle inequality of that proposition's item 7:

$$\left|\int_{-\infty}^{y}f_{(k)}(x)dx-\int_{-\infty}^{y}\varphi(x)dx\right| \leq \int_{-\infty}^{y}\left|f_{(k)}(x)-\varphi(x)\right|dx \to 0.$$

This proves that $F_{(k)}(y) \to \Phi(y)$ and thus (6.19). ∎

Remark 6.16 (Scheffé's Theorem) *The last paragraphs of the above proof derive a special case of **Scheffé's Theorem**, named for a 1947 result of **Henry Scheffé** (1907–1977). This theorem states that pointwise convergence of density functions assures pointwise convergence of the associated distribution functions, and thus the weak convergence of distributions.*

It is more generally true for density functions defined on arbitrary measure spaces and will be proved in Book VI using the general version of Lebesgue's dominated convergence theorem of Book V. Otherwise, the general proof will be identical with that above.

In the special case where $k_n/n \to q$, the qth quantile of the uniform distribution for $0 < q < 1$, Smirnov's limit theorem can be stated in a simpler way. The proof is an application of Proposition II.9.16.

Corollary 6.17 (Smirnov's Limit Theorem) *Let $\{U_{(k)}\}_{k=1}^n$ be the order statistics from a continuous, uniform distribution on $[0,1]$, and k_n a sequence so that $k_n/n \to q$ for $0 < q < 1$. Define:*

$$U''_{(k_n)} \equiv \frac{U_{(k_n)} - q}{\sqrt{q(1-q)/n}},$$

and let $G_{(k_n)}$ denote the distribution function of $U''_{(k_n)}$.
Then as $n \to \infty$,

$$G_{(k_n)} \Rightarrow \Phi, \tag{6.20}$$

where $\Phi(x)$ is the distribution function of the standard normal.
Proof. *Since $k_n/n \to q \in (0,1)$ implies that both $k_n \to \infty$ and $n - k_n \to \infty$ as $n \to \infty$, the above proposition assures that $F_{(k)}(y) \Rightarrow \Phi(y)$, where $F_{(k)}(y)$ is the distribution function of the variable:*

$$U'_{(k_n)} \equiv \frac{U_{(k_n)} - b_n}{a_n},$$

with a_n and b_n defined as above in terms of k_n.
By Proposition II.9.16, if c_n and d_n are sequences that satisfy:

$$c_n \to c, \qquad d_n \to d,$$

then:

$$F_{(k)}(c_n y + d_n) \Rightarrow \Phi(cy + d),$$

or equivalently:

$$\Pr\left[U'_{(k_n)} \le c_n y + d_n\right] \to \Pr[Z \le cy + d], \tag{1}$$

Define:

$$c_n = \frac{\sqrt{q(1-q)/n}}{a_n}, \qquad d_n = \frac{q - b_n}{a_n},$$

and note that:

$$\Pr\left[U'_{(k_n)} \le c_n y + d_n\right] = \Pr\left[U''_{(k_n)} \le y\right] = G_{(k_n)}(y). \tag{2}$$

Thus by (1) and (2), the proof will be complete by showing that $c_n \to 1$ and $d_n \to 0$.
Now $k_n/n - q \to 0$ by assumption, and thus $(k_n - 1)/(n-1) \to q$ and $(n - k_n)/(n-1) \to 1 - q$. This obtains:

$$c_n = \frac{\sqrt{q(1-q)/n}}{a_n} = \sqrt{\frac{n-1}{k_n - 1}\frac{n-1}{n - k_n}\frac{n-1}{n}q(1-q)} \to 1,$$

and

$$d_n = \frac{q - b_n}{a_n} = \left(q - \frac{k_n - 1}{n - 1} \right) \left(\frac{(k_n - 1)(n - k_n)}{(n - 1)^2} \right)^{-1/2} \to 0.$$

∎

6.2.7 A Limit Theorem on General Quantiles

In this section, we generalize Corollary 6.17 to Smirnov's limit theorem from the uniform distribution to the order statistics of any distribution function $F(x)$ which is differentiable at $F^{-1}(q)$ for $q \in (0,1)$ with $F'(F^{-1}(q)) \neq 0$, and is continuous and strictly increasing in an interval about $F^{-1}(q)$. This generalization will be proved with the aid of the Δ-Method of Proposition II.8.40.

Proposition 6.18 (General quantile limits) *Let $\{X_{(k)}\}_{k=1}^n$ be the order statistics from a distribution function F which is continuous and strictly increasing in a neighborhood of $F^{-1}(q)$ for $0 < q < 1$, and differentiable at $F^{-1}(q)$ with $F'(F^{-1}(q)) \neq 0$. Given a sequence $\{k_n\}$ with $k_n/n \to q$ as $n \to \infty$, define:*

$$X'_{(k_n)} \equiv \frac{X_{(k_n)} - F^{-1}(q)}{\left[\sqrt{q(1-q)/n} \right] \Big/ F'(F^{-1}(q))},$$

and let $F_{(k_n)}$ denote the distribution function of $X'_{(k_n)}$.
 Then as $n \to \infty$:

$$F_{(k_n)} \Rightarrow \Phi, \tag{6.21}$$

where $\Phi(x)$ is the distribution function of the standard normal.
Proof. *Given continuous, uniform U on $(0,1)$, define $X = F^*(U)$ where F^* is the left-continuous inverse of F, and recall Proposition II.4.9 in the Chapter 5 introduction that X has distribution function F. If $\{U_{(k)}\}_{k=1}^n$ are order statistics from this uniform distribution, define $X_{(k)} = F^*(U_{(k)})$, order statistics for X since F^* is increasing by Proposition II.3.16.*
 If k_n is a sequence as defined above, then Corollary 6.17 applies. In the notation of weak convergence of random variables (Definition II.5.19):

$$c_n \left(U_{(k_n)} - q \right) \Rightarrow Z,$$

where $c_n \equiv 1/\sqrt{q(1-q)/n}$, and Z is a standard normal random variable. Because $c_n \to \infty$, the Δ-Method of Proposition II.8.40 obtains that if $F^*(y)$ is differentiable at $y = q$, then:

$$c_n \left[F^*(U_{(k_n)}) - F^*(q) \right] \Rightarrow (F^*)'(q)Z. \tag{1}$$

By assumption, $F(x)$ is continuous and strictly increasing in a neighborhood of $F^{-1}(q)$, say $(F^{-1}(q) - \epsilon, F^{-1}(q) - \epsilon) \equiv (a, b)$. Define the function G :

$$G(x) = \begin{cases} F(a), & x \in (-\infty, a], \\ F(x), & x \in (a, b), \\ F(b), & x \in [b, \infty). \end{cases}$$

As $G(x)$ is continuous and strictly increasing on $D \equiv \{x | F(a) < G(x) < F(b)\} = (a, b)$, Corollary II.3.23 obtains that $G^* = G^{-1}$ on (a, b). But $G = F$ on (a, b), and thus $F^* = F^{-1}$ on (a, b), and in particular, $F^*(q) = F^{-1}(q)$.

Now by assumption $F'\left(F^{-1}(q)\right) \neq 0$, and so:

$$\left(F^{-1}\right)'(q) = 1/F'(F^{-1}(q)).$$

Thus with $F^*(U_{(k_n)}) \equiv X_{(k_n)}$, (1) obtains:

$$c_n \left[X_{(k_n)} - F^{-1}(q)\right] \Rightarrow Z/F'(F^{-1}(q)),$$

which is (6.21). ∎

Remark 6.19 (On $F'(F^{-1}(q))$) *It seems natural to expect that $F'(F^{-1}(q))$ should be expressible in terms of the density function associated with the distribution function $F(x)$. As was seen in Book III, not every distribution function has an associated density function, meaning a measurable function $f(x)$ so that for all x :*

$$F(x) = (\mathcal{L}) \int_{-\infty}^{x} f(y)dy,$$

defined as a Lebesgue integral. This is true even if $F(x)$ is continuous as assumed above.

*Existence of a density function $f(x)$ requires that $F(x)$ be **absolutely continuous**, introduced in Definition III.3.54. Absolute continuity is a stronger condition than either uniform continuity or of bounded variation, and is weaker than continuously differentiable. Interestingly, $F(x)$ in the above result need not be absolutely continuous even in the interval $(F^{-1}(q) - \epsilon, F^{-1}(q) - \epsilon)$ of the proof, where it is strictly increasing and continuous. As an example, let $F(x) = x + F_C(x)$ on $(a, b) \subset (0, 1)$ where $F_C(x)$ is the Cantor function of Definition III.3.51, which is continuous, increasing, and not absolutely continuous by Example III.3.57. The function $F(x)$ is strictly increasing and continuous on (a, b) by construction, and not absolutely continuous.*

When $F(x)$ is absolutely continuous, then $F'(x)$ exists almost everywhere and is Lebesgue measurable by Proposition III.3.59. The Lebesgue integral above is then satisfied with $f(x) \equiv F'(x)$ by Proposition III.3.62, generalized as before since $F(a) \to 0$ as $a \to -\infty$ by Proposition 1.3. This representation of $F(x)$ is also satisfied with any function $g(x)$ with $g(x) = F'(x)$ a.e., meaning outside a set of Lebesgue measure 0.

Thus even in this case of absolutely continuous $F(x)$, the derivative $F'(x)$ is not uniquely defined in terms of the density function $f(x)$, since this density is not unique. This is also seen in Proposition III.3.39, that if $F(x)$ is expressible as above with measurable $f(x)$, then $F'(x)$ exists almost everywhere and $F'(x) = f(x)$ a.e.

*In the very special case of **continuous probability theory**, where $F(x)$ is continuously differentiable and $f(x)$ is continuous, we obtain an affirmative conclusion on the inquiry of this remark. Then $F'(x) = f(x)$ everywhere by Proposition III.1.33. And in this special case, $F^{-1}(q) = x_q$, the qth quantile of F, which is well defined since F is locally strictly increasing. Thus:*

$$F'(F^{-1}(q)) = f(x_q).$$

Example 6.20 (Estimating $F^{-1}(q)$) *It follows from Proposition 6.18 that $X'_{(k_n)}$ is approximately standard normally distributed for n large. Hence $\left\{\left|X'_{(k_n)}\right| \leq 1.96\right\}$ defines an approximate 95% confidence interval for this variate, recalling that the 97.5th quantile of the standard normal is $z_{0.975} = 1.96$.*

Defining $k_n = \lfloor qn \rfloor$, the greatest integer less than or equal to n, it follows that $k_n/n \to q$ as $n \to \infty$. This then obtains a confidence interval for the quantile $F^{-1}(q)$ of X given a random sample ordered variate $\widehat{X}_{(\lfloor qn \rfloor)}$:

$$\left|\widehat{X}_{(\lfloor qn \rfloor)} - F^{-1}(q)\right| \leq \frac{1.96\sqrt{q(1-q)}}{\sqrt{n}F'(F^{-1}(q))}.$$

Of course, F and F' are unknown generally, but assuming F(x) has a continuous density,
$F'(F^{-1}(q)) \approx f(\widehat{X}_{(\lfloor qn \rfloor)})$ *would need to be estimated.*

*Using the normal density $\phi(x)$ of (1.66), it is reasonable to assume that $\varphi(\widehat{X}_{(\lfloor qn \rfloor)}) <$
$f(\widehat{X}_{(\lfloor qn \rfloor)})$ since $\phi(x)$ has "skinny" tails. Thus replacing $F'(F^{-1}(q))$ by $\varphi(\widehat{X}_{(k_n)})$ provides
an upper bound for this inequality and a conservative confidence interval:*

$$\left| \widehat{X}_{(\lfloor qn \rfloor)} - F^{-1}(q) \right| \leq \frac{1.96\sqrt{q(1-q)}}{\sqrt{n}\varphi(\widehat{X}_{(k_n)})}.$$

6.2.8   A Limit Theorem on Exponential Order Statistics

In this section, we prove a limiting result on the average "gap" of the higher order statistics
of the standard exponential distribution. This result is essentially a corollary of the central
limit theorem and the Rényi representation theorem. This result will prove to be important
in Section 7.2.5 on extreme value theory, in the investigation of the **Hill estimator**.

For the result below, let $\{X_i\}_{i=1}^n$ be a sample of **standard exponential variates**,
meaning $\lambda = 1$, with associated order statistics $\{X_{(j)}\}_{j=1}^n$. Define the random variable:

$$Y_{k,n} = \frac{1}{k} \sum_{j=0}^{k-1} \left[X_{(n-j)} - X_{(n-k)} \right].$$

In other words, $Y_{k,n}$ equals the average of the k "gaps" between $X_{(n-k)}$ and higher order
variates $X_{(n-j)}$ for $j = 0$ to $k - 1$.

This result states that if properly normalized, $Y_{k,n}$ is asymptotically normal as $n \to$
∞ and $k \to \infty$.

Example 6.21 *Though the proof does not require that k and n increase proportionately, if
$k = \lfloor (1 - q) n \rfloor$ for $0 < q < 1$, then $k \to \infty$ as $n \to \infty$. Recall that $\lfloor x \rfloor$ denotes the greatest
integer less than or equal to x, where the **greatest integer function**, also called the **floor
function**, is defined by:*

$$\lfloor x \rfloor \equiv \max\{m \in \mathbb{N} | m \leq x\}.$$

Then:

$$X_{(n-k)} = X_{(n-\lfloor (1-q)n \rfloor)},$$

*and since $n - \lfloor (1 - q) n \rfloor$ is within 1 of $\lfloor qn + 1 \rfloor$, $X_{(n-k)}$ is essentially the $(qn + 1)$ st order
statistic.*

*Thus $Y_{k,n}$ above is the average of the gaps between this $(qn + 1)$ st order statistic and
all higher order statistics $X_{(m)}$ with $m > \lfloor qn + 1 \rfloor$.*

The following proposition identifies $Y_{k,n}'$ as a normalized version of $Y_{k,n}$. This will be
justified in the proof, where it will be seen that $E[Y_{k,n}] = 1$ and $Var[Y_{k,n}] = 1/k$.

Proposition 6.22 (Average gaps of standard exponential order statistics) *Let
$\{X_{(j)}\}_{j=1}^n$ be an ordered sample from a standard exponential distribution, meaning as in
(1.54) with $\lambda = 1$. With $Y_{k,n}$ defined above, define the normalized average variate $Y_{k,n}'$:*

$$Y_{k,n}' = \frac{Y_{k,n} - 1}{1/\sqrt{k}},$$

*and let $F_{k,n}$ denote the distribution function of $Y_{k,n}'$.
Then as $n \to \infty$ and $k \to \infty$:*

$$F_{k,n} \Rightarrow \Phi, \tag{6.22}$$

where $\Phi(x)$ is the distribution function of the standard normal.
    Further as $n \to \infty$ and $k \to \infty$:

$$Y_{k,n} \to_1 1, \tag{6.23}$$

meaning that $Y_{k,n}$ converges to 1 with probability 1.
Proof. *By (3.27) of Corollary 3.31 to Rényi's representation theorem with $\lambda = 1$:*

$$X_{(j)} = \sum_{i=1}^{j} \frac{E_i}{n-i+1},$$

where $\{E_i\}_{i=1}^{n}$ are independent, standard exponential random variables. Hence for $j < k$, and using an index substitution $l = k + i - n$:

$$
\begin{aligned}
X_{(n-j)} - X_{(n-k)} &= \sum_{i=n-k+1}^{n-j} \frac{E_i}{n-i+1} \\
&= \sum_{l=1}^{k-j} \frac{E_{l+n-k}}{k-l+1}.
\end{aligned}
$$

Thus:

$$
\begin{aligned}
\sum_{j=0}^{k-1} \left[X_{(n-j)} - X_{(n-k)} \right] &= \sum_{j=0}^{k-1} \sum_{l=1}^{k-j} \frac{E_{l+n-k}}{k-l+1} \\
&= \sum_{l=1}^{k} \sum_{j=0}^{k-l} \frac{E_{l+n-k}}{k-l+1} \\
&= \sum_{l=1}^{k} E_{l+n-k}. \tag{1}
\end{aligned}
$$

    Since the collection $\{E_j\}_{j=1}^{\infty}$ used in the Rényi representation theorem are independent standard exponentials, we can for notational simplicity replace $\{E_{l+n-k}\}_{l=1}^{k}$ in this summation with $\{E_l\}_{l=1}^{k}$. Hence by (1) and the definition of $Y_{k,n}$:

$$Y_{k,n} = \frac{1}{k} \sum_{i=1}^{k} E_i,$$

and average of k independent, standard exponentials.
    Now by $E[E_i] = Var[E_i] = 1$ by (4.72), and thus $E[Y_{k,n}] = 1$ and $Var[Y_{k,n}] = 1/k$ by (4.39) and (4.43). Since the exponential distribution has a moment generating function as given in (4.73) with $\alpha = \lambda = 1$, the result in (6.22) follows from the central limit theorem of Proposition 6.13 because $k \to \infty$ as $n \to \infty$.
    Also, $Y_{k,n} \to_1 1$ by (6.37) of Proposition 6.48, since the exponential distribution has a finite fourth moment as can be verified directly, or indirectly by Proposition 4.25. ∎

6.3   Laws of Large Numbers

In this section, we study convergence results for sums of independent random variables $\{X_n\}_{n=1}^{\infty}$ defined on a probability space $(\mathcal{S}, \mathcal{E}, \lambda)$. We begin by briefly reviewing the notion of **tail event** from Definition II.5.34, an example being the convergence set for such a series, and recall the **Kolmogorov zero-one law** of Proposition II.5.36 which states that tail events, and in particular such convergence sets, must have probability 0 or 1.

    We then study **laws of large numbers,** extending the results of Section II.5.1. The **weak laws of large numbers** will generalize **Bernoulli's theorem** of Proposition II.5.3

and give specific conditions which ensure that the summation $\sum_{n=1}^{\infty} X_n$ converges in probability. The **strong laws of large numbers** generalize **Borel's theorem** of Proposition II.5.9 and provide conditions which ensure that this series converges with probability 1.

The final section investigates a limit theorem on quantiles which is a corollary to the strong laws, and proves a result in extreme value theory identified in Remark II.9.8.

6.3.1 Tail Events and Kolmogorov's 0-1 Law

In this section, we review notions from Book II. The reader is invited to look back to the earlier development for additional details and results.

Recall Definition II.3.43:

Definition 6.23 (Sigma algebras generated by random variables) *Given a random variable X defined on a probability space $(S, \mathcal{E}, \lambda)$, the **sigma algebra generated by** X and denoted $\sigma(X)$ is defined to be the smallest sigma algebra with respect to which X is measurable. By Exercise II.3.44:*

$$\sigma(X) \equiv \{X^{-1}(A) | A \in \mathcal{B}(\mathbb{R})\}. \tag{6.24}$$

*If $\{X_n\}$ is a finite or infinite collection of random variables, **the sigma algebra generated by** $\{X_n\}$ and denoted $\sigma(X_1, X_2, ...)$ is the smallest sigma algebra with respect to which each X_n is measurable. By Proposition II.3.45:*

$$\sigma(X_1, X_2, ...) = \sigma\left(\sigma(X_1), \sigma(X_2), ...\right),$$

meaning $\sigma(X_1, X_2, ...)$ is the sigma algebra generated by the collection of sigma algebras $\{\sigma(X_n)\}$.

By measurability of random variables, $\sigma(X) \subset \mathcal{E}$ for any X and $\sigma(X_1, X_2, ...) \subset \mathcal{E}$ for any $\{X_n\}$.

We next recall the essence of Definition 1.13:

Definition 6.24 (Independent random variables) *Random variables $\{X_n\}_{n=1}^{\infty}$ defined on a probability space $(S, \mathcal{E}, \lambda)$ are said to be **independent random variables** if $\{\sigma(X_n)\}_{n=1}^{\infty}$ are independent sigma algebras. That is, given any **finite** index subcollection $J = (j(1), j(2), ..., j(n))$, and $\{B_{j(i)}\}_{i=1}^{n}$ with $B_{j(i)} \in \sigma(X_{j(i)})$:*

$$\mu\left(\bigcap_{i=1}^{n} B_{j(i)}\right) = \prod_{i=1}^{n} \mu\left(B_{j(i)}\right).$$

Equivalently by (6.24), given $\{A_{j(i)}\}_{i=1}^{n} \subset \mathcal{B}(\mathbb{R})$:

$$\mu\left(\bigcap_{i=1}^{n} X_{j(i)}^{-1}\left(A_{j(i)}\right)\right) = \prod_{i=1}^{n} \mu\left(X_{j(i)}^{-1}\left(A_{j(i)}\right)\right).$$

Finally, we recall the notion of the **tail sigma algebra** $\mathcal{T}$ associated with an arbitrary collection of random variables $\{X_n\}_{n=1}^{\infty}$. The intersection of sigma algebras is a sigma algebra by Proposition I.2.8, and note that the sigma algebras in (6.25) are nested:

$$\sigma(X_{n+1}, X_{n+2}, ...) \subset \sigma(X_n, X_{n+1}, X_{n+2}, ...).$$

Definition 6.25 (Tail sigma algebra: $\mathcal{T} \equiv \mathcal{T}(\{X_n\}_{n=1}^{\infty})$) *Given a probability space $(S, \mathcal{E}, \lambda)$ and countable collection of random variables $\{X_n\}_{n=1}^{\infty}$, the **tail sigma algebra** **associated with** $\{X_n\}_{n=1}^{\infty}$ and denoted:*

$$\mathcal{T} \equiv \mathcal{T}(\{X_n\}_{n=1}^{\infty}),$$

is defined:

$$\mathcal{T} = \bigcap_{n=1}^{\infty} \sigma(X_n, X_{n+1}, X_{n+2}, \ldots), \tag{6.25}$$

where $\sigma(X_n, X_{n+1}, X_{n+2}, \ldots)$ is the sigma algebra generated by $\{X_j\}_{j=n}^{\infty}$.
*Thus $\mathcal{T} \subset \mathcal{E}$, and a **tail event** is any set $A \in \mathcal{T}$.*

An example of a tail event, which is at the heart of this section, is the convergence set A of a countable collection of random variables $\{X_n\}_{n=1}^{\infty}$. Note that it is not apparent from this definition even that $A \in \mathcal{E}$.

Definition 6.26 (Convergence set) *Given a countable collection of random variables $\{X_n\}_{n=1}^{\infty}$ defined on $(\mathcal{S}, \mathcal{E}, \lambda)$, define the **convergence set** $A \subset \mathcal{S}$ by:*

$$A = \left\{ \sum_{n=1}^{\infty} X_n(s) \text{ converges} \right\}.$$

To prove that A is a tail event, $A \in \mathcal{T} \equiv \mathcal{T}(\{X_n\}_{n=1}^{\infty})$, recall that a series of real numbers $\sum_{n=1}^{\infty} a_n$ converges if and only if this series satisfies the **Cauchy convergence criterion**, also called the **Cauchy criterion**, and named for **Augustin-Louis Cauchy** (1789–1857).

Definition 6.27 (Cauchy convergence criterion) *A series $\sum_{n=1}^{\infty} a_n$ satisfies the **Cauchy criterion** if given any $\epsilon > 0$ there is an N so that $\left| \sum_{j=m}^{n} a_j \right| < \epsilon$ for all $n, m \geq N$.*

Example 6.28 (Convergence sets are tail events) *That the convergence set $A \in \mathcal{T} \equiv \mathcal{T}(\{X_n\}_{n=1}^{\infty})$ is intuitively plausible because the convergence of a series does not depend on any finite number of terms.*

More formally, we can by the Cauchy criterion define the convergence set as follows, using rational ϵ :

$$A = \bigcap_{\epsilon \in \mathbb{Q}} \bigcup_{N=1}^{\infty} \bigcap_{n \geq m \geq N} \left\{ \left| \sum_{j=m}^{n} X_j(s) \right| < \epsilon \right\}. \tag{6.26}$$

Because $\left\{ \left| \sum_{j=m}^{n} X_j(s) \right| < \epsilon \right\} \subset \sigma(X_m, X_{m+1}, X_{m+2}, \ldots)$, it follows that $A \subset \sigma(X_m, X_{m+1}, X_{m+2}, \ldots)$ for all m and hence $A \in \mathcal{T}$.

Exercise 6.29 *Confirm (6.26). Hint: Given N, either:*

$$\bigcap_{n \geq m \geq N} \left\{ \left| \sum_{j=m}^{n} X_j(s) \right| < \epsilon \right\} = \emptyset,$$

or N works for the Cauchy criterion for given ϵ.

Kolmogorov's zero-one law is named for **Andrey Kolmogorov** (1903–1987). It states that if $\{X_n\}_{n=1}^{\infty}$ are **independent random variables**, then the measure of any set $A \in \mathcal{T}$ is predictable, but not precisely predictable. The proof is found in Proposition II.5.36.

Proposition 6.30 (Kolmogorov's zero-one law) *Given independent random variables $\{X_n\}_{n=1}^{\infty}$ defined on a probability space $(\mathcal{S}, \mathcal{E}, \lambda)$, then for any $A \in \mathcal{T} \equiv \mathcal{T}(\{X_n\}_{n=1}^{\infty})$:*

$$\lambda(A) \in \{0, 1\}. \tag{6.27}$$

We now have an immediate corollary by Example 6.28.

Corollary 6.31 (Kolmogorov's zero-one law) *Given independent random variables $\{X_n\}_{n=1}^{\infty}$ defined on a probability space $(\mathcal{S}, \mathcal{E}, \lambda)$, then:*

$$\lambda \left[\left\{ \sum_{n=1}^{\infty} X_n(s) \text{ converges} \right\} \right] \in \{0, 1\}. \tag{6.28}$$

In the next two sections we continue the study of convergence of series. The weak laws of large numbers will provide additional information on the weaker notion of **convergence in probability**, while the strong laws of large numbers will provide additional information on when such series **converge with probability one**, meaning:

$$\lambda\left[\left\{\sum_{n=1}^{\infty} X_n(s) \text{ converges}\right\}\right] = 1.$$

6.3.2   Weak Laws of Large Numbers

Weak laws of large numbers, often abbreviated WLLNs, identify conditions on a random variable sequence $\{X_n\}_{n=1}^{\infty}$ which assure **converge in probability** to a specified random variable. Recall Definition II.5.11:

Definition 6.32 (Convergence in probability: $Y_n \to_P Y$) *Given random variables Y and $\{Y_n\}_{n=1}^{\infty}$ on a probability space $(S, \mathcal{E}, \lambda)$, we say that Y_n **converges to Y in probability**, denoted $Y_n \to_P Y$, if for every $\epsilon > 0$:*

$$\lim_{n\to\infty} \Pr(|Y_n - Y| \geq \epsilon) = 0. \tag{6.29}$$

Notation 6.33 (Pr statements) *Using the probability notation $\Pr$ is often preferred in probability theory to using the measure-theoretic notation with λ. By definition:*

$$\Pr(|Y_n - Y| \geq \epsilon) = \lambda\{s \mid |Y_n(s) - Y(s)| \geq \epsilon\}.$$

Denoting $X_n \equiv Y_n - Y$ and $I_\epsilon = (-\infty, \epsilon] \bigcup [\epsilon, \infty\}$, this can also be expressed:

$$\Pr(|Y_n - Y| \geq \epsilon) = \lambda\{X_n^{-1}(I_\epsilon)\},$$

noting that this measure is well-defined since $X_n^{-1}(I_\epsilon) \in \mathcal{E}$.

Remark 6.34 *Since for any ϵ:*

$$\Pr(|Y_n - Y| < \epsilon) + \Pr(|Y_n - Y| \geq \epsilon) = 1,$$

$Y_n \to_P Y$ if and only if for every $\epsilon > 0$:

$$\lim_{n\to\infty} \Pr(|Y_n - Y| < \epsilon) = 1.$$

Weak laws are most often stated in the context of:

$$Y_n = \frac{1}{n} S_n \equiv \frac{1}{n} \sum_{j=1}^{n} X_j,$$

with $\{X_j\}_{j=1}^{\infty}$ a sequence of independent random variables. When random variables are assumed identically distributed, we will investigate convergence in probability to $\mu \equiv E[X]$, though this assumption is not necessary and there are other versions of this result. This average of the first n random variables also denoted:

$$\bar{X}_n \equiv \frac{1}{n} \sum_{j=1}^{n} X_j.$$

We begin with a version of a weak law with the simplest proof, due largely to a strong assumption. It states that when the sequence $\{X_j\}_{j=1}^{\infty}$ is independent and identically distributed, with finite mean and variance, then the associated average sequence $\{\bar{X}_n\}$ converges in probability to the mean.

Proposition 6.35 (WLLN 1) *Let $\{X_j\}_{j=1}^{\infty}$ be a sequence of independent, identically distributed random variables defined on a probability space $(S, \mathcal{E}, \lambda)$ with finite mean μ and variance σ^2. Then:*

$$\frac{1}{n}\sum_{j=1}^{n} X_j \to_P \mu. \tag{6.30}$$

Proof. *Using (4.39) and (4.43):*

$$Var\left(\frac{1}{n}\sum_{j=1}^{n} X_j\right) = \frac{\sigma^2}{n}, \qquad E\left[\frac{1}{n}\sum_{j=1}^{n} X_j\right] = \mu.$$

Thus by Chebyshev's inequality in (4.82):

$$\Pr\left(\left|\frac{1}{n}\sum_{j=1}^{n} X_j - \mu\right| \geq \epsilon\right) \leq \frac{\sigma^2}{n\epsilon^2},$$

and (6.30) follows. ∎

Example 6.36 (Bernoulli's theorem) *A special case of this version of the weak law is* **Bernoulli's theorem** *of Proposition II.5.3. This result stated that the average of independent, identically distributed binomial random variables $\{X_j^B\}_{j=1}^{\infty}$ converges in probability to the binomial probability p. The assumptions of the above proposition are satisfied for this variate since X_j^B has finite mean and variance, and unsurprisingly, so too is the conclusion of this result since $p = E[X_j^B]$.*

While the existence of σ^2 provides a very simple proof thanks to Chebyshev, the above proposition remains true with only the assumption of the existence of the first moment μ. The proof is significantly harder due principally to the need for the general **Lebesgue dominated convergence theorem** of Book V for integrals as in (4.7). This theorem is needed for a technical result on "truncations" of random variables which may appear elementary, but resist direct and more elementary validation.

In cases where $E[X]$ as defined in (4.7) can be transformed to a Lebesgue integral as in (4.14), the Lebesgue dominated convergence theorem of Proposition III.2.52 suffices.

Definition 6.37 (Truncation of X) *Let X be a random variable on $(S, \mathcal{E}, \lambda)$. Given $m > 0$, define a random variable $X^{(m)}$, called* **the truncation of X at m,** *by:*

$$X^{(m)} = \begin{cases} X, & |X| \leq m \\ 0, & |X| > m. \end{cases}$$

Proposition 6.38 (Properties of truncations, $E[X] < \infty$) *Let X be a random variable on $(S, \mathcal{E}, \lambda)$ with $E[X] < \infty$, where E is defined as in (4.7). Then:*

1. *For any m, $X^{(m)}$ has finite mean and variance:*

$$\left|E\left[X^{(m)}\right]\right| \leq E[|X|], \qquad Var\left[X^{(m)}\right] \leq mE[|X|].$$

2. *For any $\alpha > 0$ there exists $M = M(\alpha)$ so that for all $m \geq M$:*

$$E\left[\left|X - X^{(m)}\right|\right] < \alpha.$$

Proof. *For item 1, recall the integral representation of $E[X]$ in (4.7). Applying (4.12) and (4.11) obtains that $X^{(m)}$ has finite mean:*

$$\left| E\left[X^{(m)} \right] \right| \leq E\left[\left| X^{(m)} \right| \right] \leq E\left[|X| \right] < \infty,$$

since $E[X]$ exists.

Then by (4.50) and (4.10):

$$Var[X^{(m)}] \leq E[\left(X^{(m)} \right)^2] \leq mE[\left| X^{(m)} \right|] \leq mE[|X|].$$

For item 2, define $Z_m \equiv \left| X - X^{(m)} \right|$. Then $Z_m \to 0$ pointwise on S, and $|Z_m| \leq |X|$ with $|X|$ integrable by existence of $E[X]$ and (4.8). Thus by the general version of Lebesgue's dominated convergence theorem of Book V, $E[Z_m] \to 0$ as $n \to \infty$, and the result follows. ∎

Exercise 6.39 *Provide the details of item 2 above in the case where $X = f(x)$ is a Lebesgue measurable function that is integrable on $\mathbb{R}$ (or $\mathbb{R}^n$). Given m, define the truncation of $f(x)$ as above, denoting this $f_m(x)$. Use Proposition III.2.52 to complete the proof that for any $\alpha > 0$ there exists $M = M(\alpha)$ so that for all $m \geq M$:*

$$E\left[|f - f_m| \right] < \alpha.$$

Now assume that $f(x) = \phi(x)$, the standard normal density function in (1.66), and formulaically determine $M(\alpha)$ in terms of the normal distribution $\Phi(x)$.

Proposition 6.40 (WLLN 2) *Let $\{X_j\}_{j=1}^{\infty}$ be a sequence of independent, identically distributed random variables defined on $(S, \mathcal{E}, \lambda)$ with finite mean $\mu \equiv E[X]$. Then (6.30) is satisfied:*

$$\frac{1}{n} \sum_{j=1}^{n} X_j \to_P \mu.$$

Proof. *Given $\epsilon > 0$, we prove by truncation that for any $\delta > 0$ there exists N so that for $n \geq N$:*

$$\Pr\left[\left| \frac{1}{n} \sum_{j=1}^{n} X_j - E[X] \right| \geq \epsilon \right] < \delta. \tag{1}$$

The reader should verify that (1) implies the convergence in probability of (6.30).

Given m, let $X_j^{(m)}$ denote the truncation of X_j, and define $\bar{X}_n \equiv \frac{1}{n} \sum_{j=1}^{n} X_j$ and $\bar{X}_n^{(m)} \equiv \frac{1}{n} \sum_{j=1}^{n} X_j^{(m)}$. By the triangle inequality:

$$\left| \bar{X}_n - E[X] \right| \leq \left| \bar{X}_n - \bar{X}_n^{(m)} \right| + \left| \bar{X}_n^{(m)} - E[X^{(m)}] \right| + \left| E[X^{(m)}] - E[X] \right|.$$

By considering the defining sets in S and using subadditivity of the measure λ it follows that:

$$\begin{aligned}
\Pr\left[\left| \bar{X}_n - E[X] \right| \geq \epsilon \right] \leq \quad & \Pr\left[\left| \bar{X}_n - \bar{X}_n^{(m)} \right| \geq \epsilon/3 \right] \\
& + \Pr\left[\left| \bar{X}_n^{(m)} - E[X^{(m)}] \right| \geq \epsilon/3 \right] \\
& + \Pr\left[\left| E[X^{(m)}] - E[X] \right| \geq \epsilon/3 \right].
\end{aligned} \tag{2}$$

Given $\alpha > 0$ to be specified below, there exists $M = M(\alpha)$ by Proposition 6.38 so that $E\left[\left|X - X^{(m)}\right|\right] < \alpha$ for any truncation $X^{(m)}$ with $m \geq M(\alpha)$. We now prove that each of the probabilities on the right in (2) can be made arbitrarily small by making α small.

1. *By the triangle inequality, and that $\{X_j\}_{j=1}^{\infty}$ are identically distributed:*

$$E\left[\left|\bar{X}_n - \bar{X}_n^{(m)}\right|\right] \leq \frac{1}{n}\sum_{j=1}^{n} E\left[\left|X_j - X_j^{(m)}\right|\right] = E\left[\left|X - X^{(m)}\right|\right] < \alpha.$$

The Markov inequality in (4.87) obtains that for $m \geq M(\alpha)$:

$$\Pr\left[\left|\bar{X}_n - \bar{X}_n^{(m)}\right| \geq \epsilon/3\right] \leq \frac{3\alpha}{\epsilon}. \tag{3a}$$

2. *Let $g(x) \equiv \chi_{[-m,m]}(x)$, the characteristic function of the interval $[-m, m]$, defined to equal 1 on this interval and 0 elsewhere. Using Proposition II.3.56 and Borel measurable $g(x)$, independence of $\{X_j\}$ and $X_j^{(m)} = g(X_j)$ assure independence of $\{X_j^{(m)}\}$. Thus by independence, (4.43), and Proposition 6.38:*

$$Var[\bar{X}_n^{(m)}] \leq mE[|X|]/n.$$

As $E[X^{(m)}] = E[\bar{X}_n^{(m)}]$ by (4.39), Chebyshev's inequality in (4.82) obtains:

$$\Pr\left[\left|\bar{X}_n^{(m)} - E[X^{(m)}]\right| \geq \epsilon/3\right] \leq \frac{9mE[|X|]}{n\epsilon^2}. \tag{3b}$$

3. *By the triangle inequality:*

$$\left|E[X^{(m)}] - E[X]\right| \leq E\left[\left|X - X^{(m)}\right|\right] < \alpha,$$

and if $\alpha < \epsilon/3$:

$$\Pr\left[\left|E[X^{(m)}] - E[X]\right| \geq \epsilon/3\right] = 0. \tag{3c}$$

Combining $(3a) - (3c)$, if $\alpha < \epsilon/3$:

$$\Pr\left[\left|\bar{X}_n - E[X]\right| \geq \epsilon\right] \leq \frac{3\alpha}{\epsilon} + \frac{9mE[|X|]}{n\epsilon^2}.$$

Choosing $\alpha < \epsilon\delta/6$ satisfies this initial constraint and makes $3\alpha/\epsilon < \delta/2$ for $n \geq M(\alpha)$. With this α and $m = M(\alpha)$, it follows that $9M(\alpha)E[|X|]/n\epsilon^2 < \delta/2$ for $n > 18M(\alpha)E[|X|]/\delta\epsilon^2$. Defining $N = \max[M(\alpha), 18M(\alpha)E[|X|]/\delta\epsilon^2]$ completes the proof. ∎

The final generalization of Proposition 6.35 again assumes finite second moments, and applies to an independent sequence of random variables $\{X_j\}_{j=1}^{\infty}$ with arbitrary distributions. However, an assumption will be needed on the growth rates of the mean and variance of $\sum_{j=1}^{n} X_j$ as functions of n. By independence, $E\left[\sum_{j=1}^{n} X_j\right] = \sum_{j=1}^{n} E[X_j]$ by (4.39) and $Var\left(\sum_{j=1}^{n} X_j\right) = \sum_{j=1}^{n} Var(X_j)$ by (4.43).

Note that the requirements of the next result are automatically satisfied when $\{X_j\}_{j=1}^{\infty}$ are also identically distributed.

Proposition 6.41 (WLLN 3) *Let $\{X_j\}_{j=1}^{\infty}$ be a sequence of independent random variables defined on $(\mathcal{S}, \mathcal{E}, \lambda)$ with finite means $\{\mu_j\}_{j=1}^{\infty}$ and variances $\{\sigma_j^2\}_{j=1}^{\infty}$. Denote:*

$$m_n = \sum\nolimits_{j=1}^{n} \mu_j, \qquad s_n^2 = \sum\nolimits_{j=1}^{n} \sigma_j^2.$$

If $s_n^2/n^2 \to 0$, and there exists μ so that $m_n/n \to \mu$, then (6.30) is satisfied:

$$\frac{1}{n} \sum\nolimits_{j=1}^{n} X_j \to_P \mu.$$

Proof. *Letting $\bar{X}_n \equiv \frac{1}{n}\sum_{j=1}^{n} X_j$, the triangle inequality obtains:*

$$\left|\bar{X}_n - \mu\right| \le \left|\bar{X}_n - \frac{m_n}{n}\right| + \left|\frac{m_n}{n} - \mu\right|.$$

By consideration of the defining sets in $\mathcal{S}$:

$$\Pr\left(\left|\bar{X}_n - \mu\right| < \epsilon\right) \ge \Pr\left(\left|\bar{X}_n - \frac{m_n}{n}\right| < \epsilon - \left|\frac{m_n}{n} - \mu\right|\right). \tag{1}$$

Applying Chebyshev's inequality with $E\left[\bar{X}_n\right] = m_n/n$ and $Var\left[\bar{X}_n\right] = s_n^2/n^2$:

$$\Pr\left(\left|\bar{X}_n - \frac{m_n}{n}\right| < \epsilon - \left|\frac{m_n}{n} - \mu\right|\right) \ge 1 - \frac{s_n^2/n^2}{\left[\epsilon - |m_n/n - \mu|\right]^2}.$$

By assumption, $|m_n/n - \mu| \to 0$ and $s_n^2/n^2 \to 0$, and thus as $n \to \infty$:

$$\Pr\left(\left|\bar{X}_n - \frac{m_n}{n}\right| < \epsilon - \left|\frac{m_n}{n} - \mu\right|\right) \to 1.$$

Hence by (1) :
$$\Pr\left(\left|\bar{X}_n - \mu\right| < \epsilon\right) \to 1.$$

■

Remark 6.42 (On further generalizations) *In the above result, the assumption that $m_n/n \to \mu$ can be eliminated to obtain the result:*

$$\frac{1}{n}\sum\nolimits_{j=1}^{n} (X_j - \mu_j) \to_P 0.$$

Similarly, we can replace the independence assumption in any version of the weak law by the assumption that correlations in (4.45) satisfy $\rho_{ij} \le 0$ for all i, j. Then $Var\left[\sum_{j=1}^{n} X_j/n\right] \le s_n^2/n^2$ by (4.47) and the proof goes through without change.

A positive result is possible even in the nonnegative correlation case if we assume for example that for some $0 < r < 1$, $\rho_{ij} \le r^{|i-j|}$ and $\sigma_j^2 \le B$ for all j.

Exercise 6.43 *Let $\{X_j\}_{j=1}^{\infty}$ be a sequence of random variables with the same finite mean μ, and variances σ_j^2 with $\sigma_j^2 \le B$ for all j. Show that if $\rho_{ij} \le r^{|i-j|}$ for some $0 < r < 1$, then (6.30) remains true. Hint: Repeat the above proof with the new variance estimate from (4.47).*

6.3.3 Strong Laws of Large Numbers

Strong laws of large numbers, often abbreviated SLLNs, identify conditions on a random variable sequence which ensure **convergence with probability 1** to a specified random variable. This notion is also called **convergence almost surely**. Recall Definition II.5.15:

Definition 6.44 (Convergence with probability 1 : $Y_n \to_1 Y$) *Given random variables Y and $\{Y_n\}_{n=1}^\infty$ on a probability space $(\mathcal{S}, \mathcal{E}, \lambda)$, we say that Y_n **converges to Y with probability** 1, denoted $Y_n \to_1 Y$, if:*

$$\lambda\left[\left\{\lim_{n\to\infty} \sum_{j=1}^n Y_n = Y\right\}\right] = 1. \tag{6.31}$$

*This notion is also called **convergence almost surely**, and then denoted $Y_n \to_{a.s.} Y$.*

As was the case for weak laws, strong laws are most often stated in the context of:

$$Y_n = \frac{1}{n} S_n \equiv \frac{1}{n} \sum_{j=1}^n X_j,$$

with $\{X_j\}_{j=1}^\infty$ a sequence of independent random variables. When identically distributed, we will investigate convergence with probability 1 to $\mu \equiv E[X]$, though this assumption is not necessary and there are other versions of this result. The average of the first n random variables is also denoted:

$$\bar{X}_n \equiv \frac{1}{n} \sum_{j=1}^n X_j.$$

Weak laws and strong laws can be related as follows. Let $\{X_j\}_{j=1}^\infty$ be a sequence of independent, identically distributed random variables defined on a probability space $(\mathcal{S}, \mathcal{E}, \lambda)$. Define the set $A_n(\epsilon) \subset \mathcal{S}$ as in II.(5.3):

$$A_n(\epsilon) \equiv \left\{ \left| \frac{1}{n} \sum_{j=1}^n X_j - \mu \right| \geq \epsilon \right\}.$$

It is an exercise to check that $A_n(\epsilon) \in \mathcal{E}$ and thus is measurable. In the first two versions of the weak law in Propositions 6.35 and 6.40, it was proved that for such independent and identically distributed random variables and any $\epsilon > 0$:

$$\lambda[A_n(\epsilon)] \to 0 \text{ as } n \to \infty.$$

As noted in Section II.5.1.1, the sets $\{A_n(\epsilon)\}_{n=1}^\infty$ are not necessarily nested and it is possible to have both $A_{n+1}(\epsilon) - A_n(\epsilon) \neq \emptyset$ and $A_n(\epsilon) - A_{n+1}(\epsilon) \neq \emptyset$. Hence there is no apparent event in $\mathcal{S}$ associated with weak laws that is definable in terms of some limit of $A_n(\epsilon)$ as $n \to \infty$.

For the strong laws, the goal is to determine the measure of the **strong convergence set** C_S on which this averaging series converges to μ:

$$C_S \equiv \left\{ \frac{1}{n} \sum_{j=1}^n X_j \to \mu \right\} = \left\{ \frac{1}{n} \sum_{j=1}^n (X_j - \mu) \to 0 \right\}.$$

This definition was introduced in II.(5.8) in the context of binomial $\{X_j\}_{j=1}^\infty$, and called the "convergence set," but here we add the qualifier "strong" to distinguish this set from the general convergence set A of Definition 6.26.

We prove below in Proposition 6.46 that $C_S \in \mathcal{E}$ and is thus measurable. Strong laws identify conditions under which $\lambda[C_S] = 1$.

The strong convergence set C_S is closely related to the $A_n(\epsilon)$-sets, but we must first recall Definition II.2.1 for the notion of the **limit superior** of a sequence of sets $\{A_n\}_{n=1}^\infty$. We include the other limit notions for completeness.

Definition 6.45 (lim sup A_n, lim inf A_n, lim A_n) *Given a measure space* $(S, \mathcal{E}, \lambda)$ *and a countable collection of sets* $\{A_n\}_{n=1}^{\infty} \subset \mathcal{E}$, *define:*

1. **Limit superior:**

$$\limsup_n A_n = \bigcap_{n=1}^{\infty} \bigcup_{k=n}^{\infty} A_k.$$ (6.32)

2. **Limit inferior:**

$$\liminf_n A_n = \bigcup_{n=1}^{\infty} \bigcap_{k=n}^{\infty} A_k.$$ (6.33)

3. **Limit:** *If* $\limsup A_n = \liminf A_n = A$, *define:*

$$\lim_n A_n \equiv A.$$ (6.34)

It common notationally to omit the subscript n when clear from the context.

Generalizing Proposition II.5.8 for binomial sequences:

Proposition 6.46 ($\lambda[C_S]$ from $\{A_n(\epsilon)\}_{n=1}^{\infty}$) *With the notation above,* $C_S \in \mathcal{E}$ *and:*

$$\lambda\{C_S\} = 1 \iff \lambda[\limsup A_n(\epsilon)] = 0 \text{ for all } \epsilon > 0.$$ (6.35)

Proof. *Modifying (6.26),* C_S *can be expressed:*

$$\begin{aligned} C_S &= \bigcap_{j=1}^{\infty} \bigcup_{N=1}^{\infty} \bigcap_{n=N}^{\infty} \left\{ \left| \frac{1}{n} \sum_{j=1}^{n} X_j - \mu \right| < 1/j \right\} \\ &= \bigcap_{j=1}^{\infty} \bigcup_{N=1}^{\infty} \bigcap_{n=N}^{\infty} \tilde{A}_n(1/j), \end{aligned}$$

where $\tilde{A}_n(1/j)$ *denotes the complement of* $A_n(1/j)$. *Thus* $C_S \in \mathcal{E}$ *since* $A_n(\epsilon) \in \mathcal{E}$ *for all* n, ϵ.

Using De Morgan's laws of Exercise I.2.2 and (6.32), the complement of C_S *is given by:*

$$\tilde{C}_S = \bigcup_{j=1}^{\infty} \limsup A_n(1/j).$$ (1)

If $\lambda[C_S] = 1$ *then* $\lambda\left[\tilde{C}_S\right] = 0$, *and thus* $\lambda[\limsup_n A_n(\epsilon)] = 0$ *for all* $\epsilon = 1/j$ *by (1). This is then true for all* $\epsilon > 0$ *since if* $\epsilon > \epsilon'$:

$$\limsup A_n(\epsilon) \subset \limsup A_n(\epsilon').$$ (2)

Conversely, by the nested property in (2) and continuity from below (Proposition I.2.45) of the measure λ :

$$\lambda\left[\tilde{C}_S\right] = \lim_{j \to \infty} \lambda[\limsup A_n(1/j)].$$

If $\lambda[\limsup A_n(1/j)] = 0$ *for all* j, *then* $\lambda\left[\tilde{C}_S\right] = 0$ *and thus* $\lambda[C_S] = 1$. ■

Remark 6.47 (SLLN proof strategy) *To prove a strong law we must by the above proposition show that for any* $\epsilon > 0$,

$$\lambda[\limsup A_n(\epsilon)] = 0.$$

*A powerful tool for proving such a statement is the first result of the **Borel-Cantelli lemma** of Proposition II.2.6, due to Cantelli:*

Cantelli: *Given a measure space* $(S, \mathcal{E}, \lambda)$ *and a countable collection of sets* $\{A_n\}_{n=1}^{\infty} \subset \mathcal{E}$:

$$\sum_{n=1}^{\infty} \lambda(A_n) < \infty \implies \lambda(\limsup A_n) = 0. \tag{6.36}$$

Thus if we can show that for any $\epsilon > 0$:

$$\sum_{n=1}^{\infty} \lambda(A_n(\epsilon)) < \infty,$$

then $\lambda(\limsup A_n(\epsilon)) = 0$ *for all* $\epsilon > 0$, *and the strong convergence set* C_S *then has probability 1 by Proposition 6.46.*

We now derive two strong laws. As for the weak law development, the first result will have excess assumptions to more easily highlight the application of Borel-Cantelli. Much like the assumption of the existence of σ^2 in the first version of the weak law, we will require the existence of the fourth central moment μ_4 to facilitate the application of Chebyshev's inequality.

The existence of μ_4 assures the existence of all lower order moments as noted in the introduction to Section 4.2.5. Consequently, we do not have to separately assume the existence of the mean μ, and this is actually part of the conclusion.

Proposition 6.48 (SLLN 1) *Let* $\{X_j\}_{j=1}^{\infty}$ *be a sequence of independent, identically distributed random variables on a probability space* $(S, \mathcal{E}, \lambda)$ *with fourth central moment* μ_4. *Then:*

$$\frac{1}{n} \sum_{j=1}^{n} X_j \to_1 \mu.$$

In other words, $\frac{1}{n} \sum_{j=1}^{n} X_j$ *converges to the mean* μ *with probability 1:*

$$\Pr\left\{\frac{1}{n} \sum_{j=1}^{n} X_j \to \mu\right\} = 1. \tag{6.37}$$

Proof. *Let:*

$$S_n \equiv \sum_{j=1}^{n} (X_j - \mu).$$

A calculation using (4.39) and (4.38) obtains:

$$
\begin{aligned}
E\left[S_n^4\right] &= E\left[\left[\sum_{j=1}^{n} (X_j - \mu)\right]^4\right] \\
&= \sum_{j=1}^{n} E\left[(X_j - \mu)^4\right] + \binom{4}{2} \sum_{i=1}^{n} \sum_{j=i+1}^{n} E\left[(X_j - \mu)^2\right] E\left[(X_i - \mu)^2\right].
\end{aligned}
$$

This follows because all other terms have at least one factor such as $E[(X_j - \mu)]$, *which equals 0.*

By Lyapunov's inequality in (4.104):

$$E\left[(X_j - \mu)^2\right] E\left[(X_i - \mu)^2\right] \leq E(X_j - \mu)^4,$$

and hence:

$$E\left[S_n^4\right] \leq \left(3n^2 - 2n\right) \mu_4.$$

By Chebyshev's inequality in (4.83):

$$
\begin{aligned}
\Pr\left[A_n(\epsilon)\right] &\equiv \Pr\{S_n \geq n\epsilon\} \\
&\leq \frac{\left(3n^2 - 2n\right) \mu_4}{(n\epsilon)^4}.
\end{aligned}
$$

Thus $\sum_{n=1}^{\infty} \Pr[A_n(\epsilon)] < \infty$ *for any* $\epsilon > 0$, *and by the Cantelli theorem:*

$$\lambda \left[\limsup A_n(\epsilon) \right] = 0 \text{ for all } \epsilon > 0.$$

The result now follows from Proposition 6.46. ∎

Example 6.49 (Borel's theorem) *A special case of this version of the strong law is* **Borel's theorem** *presented in Proposition II.5.9. This stated that the average of indepen- dent, identically distributed binomial random variables* $\{X_j^B\}_{j=1}^{\infty}$, *converged with probability 1 to the binomial probability* $p = E[X_j^B]$ *by (4.62).*

All moments are finite for the binomial by Proposition 4.25, since this variate has a moment generating function in (4.63). Thus the above assumption that $\mu_4 < \infty$ *is satisfied.*

Like the weak law, it is also true that the strong law is valid for independent and identically distributed random variables assuming only the existence of a first moment. See for example **Feller** (1968) or **Billingsley** (1995).

We provide a version of this result that is somewhat weaker because it requires a sec- ond moment. But this version is also somewhat more general as it is also applicable to independent random variables which need not be identically distributed.

For the proof, we will use Kolmogorov's inequality in (4.94), and modify the definition of $A_n(\epsilon)$ to a related set $A'_n(\epsilon)$.

Proposition 6.50 (SLLN 2) *Let* $\{X_j\}_{j=1}^{\infty}$ *be a sequence of independent random variables on a probability space* $(\mathcal{S}, \mathcal{E}, \lambda)$ *with means* $\{\mu_j\}_{j=1}^{\infty}$, *and variances* $\{\sigma_j^2\}_{j=1}^{\infty}$ *with* $\sum_{j=1}^{\infty} \frac{\sigma_j^2}{j^2} < \infty$. *Then:*

$$\frac{1}{n} \sum_{j=1}^{n} (X_j - \mu_j) \to_1 0.$$

In other words:

$$\Pr \left\{ \frac{1}{n} \sum_{j=1}^{n} (X_j - \mu_j) \to 0 \right\} = 1. \tag{6.38}$$

Proof. *Given* $\epsilon > 0$ *and* $Y_j \equiv X_j - \mu_j$, *define:*

$$A'_n(\epsilon) \equiv \bigcup_{j=2^{n-1}+1}^{2^n} A_j(\epsilon)$$

$$= \left\{ \left| \frac{1}{k} \sum_{j=1}^{k} Y_j \right| \geq \epsilon \text{ for at least one } k \text{ with } 2^{n-1} < k \leq 2^n \right\}.$$

Thus $A'_n(\epsilon) \in \mathcal{E}$ *as a union of events.*

To prove that:

$$\sum_{n=1}^{\infty} \Pr[A'_n(\epsilon)] < \infty,$$

we first claim that $A'_n(\epsilon) \subset A''_n(\epsilon) \in \mathcal{E}$ *with:*

$$A''_n(\epsilon) = \left\{ \max_{2^{n-1} < k \leq 2^n} \left| \sum_{j=1}^{k} Y_j \right| \geq 2^{n-1} \epsilon \right\}.$$

It is an exercise to show that $A''_n(\epsilon) \in \mathcal{E}$ *(Hint: Proposition I.3.47). The inclusion* $A'_n(\epsilon) \subset A''_n(\epsilon)$ *follows because if* $\left| \sum_{j=1}^{k} Y_j / k \right| \geq \epsilon$ *for any* k *with* $2^{n-1} < k \leq 2^n$, *then* $\max_{2^{n-1} < k \leq 2^n} \left| \sum_{j=1}^{k} Y_j \right| \geq 2^{n-1} \epsilon$ *by definition.*

By Kolmogorov's inequality in (4.94), the probability of the event $A''_n(\epsilon)$ *is bounded:*

$$\Pr[A''_n(\epsilon)] \leq \frac{1}{\epsilon^2 2^{2n-2}} \sum_{j=1}^{2^n} \sigma_j^2.$$

Hence since $\Pr[A'_n(\epsilon)] \leq \Pr[A''_n(\epsilon)]$:

$$
\begin{aligned}
\sum_{n=1}^{\infty} \Pr[A'_n(\epsilon)] &\leq \frac{4}{\epsilon^2} \sum_{n=1}^{\infty} \frac{1}{2^{2n}} \sum_{j=1}^{2^n} \sigma_j^2 \\
&= \frac{4}{\epsilon^2} \sum_{j=1}^{\infty} \sigma_j^2 \sum_{2n \geq j} \frac{1}{2^{2n}} \\
&\leq \frac{4}{\epsilon^2} \sum_{j=1}^{\infty} \frac{\sigma_j^2}{j^2}.
\end{aligned}
$$

For the last step, note that for $j \geq 4$:

$$
\sum_{2n \geq j} \frac{1}{2^{2n}} \leq \sum_{k=j}^{\infty} \frac{1}{2^k} = \frac{1}{2^{j-1}} \leq \frac{2}{j^2}.
$$

Hence $\sum_{n=1}^{\infty} \Pr[A'_n(\epsilon)] < \infty$, *and by the Cantelli theorem this implies that for all* $\epsilon > 0$:

$$
\Pr[\limsup A'_n(\epsilon)] = 0,
$$

or by Definition 6.45:

$$
\Pr\left[\bigcap_{N=1}^{\infty} \bigcup_{n=N}^{\infty} A'_n(\epsilon)\right] = 0.
$$

Since $A'_n(\epsilon) \equiv \bigcup_{j=2^{n-1}+1}^{2^n} A_j(\epsilon)$, *this implies:*

$$
\Pr\left[\bigcap_{N=1}^{\infty} \bigcup_{n=N}^{\infty} A_n(\epsilon)\right] = 0,
$$

and so $\Pr[\limsup_n A_n(\epsilon)] = 0$.
Proposition 6.46 then obtains (6.38). ∎

Remark 6.51 *Two comments on the above result:*

1. *If* $\sum_{j=1}^{n} \mu_j/n \to \mu$ *or if* $\mu_j = \mu$ *for all* j, *then (6.38) can be restated as (6.37).*

2. *The assumption in this Strong Law, that* $\sum_{j=1}^{\infty} \frac{\sigma_j^2}{j^2} < \infty$, *is apparently an assumption about the growth rate of* σ_j^2 *as* $j \to \infty$.

   For example, if $\sigma_j^2 = \sigma^2$, *the assumption of no growth, then this strong law applies since* $\sum_{j=1}^{\infty} 1/j^2 < \infty$. *This is the case with independent, identically distributed random variables. This strong law also applies if* $\sigma_j^2 = j^a \sigma^2$ *for any* $a < 1$. *On the other hand, if* $\sigma_j^2 = j\sigma^2$, *this strong law does not apply since* $\sum_{j=1}^{N} 1/j \to \infty$ *with* N.

   Consequently, linear variance growth, or equivalently, square root growth in standard deviation, is a bit too fast for this strong law to apply.

6.3.4 A Limit Theorem in EVT

In this section, we present a result that is a corollary to the strong law of large numbers. It proves a statement made in Remark II.9.8 on extreme value theory, concerning the probability 1 convergence of a sequence of high quantile order statistics.

Let $\{X_j\}_{j=1}^{n}$ be a collection of independent, identically distributed random variables on a probability space $(\mathcal{S}, \mathcal{E}, \lambda)$ with distribution function $F(x)$, with associated order statistics $\{X_{(k)}\}_{k=1}^{n}$, where as usual $X_{(k)} \leq X_{(k+1)}$. Proposition 6.18 addressed the case where F is continuous and strictly increasing in a neighborhood of $F^{-1}(q)$ for $0 < q < 1$, and differentiable at $F^{-1}(q)$ with $F'(F^{-1}(q)) \neq 0$. Then given a sequence k_n with $k_n/n \to q$ as $n \to \infty$, and $X'_{(k_n)}$ defined:

$$X'_{(k_n)} \equiv \frac{X_{(k_n)} - q}{\sqrt{q(1-q)/n}},$$

the distribution function $F_{(k_n)}$ of $X'_{(k_n)}$ was proved to be asymptotically standard normal.

In this section, we address the case where $k_n/n \to 1$, and do so in the context of the strong law, meaning that we seek to prove this limiting result with probability 1. In the notational conventions of extreme value theory and the development of the Hill estimator in Chapter 7, what is called k_n here will correspond to the order variate $n - k_n$ there. Thus the assumption of this section that $k_n/n \to 1$ will be seen to be interpreted there as $k_n/n \to 0$.

As defined in II.(9.7), given a distribution function $F(x)$:

$$x^* \equiv \inf\{x|F(x) = 1\}, \qquad x^* \equiv \infty \text{ if } F(x) < 1 \text{ for all } x. \tag{6.39}$$

We now have the following result.

Proposition 6.52 $(X_{(k_n)} \to_1 x^*)$ *Let* $\{X_j\}_{j=1}^{\infty}$ *be a sequence of independent, identically distributed random variables on a probability space* $(\mathcal{S}, \mathcal{E}, \lambda)$ *with continuous distribution function F. For each n, let* $\{X_{(k)}\}_{k=1}^{n}$ *be the order statistics associated with* $\{X_j\}_{j=1}^{n}$, *and* $X_{(k_n)}$ *a given variate from this collection.*

If $k_n/n \to 1$, *then* $X_{(k_n)} \to x^*$ *with probability 1 :*

$$X_{(k_n)} \to_1 x^*. \tag{6.40}$$

Proof. *For* $r < x^*$, *define the random variable* $Z_j = \chi_{(r,\infty)}(X_j)$. *Each* Z_j *is binomially distributed with:*

$$p \equiv \Pr[Z_j = 1] = 1 - F(r).$$

Thus by (4.62), $E[Z_j] = 1 - F(r)$ *and* $Var[Z_j] = F(r)(1 - F(r))$.

By Proposition 6.48 on the strong law, as $n \to \infty$:

$$\frac{1}{n}\sum_{j=1}^{n} Z_j \to_1 1 - F(r). \tag{1}$$

Since F is continuous and $F(r) \to 1$ *as* $r \to x^*$, *it follows from (1) and the definition of* Z_j *that for any* $r < x^*$:

$$\frac{1}{n}\sum_{j=1}^{n} \chi_{(r,\infty)}(X_j) \to_1 1 - F(r) > 0. \tag{2}$$

Assume that for some $r < x^*$ *that:*

$$\lambda\left[\limsup_n\{X_{(k_n)} \leq r\}\right] > 0.$$

Thus there are infinitely many values of k_n *and n for which:*

$$\frac{n - k_n}{n} \equiv \frac{1}{n}\sum_{j=1}^{n} \chi_{(X_{(k_n)},\infty)}(X_j) \geq \frac{1}{n}\sum_{j=1}^{n} \chi_{(r,\infty)}(X_j).$$

But $\frac{k_n}{n} \to 1$ *by assumption, and this implies that the right hand summation converges to 0 with probability 1. This contradicts (2), that this summation converges to* $1 - F(r) > 0$.

Thus for all $r < x^*$:

$$\lambda\left[\limsup_n\{X_{(k_n)} \leq r\}\right] = 0,$$

and by complementarity:

$$\lambda\left[\liminf_n\{X_{(k_n)} > r\}\right] = 1. \tag{3}$$

If $\{r_j\}$ is an enumeration of rationals with $r_j \to x^*$, and:

$$B_j \equiv \liminf_n \{X_{(k_n)} > r_j\},$$

then

$$\lambda\left[\bigcap_{j=1}^{\infty} B_j\right] = 1, \tag{4}$$

where by Definition 6.45:

$$\bigcap_{j=1}^{\infty} B_j = \bigcap_{j=1}^{\infty} \bigcup_{n=1}^{\infty} \bigcap_{m=n}^{\infty} \{X_{(k_m)} > r_j\}.$$

If $s \in \bigcap_{j=1}^{\infty} B_j$, then for for every j there exists n so that $X_{(k_m)}(s) > r_j$ for all $m \geq n$, and thus $\lim_{m \to \infty} X_{(k_m)}(s) > r_j$. As this is true for all r_j, it follows from $\lim_{m \to \infty} X_{(k_m)}(s) \leq x^*$ and $r_j \to x^*$ that $\lim_{m \to \infty} X_{(k_m)}(s) = x^*$. Hence:

$$\bigcap_{j=1}^{\infty} B_j \subset \left\{\lim_{m \to \infty} X_{(k_m)} = x^*\right\},$$

and the proof is complete by (4). ∎

Corollary 6.53 ($X_{(k_n)} \to_1 x_*$) Let $\{X_j\}_{j=1}^{\infty}$ be a sequence of independent, identically distributed random variables on a probability space $(S, \mathcal{E}, \lambda)$ with a continuous distribution function F. For each n let $\{X_{(k)}\}_{k=1}^{n}$ be the order statistics associated with $\{X_j\}_{j=1}^{n}$.
If $k_n/n \to 0$, then $X_{(k_n)} \to x_*$ with probability 1 :

$$X_{(k_n)} \to_1 x_*, \tag{6.41}$$

where:

$$x_* \equiv \sup\{x | F(x) = 0\}, \qquad x_* \equiv -\infty \text{ if } F(x) > 0 \text{ for all } x.$$

Proof. Left as an exercise. Hint: Define the random variable $Y = -X$. Relate $F_X(x)$ and $F_Y(y)$, and show that $x_* = y^*$. Use the above result by relating k_n for X with k'_n for Y. ∎

6.4 Convergence of Empirical Distributions

Chapter 5 focused on simulating samples from a given and thus known distribution function $F(x)$. In empirical work, one may be confronted with a "random sample" of real world observations, deemed to be independent realizations of some experiment or event, and be interested in identifying the underlying distribution function or at least some of its properties.

For example, such observations could be on the IQs of voters, or non-voters, or the weekly total rainfall in a given rain forest, as well as examples found everywhere in finance. In the investment markets one finds the daily and other period returns of a variety of investment classes or individual securities, as well as data on financial variables such as inflation rates, interest rates on various fixed income securities, default rates, foreign currency exchange rates, commodity prices, and so forth. In the insurance markets one observes mortality rates by age, as well as various morbidity rates related to disability, hospitalization, and disability, or claims rates on various categories of automobile, homeowners and commercial insurance policies.

Any such quest to identify an underlying distribution function or some of its properties is based on the necessary **assumption** that the given data in fact represent observations of a random variable with a given distribution function. In particular it reflects the necessary assumption that this data is more or less stable, or **stationary**, over the time period observed. This issue is often addressed informally based on an intuitive understanding of the selection process for the given data, rather than by a formal analysis which seeks to prove that such a distribution function must exist.

For example, if one looked at a series of observations of the month-end value of a given equity index over the last 25 years, it would hardly seem logical to define this as a random variable and then attempt to identify the associated distribution function. Outside of inherent volatility and market corrections, most everyone's expectation is that the value of this index will increase over time, due at least to inflation and productivity. So while we can formally identify a "distribution function" for the given historical data, it is not clear that such an effort will produce something of value, if by "of value" we mean, providing some predictive insights for the future.

On the other hand, if we instead convert this price series into a monthly return series, there would seem to be a better argument that the distribution of returns over time had some stability, and hence some predictability. This argument is then challenged by the fact that such observations can potentially provide very different insights when grouped into various subperiods.

This forces the analyst to choose between several competing interpretations, a few being:

1. There is no single underlying distribution function for the given data because, for example, different periods have different distributions and the timing and magnitude of the change between distributions is unpredictable.

2. There is an underlying distribution function but no single period reveals all of its qualities; thus more data over longer periods is needed to reveal the ultimate distribution.

3. There may have been an underlying distribution function, but due to a significant event, the distribution in the future can be expected to be different.

Unfortunately, there is no universally accepted approach to resolving which of these or other interpretations is correct in a given situation. Ultimately, such a data analysis and the assumptions made to justify this analysis are part of the quantitative analyst's model building process, within which many other assumptions will also likely be made.

Are the assumptions valid? Is the model correct? What questions can the model answer?

Perhaps the best summary comment on this matter is one attributed to the statistician **George E. P. Box** (1919–2013). While there are many versions of his dictum, a commonly cited version is the quote:

"All models are wrong, but some are useful."

In this section, we study **empirical distribution functions** constructed from **samples** from a given fixed distribution, and investigate convergence of this empirical distribution to the underlying theoretical distribution function as the sample size increases.

Thus nothing in this section will ease the plight of the analyst in deciding if the random samples obtained are indeed from a given distribution. What these results confirm, is that IF the random sample has such an underlying distribution function, then the empirical distributions constructed will provide insights to it.

6.4.1   Definition and Basic Properties

Let $\{X_j\}_{j=1}^n$ denote a sample from a random variable X defined on a probability space $(\mathcal{S}, \mathcal{E}, \lambda)$ with an unknown distribution function F. Such a collection is also called a **realization of X**, or a **sample of X**. As discussed in this chapter's introduction, $\{X_j\}_{j=1}^n$

can be constructed and defined in terms of the first n-components of a point in an infinite dimensional space $(\mathbb{R}^\mathbb{N}, \sigma(\mathbb{R}^\mathbb{N}), \mu_\mathbb{N})$, and this allows one to investigate various results as $n \to \infty$.

That $\{X_j\}_{j=1}^n$ is a "sample" means that this collection is **i.i.d.**, or **independent and identically distributed.**

But there are two interpretations for this collection in this section:

1. Consistent with Chapter 6, these are independent, identically distributed random variables for any n, constructed on a probability space such as $(\mathbb{R}^\mathbb{N}, \sigma(\mathbb{R}^\mathbb{N}), \mu_\mathbb{N})$.

2. Consistent with empirical analysis, these are a **random sample** of the variate X, which can be envisioned as the numerical values $\{X_j(s)\}_{j=1}^n$ for some $s \in \mathbb{R}^\mathbb{N}$. What makes this numerical sample "random" is that these are deemed to have been obtained from a process, that if repeated many times, would produce collections which are approximately i.i.d.

While all empirical analysis is based on the random samples of item 2, one can only make probabilistic statements on the resulting empirical distributions by interpreting such samples within the context of item 1. Hence this section alternates between these perspectives.

Given a random sample, one approach to visualizing the underlying assumed-to-exist distribution function $F(x)$ is the construction of a **histogram,** which focuses on the underlying assumed-to-exist **density function** $f(x)$. In this construction one assigns a probability of $1/n$ to each observed variate, and this can be formally justified by Proposition II.4.9 of the Chapter 5 introduction, as follows.

If $F(x)$ is a given distribution function, such a sample can be produced as in Chapter 5 by $\{F^*(U_j)\}_{j=1}^n$, where $\{U_j\}_{j=1}^n$ is a numerical sample from the continuous, uniform distribution on $(0,1)$, and $F^*(y)$ denotes the left-continuous inverse of $F(x)$. Each such U_j is independent and uniformly distributed, so for any interval $[c,d] \subset (0,1)$:

$$\Pr\{U_j \in [c,d]\} = d - c.$$

By continuity of measures, this same result is true for $(c,d) \subset (0,1)$ or semi-closed intervals.

By independence:

$$\Pr\left\{(U_1, ..., U_n) \in \prod_{j=1}^n [c_j, d_j]\right\} = \prod_{j=1}^n (d_j - c_j).$$

Thus the value of each variate U_j is deemed equally likely, as is the value of each observed variate $X_j \equiv F^*(U_j)$, and one logically assigns a probability of $1/n$ to each value of X_j.

A **histogram** is then a graphical depiction of the sample $\{X_j\}_{j=1}^n$ with each observation given probability $1/n$, and this provides visual clues on the underlying density function $f(x)$.

There are two approaches for creating a histogram.

1. If $\{X_{(j)}\}_{j=1}^n$ are the associated order statistics of the sample, then for each j we assume $X_{(j)} \in [a_j, b_j)$ where $\{[a_j, b_j)\}_{j=1}^n$ are disjoint intervals with union equal to the assumed range of X and:

$$\int_{a_j}^{b_j} f(x)dx = 1/n.$$

For example, one could define $b_j = \left(X_{(j)} + X_{(j+1)}\right)/2 = a_{j+1}$, and extend symmetrically to a_1 and b_n. Alternatively, one could define $a_j = X_{(j)} = b_{j-1}$.

In this representation, the **empirical density function** $f(x)$ is assumed constant on each interval:

$$f(x) = \frac{1}{n(b_j - a_j)}, \quad x \in [a_j, b_j]. \tag{6.42}$$

2. An alternative approach to a histogram and the empirical density function $f(x)$ is to group sample variates into disjoint intervals of equal length, called **bins.** Specifically, one defines such disjoint intervals $\{[c_k, d_k]\}_{k=1}^m$, and defines the **empirical density function** $f(x)$:

$$f(x) = \frac{n_k}{n}, \qquad x \in [c_k, d_k]. \tag{6.43}$$

Here n_k equals the number of variates in bin $[c_k, d_k]$, and $\bigcup_{k=1}^m [c_k, d_k]$ is the assumed range of X.

While there is no unique approach to the determination of the best binning structure, many methods exist based on various assumptions on the underlying density f.

Thus a histogram is not a discrete probability density function, but a piecewise continuous approximation to the unknown density $f(x)$. The goal of these constructions is to provide some information on the "shape" of $f(x)$. Requiring $a_{j+1} = b_j$ and $c_{k+1} = d_k$ reflects the assumption that the range of X is connected. In the first approach, the various intervals will have different lengths and this uniform probability assumption will get translated into the various values for $f(x)$ by (6.42). In the second approach the bin intervals have fixed length, and the various values for $f(x)$ given by (6.43) provide a frequency count for that interval.

An alternative approach which we study in this section, reflects the **empirical distribution function** defined as follows.

Definition 6.54 (Empirical distribution function) *Given a sample $\{X_j\}_{j=1}^n$ of a random variable X defined on a probability space $(\mathcal{S}, \mathcal{E}, \lambda)$, the associated **empirical distribution function,** denoted $F_n(x) \equiv F_n(x| \{X_j\}_{j=1}^n)$, is defined:*

$$F_n(x) = \frac{1}{n} \sum_{j=1}^n \chi_{(-\infty, x]}(X_j), \tag{6.44}$$

where $\chi_{(-\infty, x]}(y)$ is the characteristic function of $(-\infty, x]$, and defined to equal 1 for $x \in (-\infty, x]$ and 0 otherwise.

Given $\{X_j\}_{j=1}^n$, we can look at $F_n(x)$ from at least two perspectives based on the comments above:

1. For a fixed random sample $\{X_j\}_{j=1}^n$, $F_n(x) \equiv F_n(x| \{X_j\})$ is indeed a distribution function by Proposition 1.4. Specifically, $F_n(x)$ is increasing and right-continuous, and when defined as limits, $F_n(-\infty) = 0$ and $F_n(\infty) = 1$.

In addition, $F_n(x)$ is continuous as a function of x except at $\{X_j\}_{j=1}^n$ with:

$$F_n(X_j) = F_n(X_j^-) + \frac{1}{n},$$

where $F_n(X_j^-)$ denotes the left limit of $F_n(x)$ at X_j.

This interpretation is conceptually consistent with the histograms above, but $F_n(x)$ only equals the distribution function associated with these empirical density functions in case 1 with $a_j = X_{(j)} = b_{j-1}$. In the other two constructions the discontinuities of the associated distribution functions will occur at $\{a_j\}_{j=1}^n$ and $\{c_k\}_{k=1}^m$ rather than at $\{X_j\}_{j=1}^n$. In addition, the increases at the discontinuities associated with (6.43) will not in general equal $1/n$.

2. For fixed x and $\{X_j\}_{j=1}^n$ interpreted as i.i.d. random variables on the probability space $(\mathbb{R}^{\mathbb{N}}, \sigma(\mathbb{R}^{\mathbb{N}}), \mu_{\mathbb{N}})$, let $\{Y_j\}_{j=1}^n$ be defined by $Y_j \equiv \chi_{(-\infty,x]}(X_j)$:

$$Y_j : (\mathbb{R}^{\mathbb{N}}, \sigma(\mathbb{R}^{\mathbb{N}}), \mu_{\mathbb{N}}) \to (\mathbb{R}, \mathcal{B}(\mathbb{R}), m),$$

with m Lebesgue measure. In other words:

$$Y_j(s) = \chi_{(-\infty,x]}(X_j(s)).$$

Then $\{Y_j\}_{j=1}^n$ are **standard binomial** variables defined on this probability space, with:

$$p \equiv \Pr\{Y_j = 1\} = \Pr\{X_j \le x\} = F(x).$$

Further, $\{Y_j\}_{j=1}^n$ are **independent** by Proposition II.3.56, since $g(y) = \chi_{(-\infty,x]}(y)$ is Borel measurable for any x.

Thus, for each x :

$$F_n(x) \equiv \frac{1}{n} \sum_{j=1}^n Y_j, \tag{6.45}$$

is a random variable on this probability space, and in fact is a general binomial with parameters n and $p = F(x)$.

Hence, in this case we have by (4.62) that **for each x:**

$$E\left[F_n(x)\right] = F(x), \qquad Var\left[F_n(x)\right] = F(x)(1 - F(x))/n. \tag{6.46}$$

Based on the second interpretation, the following proposition presents two results which are corollaries to the earlier results of this section. For these results we fix x and consider $F_n(x)$ of (6.45) as a random variable on the probability space $(\mathcal{S}, \mathcal{E}, \lambda) \equiv (\mathbb{R}^{\mathbb{N}}, \sigma(\mathbb{R}^{\mathbb{N}}), \mu_{\mathbb{N}})$. Thus in this result, $F(x)$ is also a constant.

Proposition 6.55 (Limit results for $F_n(x)$, fixed x) *Let $\{X_j\}_{j=1}^n$ be independent, identically distributed random variables on $(\mathcal{S}, \mathcal{E}, \lambda)$ with distribution function $F(x)$, and define $F_n(x)$ as in (6.45).*
Then for each x :

1. *As $n \to \infty$, $F_n(x)$ converges with probability 1 to $F(x)$:*

$$F_n(x) \to_1 F(x).$$

In other words,

$$\lambda\{F_n(x) \to F(x)\} = 1.$$

2. *Define the normalized random variable as in (6.10):*

$$Z_n = \frac{F_n(x) - F(x)}{\sqrt{F(x)(1 - F(x))/n}},$$

and let G_n denote the distribution function of Z_n.
Then as $n \to \infty$:

$$G_n \Rightarrow \Phi,$$

where Φ denotes the distribution function of the standard normal.

Proof. *Item 1 is an immediate consequence of the strong law of large numbers of Proposition 6.48, while item 2 is an application of the central limit theorem of Proposition 6.13 using (6.46).* ∎

The above proposition provides the important insight that random samples are informative relative to the underlying distribution function F at each x, when these random samples are defined as realizations of i.i.d variates. Not only does $F_n(x)$ converge to $F(x)$ for each x with probability 1, but we can in theory make approximate probability statements about $F(x)$ based on the random sample values $\{X_j\}_{j=1}^n$ and value of $F_n(x)$.

Example 6.56 (Confidence limits for $F(x)$, fixed x) *As is typical in estimating confidence intervals about an unknown binomial parameter $p \equiv F(x)$, we do not know the exact standard deviation $\sqrt{F(x)(1 - F(x))/n}$, and must approximate this with the sample value:*

$$\hat{\sigma} \equiv \sqrt{F_n(x)(1 - F_n(x))/n}.$$

*We now use a **Student T distribution** with $n - 1$ **degrees of freedom** in place of the normal as noted in Example 2.29.*
    This then produces the familiar $100(1 - \alpha)\%$ confidence interval for $F(x)$:

$$F_n(x) + t_{\alpha/2}^{(n-1)}\sqrt{F_n(x)(1 - F_n(x))/n} \le F(x) \le F_n(x) + t_{1-\alpha/2}^{(n-1)}\sqrt{F_n(x)(1 - F_n(x))/n},$$

where $t_{\alpha/2}^{(n-1)}$ and $t_{1-\alpha/2}^{(n-1)}$ are the associated quantiles of this distribution, and $t_{\alpha/2}^{(n-1)} = -t_{1-\alpha/2}^{(n-1)}$ by symmetry.
    Once $n \approx 100$ or so, these quantiles are quite close to those of the standard normal. For smaller values of n, Student's T has fatter tails than the normal, meaning $|z_\alpha| < \left|t_\alpha^{(n-1)}\right|$ for α near 0 or 1.

Hence, **for each** x we can estimate the value of $F(x)$ from a random sample as above. Further, we have probability 1 convergence of $F_n(x) \to_1 F(x)$ were $F_n(x)$ is interpreted as a random variable, and $F(x)$ a constant. However, this theory does not immediately support any conclusions on the extent to which F_n converges to F more generally.
    Indeed, for each x the above convergence result implies existence of a measurable set $A_x \subset S$ with $\lambda(A_x) = 0$ and $F_n(x) \to F(x)$ for $s \in S - A_x$. But as there are uncountably many such x, there is no theory to suggest that the union $\bigcup_x A_x$ is measurable, and even if measurable, is it possible that $\lambda(\bigcup_x A_x) > 0$. For example, $m(\{x\}) = 0$ for all $x \in [0, 1]$, but $m(\bigcup_x \{x\}) = 1$.
    Consequently, it is possible in theory that $S - \bigcup_x A_x$, the set on which $F_n(x) \to F(x)$ for all x, is not measurable, or measurable with probability less than 1.

Remark 6.57 (On right continuity of $F(x)$) *It is tempting to think that right continuity of the distribution function $F(x)$ would simplify the above discussion. Specifically, define $A = \bigcup_{x \in \mathbb{Q}} A_x$, so A is the union of A_x for rational x. Then since $\mathbb{Q}$ is countable, it follows that $A \in S$ is measurable with $\lambda(A) = 0$. Further, for all $s \in S - A$, a set of probability 1, we have that $F_n(x) \to F(x)$ for all $x \in \mathbb{Q}$.*
    It then seems compelling by right continuity that for all $s \in S - A$, that $F_n(x) \to F(x)$ for all $x \in \mathbb{R} - \mathbb{Q}$ as well.
    What is clear is that if $x \in \mathbb{R} - \mathbb{Q}$, and if $\{x_m\} \subset \mathbb{Q}$ with $x_m > x$ and $x_m \to x$, then:

- *For every $s \in S$ and every n, $F_n(x_m) \to F_n(x)$ as $m \to \infty$ by right continuity of F_n.*

- *$F(x_m) \to F(x)$ as $m \to \infty$ by right continuity of F.*

- $F_n(x_m) \to F(x_m)$ as $n \to \infty$ for all $x_m \in \mathbb{Q}$ and $s \in \mathcal{S} - A$ by probability 1 convergence.

Can we then conclude that $F_n(x) \to F(x)$ for $x \in \mathbb{R} - \mathbb{Q}$ and $s \in \mathcal{S} - A$? Unfortunately, no.

In order to verify this deduction formally, it is necessary to make an assumption on the uniformity of convergence in x of $F_n \to F$ for each $s \in \mathcal{S} - A$. For example, if $F(x) = 1$ on $[0, 1]$ and $F_n(x) = x^{1/n}$ on $[0, 1]$, then all functions are right continuous at $x = 0$, $F_n(x) \to F(x)$ for all $x > 0$, and yet it is apparent that $F_n(0) \not\to F(0)$.

The next section develops a positive result on this convergence.

6.4.2   The Glivenko-Cantelli Theorem

While not at all apparent as noted in Remark 6.57, the next result proves that outside a set $A \in \mathcal{E}$ with $\lambda(A) = 0$, that $F_n(x) \to F(x)$ **uniformly in** x. It is known as the **Glivenko-Cantelli theorem,** named for **Valery Glivenko** (1896–1940) and **Francesco Paolo Cantelli** (1875–1966), who independently derived this result and published in the same journal in 1933. In order to state this result, it is necessary to introduce a measure of the maximum distance between $F_n(x)$ as defined in (6.45), and $F(x)$, for each $s \in \mathcal{S}$.

To this end, define for $s \in \mathcal{S}$:

$$D_n(s) = \sup_x |F_n(x) - F(x)| . \tag{6.47}$$

Thus by (6.45), for each s :

$$D_n(s) = \sup_x \left| \frac{1}{n} \sum_{j=1}^n \chi_{(-\infty, x]}(X_j(s)) - F(x) \right| .$$

Although $|F_n(x) - F(x)|$ is a random variable on $(\mathcal{S}, \mathcal{E}, \lambda) \equiv (\mathbb{R}^{\mathbb{N}}, \sigma(\mathbb{R}^{\mathbb{N}}), \mu_{\mathbb{N}})$ for every x, the definition of $D_n(s)$ requires the supremum over uncountably many x. Hence Proposition I.3.47 does not apply, and we cannot immediately assert that $D_n(s)$ is a measurable function. However, by right continuity it can be verified as an exercise that:

$$\sup_x |F_n(x) - F(x)| = \sup_{x \in \mathbb{Q}} |F_n(x) - F(x)| . \tag{6.48}$$

Hence as a supremum of countably many measurable functions, $D_n(s)$ is a random variable on $(\mathcal{S}, \mathcal{E}, \lambda)$ by this Book I result.

Exercise 6.58 *Verify (6.48). Hint: Distribution functions are monotonic.*

Remark 6.59 (Kolmogorov-Smirnov statistic) *Fixing a random sample, which is to say, fixing $s \in \mathcal{S}$, $D_n(s)$ defined above is called a **Kolmogorov-Smirnov statistic,** named for **Andrey Kolmogorov** (1903–1987) and **Nikolai Smirnov** (1900–1966). This statistic is used in the **Kolmogorov-Smirnov test** or **K-S test,** to test the "goodness of fit" between an empirical distribution and an assumed distribution, as well as to test the equivalence of two empirical distributions.*

We now prove the Glivenko-Cantelli result.

Proposition 6.60 (Glivenko-Cantelli theorem) *Let $\{X_j\}_{j=1}^n$ be independent, identically distributed random variables on $(\mathcal{S}, \mathcal{E}, \lambda)$ with distribution function F, and define $D_n(s)$ as in (6.47).*

Then as $n \to \infty$:

$$D_n(s) \to_1 0. \tag{6.49}$$

In other words, $D_n(s) \to 0$ with probability 1, and thus $F_n(x) \to F(x)$ **uniformly in** x outside a set $A \in \mathcal{E}$ with $\lambda(A) = 0$.

Proof. Define:

$$F_n(x^-) = \frac{1}{n} \sum_{j=1}^{n} \chi_{(-\infty,x)}(X_j(s)),$$

and note that $F_n(x^-)$ is again a binomial random variable on $(\mathcal{S}, \mathcal{E}, \lambda)$, now with probability $p^- = F(x^-)$, the left limit of F at x. Another application of the strong law of large numbers proves as in Proposition 6.55 that for each x, $F_n(x^-) \to F(x^-)$ outside a set $B_x \in \mathcal{E}$ of λ-measure 0.

With F^* denoting the left-continuous inverse of F defined in (1.52), let $x_{k/m} = F^*(k/m)$ for integer m, and $1 \le k \le m-1$. By Proposition II.3.19, $F(F^*(y)^-) \le y \le F(F^*(y))$ for all $y \in (0,1)$. Letting $y = k/m$ obtains for $1 \le k \le m-1$:

$$F\left(x_{k/m}^-\right) \le k/m \le F\left(x_{k/m}\right).$$

Since F is increasing, this yields that for all $s \in \mathcal{S}$:

$$F\left(x_{(k-1)/m}^-\right) \le (k-1)/m \le F\left(x_{(k-1)/m}\right) \le F\left(x_{k/m}^-\right) \le k/m \le F\left(x_{k/m}\right).$$

This obtains that for $2 \le k \le m-1$:

$$F(x_{k/m}^-) - F(x_{(k-1)/m}) \le 1/m, \tag{1a}$$

and:

$$F(x_{1/m}^-) \le 1/m, \quad (m-1)/m \le F(x_{(m-1)/m}). \tag{1b}$$

Now define:

$$D_{n,m}(s) = \max\left[\max_{1 \le k \le m} \left|F_n(x_{k/m}) - F(x_{k/m})\right|, \max_{1 \le k \le m} \left|F_n(x_{k/m}^-) - F(x_{k/m}^-)\right| \right].$$

We claim that for any m:

$$D_n(s) \le D_{n,m}(s) + 1/m, \quad \text{all } s. \tag{2}$$

To see this, let $(k-1)/m \le x < k/m$ where $2 \le k \le m-1$. Then by the monotonicity of distributions and the inequality in (1a):

$$F_n(x) - F(x) \le F_n(x_{k/m}^-) - F(x_{(k-1)/m}) \le F_n(x_{k/m}^-) - F(x_{k/m}^-) + 1/m,$$
$$F_n(x) - F(x) \ge F_n(x_{(k-1)/m}) - F(x_{k/m}^-) \ge F_n(x_{(k-1)/m}) - F(x_{(k-1)/m}) - 1/m.$$

Hence:

$$F_n(x_{(k-1)/m}) - F(x_{(k-1)/m}) - 1/m \le F_n(x) - F(x) \le F_n(x_{k/m}^-) - F(x_{k/m}^-) + 1/m.$$

This obtains that for $(k-1)/m \le x < k/m$:

$$|F_n(x) - F(x)|$$
$$\le \max\left[\left|F_n(x_{(k-1)/m}) - F(x_{(k-1)/m}) - 1/m\right|, \left|F_n(x_{k/m}^-) - F(x_{k/m}^-) + 1/m\right| \right]$$
$$\le \max\left[\left|F_n(x_{(k-1)/m}) - F(x_{(k-1)/m})\right|, \left|F_n(x_{k/m}^-) - F(x_{k/m}^-)\right| \right] + 1/m,$$

Taking a supremum over such x:

$$\sup_{1/m \leq x < (m-1)/m} |F_n(x) - F(x)| \leq D_{n,m}(s) + 1/m. \tag{3}$$

A similar analysis applies when $0 < x < 1/m$ and $(m-1)/m < x < 1$, and is left as an exercise. This now proves (2).

Recalling Proposition 6.55, let:

$$A = \bigcup_{k,m} \left(A_{x_{k/m}} \cup B_{x_{k/m}} \right),$$

where $A_{x_{k/m}} \subset \mathcal{S}$ is the exceptional set of measure zero outside of which $F_n(x_{k/m}) \to F(x_{k/m})$, and $B_{x_{k/m}} \subset \mathcal{S}$ is similarly defined relative to $F_n(x_{k/m}^-) \to F(x_{k/m}^-)$. Then A is a countable union of sets of measure zero and hence $\lambda(A) = 0$. Further, if $s \in \mathcal{S} - A$, then $D_{n,m}(s) \to 0$ as $n \to \infty$ for any m by definition.

Hence by (2), if $s \in \mathcal{S} - A$ then $\lim_{n \to \infty} D_n(s) \leq 1/m$ for every m, proving (6.49). ∎

Exercise 6.61 *Complete the proof of (2) of the above proof by extending (3) to $0 < x < 1/m$ and $(m-1)/m < x < 1$. Hint: Repeat above derivation with (1b), and recall the limits in (1.4) and (1.5).*

6.4.3 Distributional Estimates for $D_n(s)$

In this section, we provide some additional results on the empirical distribution function. These results are presented mostly without proof as the tools required for a formal development are subtle and would require too much additional theory given that we have no need for these results.

As noted above, the Glivenko-Cantelli theorem was independently derived by Valery Glivenko and Francesco Paolo Cantelli and published in the same journal in 1933. Also in that same journal, **Andrey Kolmogorov** (1903–1987) published what has come to be known as **Kolmogorov's theorem**. When F is assumed to be continuous, Kolmogorov derives an asymptotic distributional result for $D_n(s)$ as $n \to \infty$.

Proposition 6.62 (Kolmogorov's theorem) *Let $\{X_j\}_{j=1}^n$ be independent, identically distributed random variables on $(\mathcal{S}, \mathcal{E}, \lambda)$ with **continuous** distribution function F. Let $F_{D_n}(t)$ denote the distribution function of $D_n(s)$ defined as in (6.47).*

Then for all $t > 0$:

$$F_{D_n}(t/\sqrt{n}) \to 1 - 2\sum_{k=1}^{\infty} (-1)^{k+1} e^{-2k^2 t^2}, \tag{6.50}$$

as $n \to \infty$. Further, this convergence is uniform in t.

Remark 6.63 (On $F_{D_n}(t/\sqrt{n})$) *The distribution function in (6.50) is known as the **Kolmogorov distribution function** and is defined to equal 0 when $t < 0$. The limit of this function is also 0 as $t \to 0^+$, and thus the Kolmogorov distribution function is continuous.*

Also note that:

$$F_{D_n}(t/\sqrt{n}) \equiv \Pr\left[\sqrt{n} D_n \leq t\right],$$

and so this can also be understood as a statement on the the limit of $F_{\sqrt{n} D_n}(t)$.

Kolmogorov derived this result by first proving that the limiting distribution in (6.50) is independent of $F(x)$ for continuous distributions, and then explicitly derived the limiting distribution for the uniform distribution function $F_U(x)$ associated with the density function defined in (1.51). This reduction utilizes the tools already developed so we provide this detail.

Proposition 6.64 (Distribution-free property of D_n) *The distribution for $D_n(s)$ defined in (6.47) is the same for all continuous distribution functions $F(x)$, and thus in particular can be evaluated with $F = F_U$, the continuous, uniform distribution function on $(0,1)$:*

$$D_n(s) = \sup_{y \in (0,1)} |F_{Un}(y) - y|.$$

Proof. *By Proposition II.4.9 in the Chapter 5 introduction, if U has a continuous uniform distribution, then $F^*(U)$ has distribution function $F(x)$, where F^* denotes the left-continuous inverse of F defined in (1.52). Letting $x = F^*(y)$:*

$$
\begin{aligned}
D_n(s) &= \sup_{x \in \mathbb{R}} |F_n(x) - F(x)| \\
&= \sup_{y \in (0,1)} |F_n(F^*(y)) - F(F^*(y))|.
\end{aligned}
$$

From Proposition II.3.22, $F(F^*(y)) = y$ if and only if F is continuous on $D' \equiv \{F^*(y)|y \in (0,1)\}$, and thus:

$$D_n(s) = \sup_{y \in (0,1)} |F_n(F^*(y)) - y|. \tag{1}$$

From (6.45):

$$F_n(F^*(y)) = \frac{1}{n} \sum_{j=1}^{n} \chi_{(-\infty, F^*(y)]}(X_j),$$

and since F is increasing:

$$X_j \le F^*(y) \text{ if and only if } F(X_j) \le F(F^*(y)).$$

As $F(F^*(y)) = y$ as noted above:

$$F_n(F^*(y)) = \frac{1}{n} \sum_{j=1}^{n} \chi_{(-\infty, y]}(F(X_j)). \tag{2}$$

Proposition II.4.9 also obtains that if F is continuous, then the distribution function of $F(X)$ is the continuous uniform distribution:

$$F_{F(X)}(y) = y = F_U(y).$$

With $U_j \equiv F(X_j)$, if $\{X_j\}_{j=1}^{n}$ are independent then $\{U_j\}_{j=1}^{n}$ are independent uniform variates by Proposition II.4.9. Thus by (2):

$$
\begin{aligned}
F_n(F^*(y)) &= \frac{1}{n} \sum_{j=1}^{n} \chi_{(-\infty, y]}(U_j) \\
&\equiv F_{Un}(y),
\end{aligned}
$$

where $F_{Un}(y)$ is defined as in (6.45) but with respect to $\{U_j\}_{j=1}^{n}$.
Hence from (1):

$$D_n(s) = \sup_{y \in (0,1)} |F_{Un}(y) - y|,$$

completing the proof. ∎

Several years after Kolmogorov's result, **Nikolai Smirnov** (1900–1966) in 1939 derived a comparable limit theorem but for the measure:

$$D_n^+(s) = \sup_x [F_n(x) - F(x)], \tag{6.51}$$

and in the same year developed results for the difference between two empirical distributions. For example, on $D_n^+(s)$ he proved:

Proposition 6.65 (Smirnov's theorem) *Let* $\{X_j\}_{j=1}^n$ *be independent, identically distributed random variables on* $(\mathcal{S}, \mathcal{E}, \lambda)$ *with continuous distribution function* $F(x)$. *Let* $F_{D_n^+}(t)$ *denote the distribution function of* D_n^+ *as defined as in (6.51). Then for all* $t > 0$:

$$F_{D_n^+}(t/\sqrt{n}) \to 1 - e^{-2t^2}, \tag{6.52}$$

as $n \to \infty$, *and this convergence is uniform in* t.

In 1956 this earlier work was used to develop a type of large deviation estimate as discussed in the next section, but for D_n with finite n. The first result is called the **Dvoretzky-Kiefer-Wolfowitz inequality** or the **Dvoretzky-Kiefer-Wolfowitz theorem,** and named for **Aryeh (Arie) Dvoretzky** (1916–2008), **Jack Kiefer** (1924–1981) and **Jacob Wolfowitz** (1910–1981). Their original result stated the inequality below in (6.53) with an undefined coefficient C. This result was then improved in 1990 by **Pascal Massart** by the derivation of a sharp estimate of $C = 2$. By **sharp** is meant that it is the best possible bound.

Note that the following result does not require that F be continuous.

Proposition 6.66 (Dvoretzky-Kiefer-Massart-Wolfowitz theorem) *Let* $\{X_j\}_{j=1}^n$ *be independent, identically distributed random variables on* $(\mathcal{S}, \mathcal{E}, \lambda)$ *with distribution function* F. *With* D_n *defined as in (6.47):*

$$\Pr\left[D_n > t\right] \leq 2e^{-2nt^2} \quad t > 0. \tag{6.53}$$

Equivalently,

$$\Pr\left[D_n \leq t\right] \geq 1 - 2e^{-2nt^2}, \quad t > 0. \tag{6.54}$$

Example 6.67 (Confidence band for $F(x)$**)** *The DKMW theorem allows the estimation of a "confidence band" about the entire empirical distribution function which will contain the theoretical distribution function with the degree of confidence implied by (6.54).*

For example, if $n = 100$ *and a 95% confidence band for* $F(x)$ *is desired, we choose* t *so that* $1 - 2e^{-200t^2} = 0.95$ *producing* $t = 0.135\,81$. *Then with* $F_n(x)$ *denoting the empirical distribution function of (6.45) based on the given sample of 100, (6.53) implies the 95% confidence band:*

$$\max\left(F_n(x) - 0.135\,81, 0\right) \leq F(x) \leq \min\left(F_n(x) + 0.135\,81, 1\right).$$

Letting $F_n(x, s)$ *be defined as in (6.45) with* $Y_j(s) = \chi_{(-\infty, x]}(X_j(s))$, *this confidence band implies that if* $A \subset \mathcal{S}$ *is defined by:*

$$A = \{\max\left(F_n(x, s) - 0.135\,81, 0\right) \leq F(x) \leq \min\left(F_n(x, s) + 0.135\,81, 1\right)\},$$

then:

$$\lambda\left[A\right] \geq 0.95.$$

For a $100\alpha\%$ *confidence band, the solution to* $1 - 2e^{-2nt^2} = \alpha$ *is given by:*

$$t = \left[\frac{1}{2n} \ln\left(\frac{2}{1-\alpha}\right)\right]^{0.5}.$$

If k *is the largest integer with* $k/n \leq t$, *then the lower bound of 0 is achieved when* $x \leq X_{(k)}$, *the kth order statistic, and the upper bound of 1 is achieved when* $x \geq X_{(n-k)}$. *Thus the confidence interval is informative primarily for* $X_{(k)} < x < X_{(n-k)}$, *an interval that decreases as* α *increases.*

7

Estimating Tail Events 2

In this chapter, we continue the investigations initiated in Chapter II.9, now using properties of expectations and the moment generating function. As in Book II, we investigate properties of the right tail of the distribution, focusing on results as $n \to \infty$:

- **Large Deviation Theory** studies tail probabilities related to the **average** of n independent random variables;

- **Extreme Value Theory** studies the limiting distribution of the **maximum** of n independent random variables.

7.1 Large Deviation Theory 2

In Book II, large deviation theory was introduced and the main result derived can be summarized as follows:

Summary 7.1 (Book II) *If $\{X_j\}_{j=1}^n$ are independent and identically distributed on a probability space $(\mathcal{S}, \mathcal{E}, \lambda)$, define:*

$$S_n \equiv \sum_{j=1}^n X_j,$$

and for $t > 0$:

$$\pi_n \equiv \ln\left[\Pr\{S_n \geq nt\}\right].$$

Then:
    Proposition II.9.2: $\lim_{n \to \infty} \left(\frac{\pi_n}{n}\right)$ *exists and equals* $\sup_m \left(\frac{\pi_m}{m}\right)$.
    Proposition II.9.4: *With* $\pi(t) = \lim_{n \to \infty} \left(\frac{\pi_n}{n}\right)$, *then* $-\infty \leq \pi(t) \leq 0$ *and for all n :*

$$\Pr\{S_n \geq nt\} \leq e^{n\pi(t)}.$$

Proposition II.9.6: *If $\pi(t) \neq 0$, then $\Pr\{S_n \geq nt\} \to 0$ exponentially fast as $n \to \infty$.*

In this section, we develop additional insights on the bounding function $\pi(t)$. We begin with the explicit identification of another bounding function when a moment generating function exists, called the **Chernoff bound**. After developing properties of this bound, we prove the **Cramér-Chernoff theorem**, which asserts that this bound is arbitrarily close to the $\pi(t)$-bound as $n \to \infty$.

DOI: 10.1201/9781003264583-7

7.1.1 Chernoff Bound

Assume that the moment generating function $M_X(t)$ exists in an interval $(-t_0, t_0)$ with $t_0 > 0$. From Markov's inequality of (4.87):

$$\Pr[|X| \geq t] \leq \frac{E[|X|]}{t},$$

and it follows that for $\theta > 0$:

$$
\begin{aligned}
\Pr\{S_n \geq nt\} &= \Pr\{\exp(\theta S_n) \geq \exp(\theta nt)\} \\
&\leq \frac{E[\exp(\theta S_n)]}{\exp(\theta nt)}.
\end{aligned}
$$

By definition of S_n and (4.48):

$$E[\exp(\theta S_n)] = M_X^n(\theta),$$

and so:

$$\Pr\{S_n \geq nt\} \leq \exp[-n[\theta t - \ln M_X(\theta)]].$$

As this inequality is true for all $\theta > 0$, we have proved the following:

Proposition 7.2 (Bound for $\Pr\{S_n \geq nt\}$) *If* $S_n = \sum_{j=1}^n X_j$ *with* $\{X_j\}_{j=1}^n$ *independent and identically distributed on a probability space* $(\mathcal{S}, \mathcal{E}, \lambda)$, *and if* $M_X(\theta)$ *exists on* $(-t_0, t_0)$ *with* $t_0 > 0$, *then for any* $t > 0$,

$$\Pr\{S_n \geq nt\} \leq \exp\left[-n \sup_{\theta \geq 0}[\theta t - \ln M_X(\theta)]\right]. \tag{7.1}$$

Thus the exponential decay function $\pi(t)$ *of Proposition II.9.4 satisfies:*

$$\pi(t) \leq -\sup_{\theta \geq 0}[\theta t - \ln M_X(\theta)]. \tag{7.2}$$

Proof. *The bound in (7.1) was derived above from Markov's inequality, while (7.2) follows from the definition of* $\pi(t)$ *and Proposition II.9.2. In detail, the definition of* π_n *obtains that* $\pi_n/n \leq -\sup_{\theta \leq 0}[\theta t - In M_X(\theta)]$. *As this is true for all* n, *the result follows from Proposition II.9.2 and the definition of* $\pi(t)$. $\blacksquare$

Remark 7.3 (Chernoff bound) *The upper bound in (7.1) is known as the **Chernoff bound**, and named for **Herman Chernoff**.*

Definition 7.4 (Rate function for X) *Let:*

$$\Gamma(\theta) \equiv \theta t - \ln M_X(\theta),$$

and denote by $\theta^*(t) > 0$ *the solution to:*

$$\Gamma(\theta^*(t)) = \sup_{\theta \geq 0} \Gamma(\theta).$$

Then $\Gamma(\theta^*(t))$ *is called the **rate function for** X, and it follows from (7.1) that the Chernoff bound can be expressed:*

$$\Pr\{S_n \geq nt\} \leq \exp[-n\Gamma(\theta^*(t))]. \tag{7.3}$$

Remark 7.5 (Questions on the Chernoff bound) *There are two immediate questions that arise from the Chernoff bound:*

1. *When is* $\Gamma(\theta^*(t)) \equiv \sup_{\theta \geq 0}(\theta t - \ln M_X(\theta)) > 0$, *so that the bound in (7.3) provides an estimate of the exponential decay rate for* $\Pr\{S_n \geq nt\}$?

2. If $\Gamma(\theta^*(t)) > 0$, is this estimate of exponential decay close to the best possible using $\pi(t)$, recalling from (7.2) that this estimate is, at the moment, just an upper bound to $e^{n\pi(t)}$?

Before investigating, we consider two examples.

Example 7.6 *In this example, we investigate the Chernoff bound for binomial and normal variates.*

1. **Binomial X_j** :

   From (4.63), $M_B(\theta) = \left(1 + p(e^\theta - 1)\right)$, and so:

   $$\Gamma_B(\theta) = \theta t - \ln\left(1 + p(e^\theta - 1)\right).$$

   It is an exercise to check that $\Gamma_B''(\theta) < 0$ for all θ and hence $\Gamma_B(\theta)$ is concave and has a maximum when $\Gamma_B'(\theta) = 0$.

   A calculation shows that this equation is solvable for any t with $0 < t < 1$, and has a solution at $\theta^* \equiv \theta^*(t)$ defined by:

   $$\theta^* = \ln\left[\frac{t(1-p)}{p(1-t)}\right].$$

   Then $\theta^* > 0$ if and only if $t > p = E[X]$, and in this case:

   $$\Gamma_B(\theta^*(t)) = t\ln\left(\frac{t}{p}\right) + (1-t)\ln\left(\frac{1-t}{1-p}\right). \tag{1}$$

   As a function of t, the rate function $\Gamma_B(\theta^*(t))$ is strictly increasing for $p < t < 1$ since:

   $$\frac{d\Gamma_B(\theta^*(t))}{dt} = \ln\frac{t}{1-t} - \ln\frac{p}{1-p} > 0.$$

   It follows from this and $\Gamma_B(\theta^*(p)) = 0$ that $\Gamma_B(\theta^*(t)) > 0$ for such t. Further, $\Gamma_B(\theta^*(t))$ is a convex function of t.

   Hence for t with $p < t < 1$:

   $$\Pr\{S_n \geq nt\} \leq \exp\left[-n\Gamma_B(\theta^*(t))\right], \tag{7.4}$$

   and $\Pr\{S_n \geq nt\}$ decreases exponentially as $n \to \infty$. This bound in (7.4) can also be expressed using (1) :

   $$\Pr\{S_n \geq nt\} \leq \left[\left(\frac{t}{p}\right)^t \left(\frac{1-t}{1-p}\right)^{1-t}\right]^{-n},$$

   and can be compared to **Bernstein's inequality** in II.(5.5).

   With $X_j \equiv B_j - p$, a one-tail version of this earlier result is found in the proof of Proposition II.5.3 and be expressed:

   $$\Pr\{S_n \geq nt\} \leq e^{-nt^2/4}.$$

2. **Normal X_j :**

From (4.78), $M_N(\theta) = \exp\left(\mu\theta + \frac{1}{2}\sigma^2\theta^2\right)$, and so:

$$\Gamma_N(\theta) = (t - \mu)\theta - \frac{1}{2}\sigma^2\theta^2.$$

It again follows that $\Gamma_N''(\theta) < 0$ for all θ, and hence $\Gamma_N(\theta)$ is concave and has a maximum when $\Gamma_N'(\theta) = 0$.

A calculation obtains that this maximum occurs at $\theta^* \equiv \theta^*(t)$:

$$\theta^* = \frac{t - \mu}{\sigma^2},$$

and thus $\theta^* > 0$ if and only if $t > \mu = E[X]$. We then have that $\sup_{\theta \geq 0} \Gamma_N(\theta) = \Gamma_N(\theta^*)$ where:

$$\Gamma_N(\theta^*(t)) = \frac{(t - \mu)^2}{2\sigma^2}, \tag{7.5}$$

and the rate function for the normal is increasing and convex for $t > \mu$.

Thus by (7.3):

$$\Pr\{S_n \geq nt\} \leq \exp\left[-\frac{n(t - \mu)^2}{2\sigma^2}\right]. \tag{7.6}$$

Setting $n = 1$, this bound can be compared to that derived in II.(9.49) of Example II.9.48:

$$\Pr\{S_1 \geq t\} \leq \frac{1}{\sqrt{2\pi}} \frac{\sigma}{t - \mu} \exp\left[-\frac{(t - \mu)^2}{2\sigma^2}\right].$$

We are now ready to address the first question posed in Remark 7.5. The next result states that when $M_X(\theta)$ exists and $t > E[X]$, the bound in (7.1) or (7.3) is meaningful in the sense that it produces exponential decay as $n \to \infty$. We also prove that $\Gamma(\theta)$ is a concave function, a property seen in the above examples.

Perhaps ironically, the critically important first result will have a short, easy proof. The second result on the concavity of $\Gamma(\theta)$ will require a good deal more work. For it we will need to derive a special case of a Book V change of variables result. See Remark 7.8.

Proposition 7.7 ($\sup_{\theta \geq 0} \Gamma(\theta) > 0$; $\Gamma(\theta)$ **is concave**) *Assume that $M_X(\theta)$ exists on $(-\theta_0, \theta_0)$ with $\theta_0 > 0$.*

If $t > E[X]$, then $\Gamma(\theta) \equiv \theta t - \ln M_X(\theta) > 0$ for some $\theta > 0$, and hence:

$$\sup_{\theta \geq 0} (\theta t - \ln M_X(\theta)) > 0. \tag{7.7}$$

In addition, $\Gamma''(\theta) < 0$ and so $\Gamma(\theta)$ is a concave function.

Proof. *For (7.7), $M_X(\theta) > 0$ for all θ by definition, and $M_X(\theta)$ is infinitely differentiable on $(-\theta_0, \theta_0)$ by Proposition 4.26. Thus $\Gamma(\theta) = \theta t - \ln M_X(\theta)$ is infinitely differentiable on this interval. Further, since $\Gamma(0) = 0$ and $\Gamma'(0) = t - E[X]$ by (4.55), it follows that if $t > E[X]$, then $\Gamma(\theta) > 0$ for $\theta \in (0, \theta')$ and some $\theta' > 0$. This proves (7.7).*

To prove concavity of $\Gamma(\theta)$, let $\theta \in (-\theta_0, \theta_0)$ and define a function $F_\theta(y)$ by the Riemann-Stieltjes integral (Section II.4.2):

$$F_\theta(y) = \frac{1}{M_X(\theta)} \int_{-\infty}^{y} e^{\theta x} dF, \tag{7.8}$$

where F is the distribution function of X. Since $e^{\theta x}$ is continuous and F is bounded and increasing, this integral exists for all y by Proposition III.4.17. It is left as an exercise to prove that $F_\theta(y)$ is continuous, increasing, and has appropriate limits at $\pm\infty$, and thus $F_\theta(y)$ is a distribution function of a random variable Y by Proposition 1.4.

We now prove that $E[Y^n] \equiv \int_{-\infty}^{\infty} y^n dF_\theta$ exists for all n, and that:

$$E[Y^n] = \frac{M_X^{(n)}(\theta)}{M_X(\theta)}, \tag{1}$$

where $M_X^{(n)}(\theta)$ is the nth derivative of $M_X(\theta)$.

Once proved, this assures concavity of $\Gamma(\theta)$. By direct calculation of $\Gamma''(\theta)$, then (1) and (4.50):

$$\begin{aligned} \Gamma''(\theta) &= \left(\frac{M_X'(\theta)}{M_X(\theta)}\right)^2 - \frac{M_X''(\theta)}{M_X(\theta)} \\ &= -Var[Y] \\ &\leq 0. \end{aligned}$$

Thus $\Gamma'(\theta)$ is a decreasing function and $\Gamma(\theta)$ is concave by Proposition 4.40.

Turning to (1), recall (4.51) from Proposition 4.26:

$$M_X^{(n)}(\theta) = \int_{-\infty}^{\infty} y^n e^{\theta y} dF,$$

and since $M_X(\theta) > 0$:

$$\frac{M_X^{(n)}(\theta)}{M_X(\theta)} = \frac{1}{M_X(\theta)} \int_{-\infty}^{\infty} y^n e^{\theta y} dF. \tag{2}$$

We now prove that the integral in (2) equals $E[Y^n]$ as claimed in (1).

First, since $F_\theta(y)$ is increasing and y^n is continuous, $\int_a^b y^n dF_\theta$ exists for any bounded interval $[a,b]$ by Proposition III.4.17. Then by Definition III.4.3, for any $\epsilon > 0$ there exists a partition $\{[y_{j-1}, y_j]\}_{j=1}^m$ of $[a,b]$ so that with arbitrary tags $\tilde{y}_j \in [y_{j-1}, y_j]$, this integral can be approximated within ϵ by a Riemann-Stieltjes summation:

$$\left| \int_a^b y^n dF_\theta - \sum_{j=1}^m \tilde{y}_j^n \left[F_\theta(y_j) - F_\theta(y_{j-1})\right] \right| < \epsilon.$$

As $\mathbb{R}$ is a countable union of such intervals and Proposition III.4.17 and Definition III.4.3 can be applied to approximate the associated integrals with an error of $\epsilon/2^j$, it follows that for any $\epsilon > 0$ there exists a partition $\{(y_{j-1}, y_j)\}_{j=-\infty}^\infty$ of $\mathbb{R}$ and so that with arbitrary tags $\tilde{y}_j \in [y_{j-1}, y_j]$, the integral $\int_{-\infty}^{\infty} y^n dF_\theta$ can be approximated within ϵ by a Riemann-Stieltjes summation:

$$\left| \int_{-\infty}^{\infty} y^n dF_\theta - \sum_{j=-\infty}^{\infty} \tilde{y}_j^n \left[F_\theta(y_j) - F_\theta(y_{j-1})\right] \right| < \epsilon. \tag{3}$$

By Definition III.4.3, the result in (3) is valid for any refinement (Definition III.4.7) of this partition. Recalling that $\theta \in (-\theta_0, \theta_0)$ is fixed, we refine this partition as necessary so that over each such interval $[y_{j-1}, y_j]$, if M_j and m_j denote the maximum and minimum values of y^n:

$$|M_j - m_j| \leq \epsilon e^{-|\theta|y_j} M_X(\theta)/2^{j+1}. \tag{4}$$

Next by Proposition III.4.24:

$$F_\theta(y_j) - F_\theta(y_{j-1}) \equiv \frac{1}{M_X(\theta)} \int_{y_{j-1}}^{y_j} e^{\theta y} dF,$$

and Definition III.4.3 can again be applied to produce a partition $\{[y_{j_k-1}, y_{j_k}]\}_{k=1}^{n_j}$ of each $[y_{j-1}, y_j]$ so that for arbitrary tags $y'_{jk} \in [y_{j_k-1}, y_{j_k}]$:

$$\left| [F_\theta(y_j) - F_\theta(y_{j-1})] - \frac{1}{M_X(\theta)} \sum_{k=1}^{n_j} e^{\theta y'_{jk}} [F(y_{j_k}) - F(y_{j_k-1})] \right| < \epsilon / (2^{j+1} M_j^n).$$

Combining with (3) obtains:

$$\left| \int_{-\infty}^{\infty} y^n dF_\theta - \frac{1}{M_X(\theta)} \sum_{j=-\infty}^{\infty} \sum_{k=1}^{n_j} (\tilde{y}_j)^n e^{\theta y'_{jk}} [F(y_{j_k}) - F(y_{j_k-1})] \right| < 2\epsilon. \qquad (5)$$

This summation in (5) is seen to be a Riemann-Stieltjes summation for the integral in (2) with partition $\left\{ \{[y_{j_k-1}, y_{j_k}]\}_{k=1}^{n_j} \right\}_{j=-\infty}^{\infty}$ except the $\tilde{y}_j$ tags are generally not in the $[y_{j_k-1}, y_{j_k}]$-intervals. However, we can replace these tags by the above subinterval tags at a cost of at most ϵ since:

$$\frac{1}{M_X(\theta)} \left| \sum_{j=-\infty}^{\infty} \left[\sum_{k=1}^{n_j} [(\tilde{y}_j)^n - (y'_{jk})^n] e^{\theta y'_{jk}} [F(y_{j_k}) - F(y_{j_k-1})] \right] \right|$$

$$\leq \epsilon \left| \sum_{j=-\infty}^{\infty} \left[\sum_{k=1}^{n_j} e^{-|\theta| y_j} e^{\theta y'_{jk}} [F(y_{j_k}) - F(y_{j_k-1})] \right] / 2^{j+1} \right|$$

$$\leq \epsilon \sum_{j=-\infty}^{\infty} [F(y_j) - F(y_{j-1})] / 2^{j+1}$$

$$\leq \epsilon.$$

The first inequality is from (4), the second since $\theta y'_{jk} \leq |\theta| y_j$ for all k, and finally $[F(y_j) - F(y_{j-1})] \leq 1$.

In summary, Riemann-Stieltjes summations for the integral $\frac{1}{M_X(\theta)} \int_{-\infty}^{\infty} y^n e^{\theta y} dF$ can be made arbitrarily close to the integral $\int_{-\infty}^{\infty} y^n dF_\theta$ by (5) and this final calculation. However, the integral $\frac{1}{M_X(\theta)} \int_{-\infty}^{\infty} y^n e^{\theta y} dF$ exists as noted in (2), and thus such Riemann-Stieltjes summations also converge to $\int_{-\infty}^{\infty} y^n dF_\theta$.

Thus:

$$\int_{-\infty}^{\infty} y^n dF_\theta = \frac{1}{M_X(\theta)} \int_{-\infty}^{\infty} y^n e^{\theta y} dF, \qquad (6)$$

and this completes the proof of (1) by (2). ∎

Remark 7.8 (Change of variables and (6)) *It will be seen in Book V that the transition from the dF_θ-integral representation for $E[Y^n]$ to the dF-integral representation in (6) is a special case of a general change of variables result.*

From (7.8):

$$F_\theta(y) = \frac{1}{M_X(\theta)} \int_{-\infty}^{y} e^{\theta x} dF.$$

It seems compelling to wonder, as would appear "obvious" at least notationally, if:

$$dF_\theta = \frac{1}{M_X(\theta)} e^{\theta y} dF?$$

If true, then it is also "obvious" that:

$$E[Y^n] \equiv \int_{-\infty}^{\infty} y^n dF_\theta = \frac{1}{M_X(\theta)} \int_{-\infty}^{\infty} y^n e^{\theta y} dF. \tag{7.9}$$

This notational "trick" was justified in the above proof by approximating the dF_θ-integral on the left with sums, converting the ΔF_θ terms to dF-integrals which can then be approximated with ΔF terms, where the final summation also approximates the dF-integral on the right.

In Book V, this change of variable result will be justified more generally for Lebesgue-Stieltjes integrals, which will be seen to agree with Riemann-Stieltjes integrals when the integrand is continuous.

Example 7.9 (Existence of $M_Y(t)$) *As an application of this general result, or derived using the approach of the above proof, if $\theta \in (-\theta_0, \theta_0)$ then $M_Y(t)$ exists for $t \in (-t_0, t_0)$ where $(\theta - t_0, \theta + t_0) \in (-\theta_0, \theta_0)$.*

Replacing y^n with e^{ty} in (7.9) obtains:

$$M_Y(t) \equiv \int_{-\infty}^{\infty} e^{ty} dF_\theta = \frac{M_X(\theta + t)}{M_X(\theta)}. \tag{7.10}$$

Definition 7.10 (Twisted/tilted distribution) *The distribution function F_θ in (7.8) is referred to as the **twisted distribution**, or the **tilted distribution**, associated with F. Sometimes either of these names is preceded by "**exponentially**," as in **exponentially tilted distribution**.*

Proposition 7.7 assures that if $t > E[X]$, then $\sup_{\theta \geq 0} (\theta t - \ln M_X(\theta)) > 0$ and that $\Gamma(\theta) \equiv \theta t - \ln M_X(\theta)$ is a concave function. Hence since $\Gamma(\theta)$ is also differentiable, if we can solve $\Gamma'(\theta) = 0$ for given t with solution $\theta^*(t) > 0$, then

$$\sup_{\theta \geq 0} \Gamma(\theta) = \Gamma(\theta^*(t)).$$

This was the approach taken in Example 7.6.

Solving $\Gamma'(\theta) = 0$ is equivalent to solving the following for $\theta^*(t)$:

$$t = \frac{M_X'(\theta^*(t))}{M_X(\theta^*(t))}. \tag{7.11}$$

By (4.50) and (1) of the above proof, $[M_X'(\theta)/M_X(\theta)]' = Var[Y]$ and so $M_X'(\theta)/M_X(\theta)$ is an increasing function. Since $M_X'(0)/M_X(0) = \mu$, it follows that if the equation in (7.11) is solvable for $t > \mu$, then it will be solvable with $\theta^*(t) > 0$.

That said, the above proposition does not assure that we can always determine the value of $\theta^*(t)$ which achieves this supremum using this approach because $M_X'(\theta)/M_X(\theta)$ may be bounded and hence (7.11) will be unsolvable for $t > \max[M_X'(\theta)/M_X(\theta)]$.

Example 7.11 (Bounded $M_X'(\theta)/M_X(\theta)$) *Given density function $f(x) = Cx^{-3}e^{-x}$ on $x \geq 1$, where C is chosen so that f integrates to 1, then $M_X(\theta)$ exists when $\theta \leq 1$. Further, $M_X'(\theta)/M_X(\theta)$ is increasing but bounded when $\theta \leq 1$ since*

$$\begin{aligned}
\frac{M_X'(\theta)}{M_X(\theta)} &\leq \frac{M_X'(1)}{M_X(1)} \\
&= \int_1^{\infty} x^{-2} dx \Big/ \int_1^{\infty} x^{-3} dx \\
&= 2.
\end{aligned}$$

Hence, the equation in (7.11) has no solution for $t > 2$.

7.1.2   Cramér-Chernoff Theorem

We next turn to item 2 in Remark 7.5, on the tightness of the bound in (7.1) and (7.3). For this we have the **Cramér-Chernoff theorem,** named for **Harald Cramér** (1893–1985) and **Herman Chernoff**. It states that if $\theta^*(t) > 0$ exists for given $t > \mu \equiv E[X]$, meaning:

$$\sup_{\theta \geq 0} \Gamma(\theta) = \Gamma(\theta^*(t)),$$

then the bound in (7.3) is exact in the limit as $n \to \infty$. In other words, the Chernoff bound is tight.

For this result, we assume that the distribution function $F(x)$ of X is absolutely continuous, recalling the discussion in Section 1.3.1. The same proof works if $F(x)$ is a saltus function, replacing integrals below with summations. The key point of these assumptions is to assure that X has a density function, so in particular it is assumed that the decomposition of $F(x)$ in Proposition 1.18 has no singular component.

Remark 7.12 (On the proof and $\theta^\#(t)$) *A new function $\theta^\#(t)$ is introduced in the statement of this proposition which is defined in terms of $\sup_\theta \Gamma(\theta)$, while $\theta^*(t)$ of Definition 7.4 is defined in terms of $\sup_{\theta \geq 0} \Gamma(\theta)$. However, the first development in the proof will be that $\theta^\#(t) = \theta^*(t)$ when $t > \mu$.*

Also, for an absolutely continuous distribution function $F(x)$, the proof below again requires Fubini's theorem of Book V, which allows the evaluation of a multivariate Lebesgue integral in terms of iterated 1-dimensional integrals. When $F(x)$ is a saltus function with discrete density, the Book V general machinery is not needed. Similarly, if $F(x)$ is absolutely continuous with continuous density $f(x) \equiv F'(x)$, then the integrals below become Riemann integrals and Fubini's result is replaced by Corollary III.1.77.

Proposition 7.13 (Cramér-Chernoff theorem) *Let $S_n = \sum_{j=1}^n X_j$ with $\{X_j\}_{j=1}^n$ independent and identically distributed with a distribution function $F(x)$ containing no singular part, and where $M_X(\theta)$ exists on $(-\theta_0, \theta_0)$ with $\theta_0 > 0$. Assume that for given $t > \mu \equiv E[X]$, that $\theta^\#(t) > 0$ exists with $\theta^\#(t) \in (0, \theta_0)$ and:*

$$\Gamma(\theta^\#(t)) = \sup_\theta \Gamma(\theta),$$

where $\Gamma(\theta) = \theta t - \ln M_X(\theta)$.

Then for any such t and $\epsilon > 0$, there is an N so that for $n \geq N$:

$$\exp\left[-n\left(\Gamma(\theta^\#(t)) + \epsilon\right)\right] \leq \Pr\{S_n \geq nt\} \leq \exp\left[-n\Gamma(\theta^\#(t))\right]. \tag{7.12}$$

Proof. *The upper bound in (7.12) is obtained by (7.3) if we can prove that $\theta^\#(t) = \theta^*(t)$ for $t > \mu$, meaning:*

$$\sup_\theta \Gamma(\theta) = \sup_{\theta \geq 0} \Gamma(\theta). \tag{1}$$

Because $\Gamma(0) = 0$, (1) will follow by showing that $\Gamma(\theta) < 0$ for $\theta < 0$.

Since $f(x) \equiv e^{\theta x}$ is convex, $M_X(\theta) \geq e^{\theta \mu}$ by Jensen's inequality in (4.93). If $\theta < 0$ and $t > \mu$ then $\exp[-\theta(t - \mu)] > 1$, and so:

$$\exp[-\theta(t - \mu)]M_X(\theta) > e^{\theta \mu}.$$

Thus $\Gamma(\theta) < 0$ is confirmed by taking logarithms.

For the lower bound in (7.12), let $\epsilon > 0$ be given. That X has an absolutely continuous distribution function $F(x)$ implies that there is an associated measurable density function $f(x)$ by Proposition III.3.62, with $f(x) = F'(x)$ almost everywhere, meaning outside a set of Lebesgue measure 0. Independence of $\{X_j\}_{j=1}^n$ assures that the joint density function $\tilde{f}(x) = \prod_{j=1}^n f(x_j)$ by item 4 of Section 4.2.3, where $x \equiv (x_1, ..., x_n)$.

By Proposition III.3.62, with $f(x) = F'(x)$ almost everywhere:

$$(\mathcal{L}) \int_a^b f(x)dx = F(b) - F(a),$$

and thus by Proposition 1.5 this integral equals $\Pr\{X \in (a, b]\}$. It will be seen in Book V that for any measurable set A that $\Pr\{X \in A\}$ is given by $(\mathcal{L}) \int_A f(x)dx$, and that this generalizes to joint density functions.

Thus for any $\delta > 0$, and dropping the notation $(\mathcal{L})$:

$$
\begin{aligned}
\Pr\{S_n \geq nt\} &\geq \Pr\left\{nt \leq \sum_{j=1}^n X_j \leq n(t+\delta)\right\} \\
&= \int_A \tilde{f}(x)dx \\
&= \int \cdots \int_A f(x_1)...f(x_n)dx_1...dx_n,
\end{aligned}
\tag{2}
$$

where $A \equiv \left\{nt \leq \sum_{j=1}^n x_j \leq n(t+\delta)\right\}$,

*Given $t > \mu$ and $\theta^* \equiv \theta^*(t) = \theta^\#(t)$, define the **twisted density function** $g(x)$ associated with the density $f(x)$:*

$$g(x) = \frac{\exp[\theta^* x]}{M_X(\theta^*)} f(x).$$

This is indeed a density function since nonnegative, and it integrates to 1 by definition of $M_X(\theta^*)$.

Let Y be a random variable associated with the distribution function $G(x)$ defined by $g(x)$, which is assured to exist by Proposition 1.4. Then by (2) :

$$
\begin{aligned}
&\Pr\left\{\sum_{j=1}^n X_j \geq nt\right\} \\
&\geq \frac{M_X^n(\theta^*)}{\exp[n(t+\delta)\theta^*]} \int \cdots \int_A \frac{\prod_{j=1}^n \exp[\theta^* x_j]}{M_X^n(\theta^*)} f(x_1)...f(x_n)dx_1...dx_n \\
&\equiv \frac{M_X^n(\theta^*)}{\exp[n(t+\delta)\theta^*]} \int \cdots \int_A g(x_1)...g(x_n)dx_1...dx_n \\
&= \frac{M_X^n(\theta^*)}{\exp[n(t+\delta)\theta^*]} \Pr\left\{nt \leq \sum_{j=1}^n Y_j \leq n(t+\delta)\right\}.
\end{aligned}
\tag{3}
$$

In the first step, $\prod_{j=1}^n \exp[\theta^* x_j] \leq \exp[n(t+\delta)\theta^*]$ since $\sum_{j=1}^n x_j \leq n(t+\delta)$ by the definition of A. The last step reflects the above comment relating $\Pr[Y \in A]$ with the integral of the density for Y over A.

Now:

$$
\begin{aligned}
\frac{M_X^n(\theta^*)}{\exp[n(t+\delta)\theta^*]} &= \exp[n \ln M_X(\theta^*) - n(t+\delta)\theta^*] \\
&= \exp[-n\Gamma(\theta^*)] \exp[-n\theta^*\delta].
\end{aligned}
\tag{4}
$$

In addition, $E[Y] = M_X'(\theta^*)/M_X(\theta^*)$ by (1) in the proof of Proposition 7.7. Since θ^* maximizes concave and differentiable $\Gamma(\theta)$, it follows that $\Gamma'(\theta^*) = 0$. A calculation with $\Gamma(\theta) = \theta t - \ln M_X(\theta)$ then produces:

$$E[Y] = t. \tag{5}$$

Also, the standard deviation σ_Y of Y exists by (1) of the prior proof and (4.50), while $M_Y(s)$ exists by Example 7.9.

  Combining $(3) - (5)$:

$$\Pr\left\{\sum\nolimits_{j=1}^{n} X_j \geq nt\right\}$$
$$\geq \exp\left[-n\Gamma(\theta^*)\right]\exp\left[-n\theta^*\delta\right] \times \qquad\qquad (6)$$
$$\Pr\left\{0 \leq \left[\sum\nolimits_{j=1}^{n}(Y_j - E[Y])\right]\Big/(\sigma_Y\sqrt{n}) \leq \frac{\delta\sqrt{n}}{\sigma_Y}\right\}.$$

The central limit theorem of Proposition 6.13 then obtains that for any $\delta > 0$, the probability expression in (6) converges to $1/2$ as $n \to \infty$.

  Choose $0 < \delta \leq \epsilon/(2\theta^*)$. Then since $\exp\left[-n\theta^*\delta\right] \geq \exp\left[-n\epsilon/2\right]$, determine N_1 so that for $n \geq N_1$:

$$\exp\left[-n\theta^*\delta\right] \geq \exp\left[-n\epsilon/2\right] \geq 4\exp\left[-n\epsilon\right].$$

For this same δ, let N_2 be defined so that for $n \geq N_2$:

$$\Pr\left\{0 \leq \left[\sum\nolimits_{j=1}^{n}(Y_j - E[Y])\right]\Big/(\sigma_Y\sqrt{n}) \leq \frac{\delta\sqrt{n}}{\sigma_Y}\right\} \geq 1/4.$$

Then for $n \geq \max[N_1, N_2]$:

$$\Pr\left\{\sum\nolimits_{j=1}^{n} X_j \geq nt\right\} \geq \exp\left[-n\left(\Gamma(\theta^*) + \epsilon\right)\right].$$

■

Corollary 7.14 (Cramér-Chernoff theorem) *Under the assumptions of the prior proposition, for $t > \mu \equiv E[X]$:*

$$\lim_{n\to\infty} \frac{1}{n}\ln\left[\Pr\left\{S_n \geq nt\right\}\right] = -\Gamma(\theta^*(t)). \qquad\qquad (7.13)$$

Thus:

$$\pi(t) = -\Gamma(\theta^*(t)).$$

Proof. *From the above result it follows that for any $\epsilon > 0$ there is an N so that for $n \geq N$:*

$$-\left(\Gamma(\theta^\#(t)) + \epsilon\right) \leq \frac{1}{n}\ln\left[\Pr\left\{S_n \geq nt\right\}\right] \leq -\Gamma(\theta^\#(t)).$$

The result in (7.13) follows because $\theta^\#(t) = \theta^*(t)$ for $t > \mu$, and by definition of limit.
  That $\pi(t) = -\Gamma(\theta^*(t))$ follows from the definition of $\pi(t)$ and (7.13). ■

7.2   Extreme Value Theory 2

Extreme value theory was introduced in Book II, where the two central results of this theory were investigated:

- The **Fisher-Tippett theorem:** Named for the earliest developers **Ronald Fisher** (1890–1962) and **L. H. C. Tippett** (1902–1985), and also called the **Fisher-Tippett-Gnedenko theorem** in recognition of the later contributions of **Boris Gnedenko** (1912–1995). This result was developed in Proposition II.9.30 and is summarized below, while the multivariate version of this theorem can be seen in Proposition II.9.52.

- The **Pickands-Balkema-de Haan theorem:** Named for 1974-5 papers of **A. A. Balkema** and **Laurens de Haan,** and, **James Pickands III** (1931–2022), and sometimes called the **Gnedenko-Pickands-Balkema-de Haan theorem** and even the **Gnedenko theorem** in recognition of the earlier 1943 paper of **Boris Gnedenko.** The key idea underlying this result was introduced in Proposition II.9.38, and then stated but not proved in Proposition II.9.44.

The next section below summarizes the Book II results on the Fisher-Tippett-Gnedenko theorem in order to introduce the **Hill estimator** γ_H for the extreme value index $\gamma > 0$, and then investigate this estimator and some of its properties in the following sections. For finance applications, $\gamma > 0$ is the range of extreme value indexes of central interest.

The final section returns to the Gnedenko-Pickands-Balkema-de Haan theorem, and provides a proof when $\gamma > 0$.

7.2.1 Fisher-Tippett-Gnedenko Theorem

The one-dimensional version this result of Proposition II.9.30 addresses weak convergence of the distribution function of the maximum value of an independent collection of n identically distributed random variables $\{X_m\}_{m=1}^n$. In the notation of order statistics, the distribution function of this variate is $F_{(n)}(x)$, and by Proposition 3.6:

$$F_{(n)}(x) = F^n(x),$$

where $F(x)$ is the distribution function of the underlying random variable X.

We summarize this result here:

Proposition 7.15 (Fisher-Tippett-Gnedenko theorem) *Let $F(x)$ be the distribution function of a random variable X and $F^n(x)$ the distribution function of $M_n = \max_{m \leq n}\{X_m\}$ for independent $\{X_m\}_{m=1}^n$. Assume that there exists sequences $\{a_n\}_{n=1}^\infty$ and $\{b_n\}_{n=1}^\infty$ where $a_n > 0$ for all n, so that:*

$$F^n(a_n x + b_n) \Rightarrow G(x),$$

for a nondegenerate distribution function $G(x)$.

Then there are real constants $A > 0$, B, and γ so that $G(x) = G_\gamma(Ax + B)$, with G_γ for $\gamma \neq 0$ defined by:

$$G_\gamma(x) = \exp\left(-(1 + \gamma x)^{-1/\gamma}\right), \qquad 1 + \gamma x \geq 0. \tag{7.14}$$

When $\gamma = 0$, $G_\gamma(x)$ is defined as the limit of $G_\gamma(x)$ as $\gamma \to 0$:

$$G_0(x) \equiv \exp\left(-e^{-x}\right). \tag{7.15}$$

Definition 7.16 (Extreme value index; Domain of attraction) *The single family of distributions identified and parametrized by γ is called the **extreme value class of distributions,** and the distribution function $G_\gamma(ax + b)$ is called a **generalized extreme value distribution,** abbreviated **GEV.***

*The parameter $\gamma \in \mathbb{R}$ is called the **extreme value index.***

*When $F^n(a_n x + b_n) \Rightarrow G_\gamma(Ax + B)$, we say that the distribution function F is in the **domain of attraction** of G_γ, denoted $F \in \mathcal{D}(G_\gamma)$.*

Proposition II.9.45 derived the **von Mises' condition,** named for **Richard von Mises** (1883–1953), which identified how the extreme value index γ could be derived in the case of a twice continuously differentiable distribution function $F(x)$.

For its statement, recall x^* as defined in (6.39):

$$x^* \equiv \inf\{x | F(x) = 1\}, \qquad x^* \equiv \infty \text{ if } F(x) < 1 \text{ for all } x,$$

and Proposition 6.52 which assured that $M_n \to x^*$ with probability 1.

Proposition 7.17 (von Mises' condition) *Let $F(x)$ be a twice continuously differentiable distribution function with $F'(x) > 0$ for some interval (x_0, x^*). If:*

$$\lim_{x \to x^*} \left(\frac{1-F}{F'} \right)' (x) = \gamma, \tag{7.16}$$

then F is in the domain of attraction of G_γ, i.e., $F \in D(G_\gamma)$.

The von Mises' condition provides one approach to determining if a given distribution function is in the domain of attraction of G_γ for some γ.

Example 7.18 ($F(x)$ normal, or Pareto) *With Φ denoting the standard normal distribution function, it was shown that $\Phi \in D(G_0)$ in Example II.9.48 using the von Mises' condition.*

If $F(x)$ is the Pareto distribution of (7.17) with $x_0 = 1$ for simplicity, so $F_P(x) = 1 - x^{-1/\gamma}$ for $x \geq 1$, then $F(x)$ satisfies the requirements for the von Mises' result. As $(1 - F)/F' = \gamma x$, it follows from (7.16) that $F_P \in D(G_\gamma)$.

When dealing with data sets from unknown distributions in finance and other disciplines, the question naturally arises as to how one can determine if the data is consistent with that from a distribution F with $F \in D(G_\gamma)$ for some γ. As $F(x)$ is unknown in this case, certainly so too are its various derivatives. Thus the von Mises' condition can only provide a workable approach when dealing with data if an underlying distribution function is first estimated and proves to be sufficiently differentiable.

While there are many approaches to this estimation problem, a popular and frequently used approach is the **Hill estimator,** introduced in 1975 by **Bruce M. Hill.** We study this estimator in the case of greatest interest in finance applications, and that is when $\gamma > 0$.

7.2.2   The Hill Estimator, $\gamma > 0$

The Hill estimator was introduced as an approach to estimating the "tail behavior" of a given sample, without making an assumption on the global form of the distribution function. The result derived was particularly simple for estimating the exponent parameter of the **Pareto distribution**. This distribution was introduced in Remark II.9.40, named for **Vilfredo Pareto** (1848–1923), and also called the **power law** or a **distribution of Zipf type,** the latter named for **George Kingsley Zipf** (1902–1950).

Defined on $x \geq x_0$, the Pareto distribution is given by:

$$F_P(x) = 1 - \left(\frac{x}{x_0} \right)^{-1/\gamma}, \tag{7.17}$$

where $\gamma > 0$. This model is reflective of many data observations made in finance and elsewhere, and is often parametrized with $\alpha = 1/\gamma$.

Parametrized as in (7.17), this distribution function satisfies the requirements of the von Mises' result, and as noted in Example 7.18, such distributions satisfy $F_P \in \mathcal{D}(G_\gamma)$. Since $\gamma > 0$ here, it follows that Pareto, power law, or Zipf-type distributions are in the domain of attraction of the **Fréchet class of distributions,** also called **Type II extreme value distributions,** as noted in Examples II.9.10 and II.9.13.

The Hill estimator for $\gamma > 0$ is defined as follows.

Definition 7.19 (Hill estimator for $\gamma > 0$) *Let $\{X_i\}_{i=1}^n$ be a random sample of a given variate, and $\{X_{(j)}\}_{j=1}^n$ the associated order statistics.*

*The **Hill estimator** $\gamma_H \equiv \gamma_H^{(k,n)}$ is based on the $k+1$ largest variates, $\{X_{(n-j)}\}_{j=0}^k$, and defined as an average of k log ratios:*

$$\gamma_H \equiv \frac{1}{k} \sum_{j=0}^{k-1} \ln \left[\frac{X_{(n-j)}}{X_{(n-k)}} \right], \tag{7.18}$$

and equivalently expressed:

$$\gamma_H \equiv \frac{1}{k} \sum_{j=0}^{k-1} \left[\ln X_{(n-j)} - \ln X_{(n-k)} \right]. \tag{7.19}$$

Example 7.20 (Pareto distribution: $\gamma_H^{(k,n)}$ and MLE) *Given a random sample $\{X_j\}_{j=1}^n$ that is assumed to come from a Pareto distribution with unknown parameter γ, we show that the Hill estimator provides a solution that equals what is known as the maximum likelihood estimate of this parameter.*

Let $F_P(x)$ be given as in (7.17) with $x_0 = 1$ for simplicity, and recall that $F_P \in \mathcal{D}(G_\gamma)$ by Example 7.18. Given the ordered sample $\{X_{(j)}\}_{j=1}^n$ and k, define the conditional distribution function for $x \geq X_{(n-k)}$ as in Definition 1.12 (see also Example II.3.40), where we temporarily denote $\alpha = 1/\gamma$ for notational simplicity:

$$
\begin{aligned}
F_P\left(x|x \geq X_{(n-k)}\right) &= \frac{F_P(x) - F_P(X_{(n-k)})}{1 - F_P(X_{(n-k)})} \\
&= \left[X_{(n-k)}^{-\alpha} - x^{-\alpha} \right] / X_{(n-k)}^{-\alpha}.
\end{aligned}
$$

The conditional density function is then given as the derivative of this differentiable function by Proposition III.1.33:

$$f_P\left(x|x \geq X_{(n-k)}\right) = \alpha x^{-(\alpha+1)} / X_{(n-k)}^{-\alpha}.$$

*In many applications it is reasonable to **assume** that the given sample $\{X_j\}_{j=1}^n$ has density function $f(x; \alpha)$, parametrized by and therefore "conditional" on an unknown parameter α to be estimated. The **conditional likelihood function,** and sometimes the **likelihood function** of the sample, is defined:*

$$L[\{X_j\}_{j=1}^n; \alpha] \equiv \prod_{j=1}^n f(X_j; \alpha).$$

Given this assumption on the parametric form of the density, $L[\{X_j\}_{j=1}^n; \alpha]$ informally provides the "probability of this sample" given any parameter α.

*A logical objective is therefore to maximize L as a function of α, producing the **conditional maximum likelihood estimate** for α, often called the **maximum likelihood estimate/estimator** or **MLE**. By maximizing L, the given parameter α provides a model which maximizes the probability of the observed sample among the family of distributions $\{f(x; \alpha)\}$ parametrized by α.*

Applying this approach to the sample $\{X_{(n-j)}\}_{j=0}^{k}$ and conditional density $f_P(x;\alpha) \equiv \alpha x^{-(\alpha+1)}/X_{(n-k)}^{-\alpha}$ obtains the likelihood function:

$$L\left[\{X_{(n-j)}\}_{j=0}^{k};\alpha\right] = \alpha^k \prod_{j=0}^{k-1} X_{(n-j)}^{-(\alpha+1)}/X_{(n-k)}^{-\alpha}.$$

To maximize L as a differentiable function of α, or equivalently to maximize the logarithm of this function, we differentiate:

$$\frac{\partial \ln L}{\partial \alpha} = \frac{k}{\alpha} - \sum_{j=0}^{k-1} \ln X_{(n-j)} + k \ln X_{(n-k)}.$$

The maximum likelihood estimate α^{MLE} equals the value of α that solves $\frac{\partial \ln L}{\partial \alpha} = 0$ if $\frac{\partial^2 \ln L}{\partial \alpha^2} < 0$, which is verifiable in this case. A calculation yields

$$\alpha^{MLE} = 1/\gamma_H.$$

It is then checked as an exercise that if parametrized as a function of γ, one again has that $\frac{\partial \ln L}{\partial \gamma} = 0$ and $\frac{\partial^2 \ln L}{\partial \gamma^2} < 0$ when $\gamma^{MLE} = \gamma_H$.

Thus we have proved the following result:

Proposition 7.21 (Pareto distribution: $\gamma_H^{(k,n)} = \gamma^{MLE}$) *Given a sample $\{X_j\}_{j=1}^{n}$ that is assumed to follow a Pareto distribution with unknown parameter γ, if γ^{MLE} denotes the maximum likelihood estimate for γ based on $\{X_{(n-j)}\}_{j=0}^{k}$, then:*

$$\gamma^{MLE} = \gamma_H^{(k,n)}, \tag{7.20}$$

where $\gamma_H^{(k,n)}$ is the Hill estimator based on this same subsample.

Example 7.22 (Pareto distribution: $\gamma_H^{(k,n)}$ as a random variable) *The collection $\{X_j\}_{j=1}^{n}$ above denoted a random sample from a random variable X defined some probability space with given distribution function $F(x)$. By random sample is meant that these variates have the same distribution function $F(x)$ as X, and are independent. See Example II.4.6 for a discussion on random samples, and the meaning of random samples having such properties.*

*Given the constructions of Chapter II.4 and summarized in Section 6.1, one can also construct a probability space $(\mathcal{S},\mathcal{E},\lambda)$, and collection of **independent, identically distributed (i.i.d.)** random variables $\{X_j\}_{j=1}^{\infty}$ with:*

$$X_j : (\mathcal{S},\mathcal{E},\lambda) \to (\mathbb{R},\mathcal{B}(\mathbb{R}),m),$$

for all j. By i.i.d is meant:

- *$\{X_j\}_{j=1}^{\infty}$ are **identically distributed,** meaning $F_{X_j}(x) = F(x)$ for all j;*

- *$\{X_j\}_{j=1}^{\infty}$ are **independent** in the sense of Definition 1.13.*

Within this framework, $\gamma_H^{(k,n)}$ is seen to be a random variable on $(\mathcal{S},\mathcal{E},\lambda)$ given any n and k, and thus we can investigate distributional properties.

One model for such $\{X_j\}_{j=1}^{\infty}$ and $(\mathcal{S},\mathcal{E},\lambda)$ is given in Proposition II.4.13 where $X_j \equiv F^*(U_j)$, $\{U_j\}_{j=1}^{\infty}$ are i.i.d random variables with a continuous, uniform distribution, and $F^*(y)$ is the left-continuous inverse of $F(x)$ as given in Definition 1.29.

If $F(x) = F_P(x)$ in (7.17) with $x_0 = 1$ for notational simplicity, then:

$$F_P^*(y) = (1-y)^{-\gamma},$$

is a continuous, strictly increasing function. Thus given n and k, the variates in the Hill estimator $\{X_{(n-j)}\}_{j=0}^k$ are defined:

$$X_{(n-j)} \equiv \left(1 - U_{(n-j)}\right)^{-\gamma},$$

and so by (7.18):

$$
\begin{aligned}
\gamma_H^{(k,n)} &\equiv \frac{1}{k} \sum_{j=0}^{k-1} \ln \left[\frac{X_{(n-j)}}{X_{(n-k)}}\right] \\
&= \frac{1}{k} \sum_{j=0}^{k-1} \ln \left[\frac{\left(1 - U_{(n-j)}\right)^{-\gamma}}{\left(1 - U_{(n-k)}\right)^{-\gamma}}\right] \\
&= \frac{\gamma}{k} \sum_{j=0}^{k-1} \left[\left(-\ln\left(1 - U_{(n-j)}\right)\right) - \left(-\ln\left(1 - U_{(n-k)}\right)\right)\right].
\end{aligned}
$$

By Proposition 5.9, the random variable $E = -\ln(1 - U)$ has the standard exponential distribution of (1.54) with $\lambda = 1$:

$$F_E(x) = 1 - \exp(-x), \quad x \geq 0,$$

and since this transformation is increasing, the order statistics are preserved. Thus $\{E_{(n-j)}\}_{j=0}^k = \{-\ln\left(1 - U_{(n-j)}\right)\}_{j=0}^k$ are the higher order statistics of a sample $\{E_j\}_{j=1}^n$ of standard exponential variates. Since $X_j \equiv (1 - U_j)^{-\gamma}$ it follows that $E_j = \gamma \ln X_j$, and so $\{E_j\}_{j=1}^n$ are independent random variables by Proposition II.3.56.

The Hill estimator for $\{X_{(n-j)}\}_{j=0}^k$ can be thus expressed:

$$\gamma_H^{(k,n)} = \frac{\gamma}{k} \sum_{j=0}^{k-1} \left[E_{(n-j)} - E_{(n-k)}\right]. \tag{1}$$

From (1), we can now state the final results:

Proposition 7.23 (Pareto: $\gamma_H^{(k,n)}/\gamma$ is asymptotically normal; and $\gamma_H^{(k,n)} \to_1 \gamma$) *Let $\{X_j\}_{j=1}^n$ be independent Pareto random variables with parameter γ, and $\gamma_H^{(k,n)}$ the Hill estimator based on $\{X_{(n-j)}\}_{j=0}^k$.*
Define:

$$\widehat{\gamma}_H^{(k,n)} = \frac{\gamma_H^{(k,n)}/\gamma - 1}{1/\sqrt{k}},$$

and let $F_{k,n}$ denote the distribution function of $\widehat{\gamma}_H^{(k,n)}$.
Then as $n \to \infty$ and $k \to \infty$:

1. $F_{k,n}$ converges weakly to Φ:

$$F_{k,n} \Rightarrow \Phi, \tag{7.21}$$

where $\Phi(x)$ is the distribution function of the standard normal.

2. $\widehat{\gamma}_H^{(k,n)}$ converges to γ with probability 1:

$$\widehat{\gamma}_H^{(k,n)} \to_1 \gamma. \tag{7.22}$$

Proof. *Both results are an immediate application of Proposition 6.22. Since $\gamma_H^{(k,n)}$ can be expressed as in (1), $\gamma_H^{(k,n)}/\gamma$ equals the variate denoted $Y_{k,n}$ in that result. In addition, the conclusion of that result that $\widehat{\gamma}_H^{(k,n)}/\gamma \to_1 1$ is equivalent to (7.22) by definition.* ∎

For general $F \in \mathcal{D}(G_\gamma)$ with $\gamma > 0$, we cannot expect that $F(x)$ is Pareto. However, Proposition II.9.38 obtained that if $\gamma > 0$, then for all $x \geq -1/\gamma$:

$$\lim_{t \to x^*} \frac{1 - F(t + xh_a(t))}{1 - F(t)} = (1 + \gamma x)^{-1/\gamma}.$$

Recall that $h_a(t) \equiv a_c \left(\frac{1}{1 - F(t)}\right)$, where $a_c(t)$ is the normalizing function in Corollary II.9.35. This is defined by $a_c(t) \equiv ca_{\lfloor t \rfloor}$ with $\lfloor t \rfloor$ is the **greatest integer function:**

$$\lfloor t \rfloor = \max\{n | n \leq t\},$$

where $\{a_n\}_{n=1}^\infty$ is the sequence in the Fisher-Tippett-Gnedenko theorem, and c is the constant in II.(9.32) of Proposition II.9.34. And as defined in (6.39), $x^* = \inf\{x | F(x) = 1\}$, with $x^* \equiv \infty$ if $F(x) < 1$ for all x.

To investigate the implications of this general result, note that:

$$1 - \frac{1 - F(t + xh_a(t))}{1 - F(t)} = \frac{F(t + xh_a(t)) - F(t)}{1 - F(t)}$$

$$= F(t + xh_a(t) | X > t),$$

where the conditional distribution function on the right is as defined in Definition 1.12 (see also Example II.3.40).

Thus for $F \in \mathcal{D}(G_\gamma)$ with $\gamma > 0$, Proposition II.9.38 asserts that the conditional distribution function $F(t + xh_a(t) | X > t)$ has limit as $t \to x^*$:

$$\lim_{t \to x^*} F(t + xh_a(t) | X > t) = 1 - (1 + \gamma x)^{-1/\gamma}.$$

With a reparametrization of $y = 1 + \gamma x$, this implies that the following conditional distribution is **asymptotically Pareto**, meaning:

$$\lim_{t \to x^*} F\left(t + \left(\frac{y - 1}{\gamma}\right) h_a(t) \,\middle|\, X > t\right) = 1 - y^{-1/\gamma}.$$

Written in terms of the Pr notation:

$$\lim_{t \to x^*} \Pr\left(X \leq t + \left(\frac{y - 1}{\gamma}\right) h_a(t) \,\middle|\, X > t\right) = 1 - y^{-1/\gamma}.$$

In summary, while general $F \in \mathcal{D}(G_\gamma)$ with $\gamma > 0$ is not a Pareto distribution, it is asymptotically Pareto in the above sense. Put another way, the conditional asymptotic tail of $F \in \mathcal{D}(G_\gamma)$ is Pareto. Thus there is some hope that the properties of the Hill estimator for the Pareto distribution will imply similar properties for such F.

In the next three sections, we develop the Hill estimator result in three steps:

1. **If $F \in \mathcal{D}(G_\gamma)$ with $\gamma > 0$, then $F(x)$ is conditionally asymptotically Pareto.**

    We will refine the above analysis of the behavior of $F(x)$ and show that $\gamma > 0$ if and only if $x^* = \infty$, and then in (7.32) that:

$$\lim_{t \to \infty} \Pr[X \leq tx | X > t] = 1 - x^{-1/\gamma}.$$

    In other words, a much simpler conditional distribution of $F(x)$ will be shown to be asymptotically Pareto.

    From this result we will then derive an integral formula for the parameter γ associated with such a function. This formula will generalize von Mises' condition by not requiring differentiability of $F(x)$, but will be more restrictive in that it will require that $\gamma > 0$.

2. **If $F \in D(G_\gamma)$ with $\gamma > 0$, then $\gamma_H^{(k,n)} \approx \gamma$.**

The integral formula for γ derived in item 1 will be approximated, and will produce the estimate $\gamma_H^{(k,n)}$.

3. **If $F \in D(G_\gamma)$ with $\gamma > 0$, then $\gamma_H^{(k,n)} \to_1 \gamma$ as $n \to \infty$.**

This is the main result for the Hill estimator. For this convergence with probability 1, it will be required that both $k, n \to \infty$ and $k/n \to 0$ as $n \to \infty$. In other words, the Hill estimator $\gamma_H^{(k,n)}$ must of necessity be based on increasingly high quantiles of X since the estimator uses ordered data in the $[n - k, n]$ range, which is equivalent to the $\left[1 - \frac{k}{n}, 1\right]$ quantile range.

The final section will address asymptotic normality of the Hill estimator, generalizing Proposition 7.23.

7.2.3 $F \in D(G_\gamma)$ is Asymptotically Pareto for $\gamma > 0$

The goals of this section are to prove that the above conditional distribution of $F \in D(G_\gamma)$ is asymptotically Pareto for $\gamma > 0$, and to derive an integral formula for this parameter. We begin by recalling Corollary II.9.35.

If $F \in D(G_\gamma)$ for $\gamma \neq 0$, then for all $x > 0$:

$$\lim_{t \to \infty} \frac{U(tx) - U(t)}{a_c(t)} = \frac{x^\gamma - 1}{\gamma}. \tag{7.23}$$

Here as above, $a_c(t) \equiv ca_{\lfloor t \rfloor}$, where $\lfloor t \rfloor$ is the **greatest integer function**, $\{a_n\}_{n=1}^\infty$ is the sequence in Fisher-Tippett-Gnedenko theorem, and c is the constant in II.(9.32) of Proposition II.9.34.

The function $U(t)$ is the left-continuous inverse of $1/(1 - F(x))$:

$$U(t) \equiv \left(\frac{1}{1 - F}\right)^*(t), \tag{7.24}$$

and defined on $t > 1$. An alternative and perhaps more intuitive representation for $U(t)$ is given in Exercise II.9.26:

$$U(t) \equiv F^*(1 - 1/t). \tag{7.25}$$

We begin by investigating and refining the limit in (7.23). The result is simple, but the derivation is subtle.

Proposition 7.24 (Asymptotics of $\frac{U(t)}{a_c(t)}$ and $\frac{U(tx)}{U(t)}$) *If $F \in D(G_\gamma)$ with $\gamma > 0$, then $U(t) \to \infty$ as $t \to \infty$ and:*

$$\lim_{t \to \infty} \frac{U(t)}{a_c(t)} = \frac{1}{\gamma}. \tag{7.26}$$

Also, for $x > 0$:

$$\lim_{t \to \infty} \frac{U(tx)}{U(t)} = x^\gamma. \tag{7.27}$$

Proof. *Simplifying notation, we write $a_c(t) = ca(t)$, and thus $a(t) \equiv a_{\lfloor t \rfloor}$.*
Defining:

$$V_t(x) \equiv \frac{U(tx) - U(t)}{a_c(t)},$$

it follows that since $\frac{a_c(tx)}{a_c(t)} = \frac{a(tx)}{a(t)}$:

$$\frac{a(tx)}{a(t)} = \frac{V_t(xy) - V_t(x)}{V_{tx}(y)}.$$

Applying (7.23) to each V-term obtains:

$$\lim_{t \to \infty} \frac{a(tx)}{a(t)} = x^\gamma. \qquad (7.28)$$

Then since:

$$\frac{U(tx)}{U(t)} = 1 + \frac{U(tx) - U(t)}{a_c(t)} \frac{a_c(t)}{U(t)}, \qquad (1)$$

it follows from (7.23) that a limit exists in (7.27) if and only if a limit exists in (7.26) and this limit is nonzero.

First, we claim that if the limit in (7.26) exists, this limit must be nonzero. To see this note that:

$$\frac{U(tx) - U(t)}{a_c(t)} = \frac{U(tx)}{a_c(tx)} \frac{a_c(tx)}{a_c(t)} - \frac{U(t)}{a_c(t)}.$$

Since $U(tx)/a_c(tx)$ *and* $U(t)/a_c(t)$ *must have the same limit as* $t \to \infty$ *if such limit exists, the limit in (7.26) must be nonzero by (7.28) and (7.23).*

Finally, if a limit exists in (7.26) and is thus nonzero, the identity in (1) and (7.23) obtain:

$$\lim_{t \to \infty} \frac{U(tx)}{U(t)} = 1 + \frac{x^\gamma - 1}{\gamma} \left[\frac{1}{\lim_{t \to \infty} \frac{U(t)}{a_c(t)}} \right]. \qquad (2)$$

Thus (7.27) is true if and only if (7.26) is true.

The final challenge is thus to prove either limit in (7.26) or (7.27), and we verify the latter limit after proving that $U(t) \to \infty$ as $t \to \infty$.

Letting $Z > 1$, an application of (7.28) with $t = Z^{k-1}$ yields $a(Z^k)/a(Z^{k-1}) \to Z^\gamma$. A similar application of (7.23) and defining $t = Z^{k-1}$ or $t = Z^k$ as appropriate obtains:

$$\lim_{k \to \infty} \frac{U(Z^{k+1}) - U(Z^k)}{U(Z^k) - U(Z^{k-1})}$$

$$= \lim_{k \to \infty} \frac{U(Z^{k+1}) - U(Z^k)}{a_c(Z^k)} \bigg/ \frac{U(Z^k) - U(Z^{k-1})}{Z^\gamma a_c(Z^{k-1})}$$

$$= Z^\gamma.$$

Hence for any $\epsilon > 0$ *there is an* $N \equiv N(\epsilon)$ *so that for* $k \geq N$:

$$Z^\gamma(1 - \epsilon) \leq \frac{U(Z^{k+1}) - U(Z^k)}{U(Z^k) - U(Z^{k-1})} \leq Z^\gamma(1 + \epsilon). \qquad (3)$$

For $n > N$:

$$U(Z^{n+1}) - U(Z^N) = [U(Z^N) - U(Z^{N-1})] \sum_{j=N}^{n} \prod_{k=N}^{j} \frac{U(Z^{k+1}) - U(Z^k)}{U(Z^k) - U(Z^{k-1})},$$

and so from (3) :

$$\lim_{n \to \infty} [U(Z^{n+1}) - U(Z^N)] \geq [U(Z^N) - U(Z^{N-1})] \lim_{n \to \infty} \sum_{j=N}^{n} \prod_{k=N}^{j} Z^\gamma(1 - \epsilon)$$

$$= [U(Z^N) - U(Z^{N-1})] \lim_{n \to \infty} \sum_{j=N}^{n} [Z^\gamma(1 - \epsilon)]^{j-N+1}.$$

Hence $\lim_{n\to\infty} U\left(Z^{n+1}\right) = \infty$ *if* $Z^{\gamma}(1-\epsilon) > 1$, *which is to say,* $\epsilon < 1 - Z^{-\gamma}$. *This proves that* $U(t) \to \infty$ *as* $t \to \infty$ *since* $U(t)$ *is increasing.*

Turning next to (7.27), since $U(t) \to \infty$ *as* $t \to \infty$:

$$\lim_{n\to\infty}\sum_{k=N}^{n}\left[\frac{U\left(Z^{k+1}\right)-U\left(Z^{k}\right)}{U\left(Z^{n}\right)}\right] = \lim_{n\to\infty}\frac{U\left(Z^{n+1}\right)-U\left(Z^{N}\right)}{U\left(Z^{n}\right)} \tag{4}$$

$$= \lim_{n\to\infty}\frac{U\left(Z^{n+1}\right)}{U\left(Z^{n}\right)}.$$

Using the inequalities in (3) applied to each term in the first summation yields:

$$Z^{\gamma}(1-\epsilon)\lim_{n\to\infty}\sum_{k=N}^{n}\left[\frac{U\left(Z^{k}\right)-U\left(Z^{k-1}\right)}{U\left(Z^{n}\right)}\right]$$

$$\leq \lim_{n\to\infty}\frac{U\left(Z^{n+1}\right)}{U\left(Z^{n}\right)}$$

$$\leq Z^{\gamma}(1+\epsilon)\lim_{n\to\infty}\sum_{k=N}^{n}\left[\frac{U\left(Z^{k}\right)-U\left(Z^{k-1}\right)}{U\left(Z^{n}\right)}\right].$$

The limits of the summations in these bounds are seen to equal 1, recalling that $U(t) \to \infty$ *as* $t \to \infty$. *Thus for any* $\epsilon < 1 - Z^{-\gamma}$:

$$Z^{\gamma}(1-\epsilon) \leq \lim_{n\to\infty}\frac{U\left(Z^{n+1}\right)}{U\left(Z^{n}\right)} \leq Z^{\gamma}(1+\epsilon),$$

and this limit therefore equals Z^{γ} *for any* $Z > 1$:

$$\lim_{n\to\infty}\frac{U\left(Z^{n+1}\right)}{U\left(Z^{n}\right)} = Z^{\gamma}. \tag{5}$$

Letting $t = Z^n$ *in (5) then suggests that* $\lim_{t\to\infty} U\left(tx\right)/U\left(t\right)$ *exists for all* $x > 1$ *and equals the limit in (7.27), but this must be formalized.*

Given $Z > 1$ *and* $y > 1$, *let* $n(y)$ *be defined as the integer such that* $Z^{n(y)} \leq y < Z^{n(y)+1}$. *Then because* U *is increasing, if* $t, x > 1$:

$$\frac{U(Z^{n(x)}Z^{n(t)})}{U(Z^{n(t)+1})} \leq \frac{U\left(tx\right)}{U\left(t\right)} \leq \frac{U(Z^{n(x)+1}Z^{n(t)+1})}{U(Z^{n(t)})}. \tag{6}$$

Using (5) for the left bound in (6) :

$$\lim_{t\to\infty}\frac{U(Z^{n(x)}Z^{n(t)})}{U(Z^{n(t)+1})} = \lim_{t\to\infty}\frac{U(Z^{n(x)}Z^{n(t)})}{U(Z^{n(t)})}\frac{U(Z^{n(t)})}{U(Z^{n(t)+1})}$$

$$= Z^{(n(x)-1)\gamma}$$

$$\geq (x/Z^{2})^{\gamma}.$$

For this calculation, the substitution $W \equiv Z^{n(x)}$ *and* $m(t) \equiv \frac{n(t)}{n(x)}$ *obtains:*

$$\frac{U(Z^{n(x)}Z^{n(t)})}{U(Z^{n(t)})} = \frac{U(W^{m(t)+1})}{U(W^{m(t)})} \to W^{\gamma} = Z^{n(x)\gamma}.$$

Similarly, for the right bound, let $W \equiv Z^{n(x)+2}$ and $m(t) \equiv \frac{n(t)}{n(x)+2}$, and this obtains:

$$\lim_{t \to \infty} \frac{U(Z^{n(x)+2} Z^{n(t)})}{U(Z^{n(t)})} \leq (xZ^2)^\gamma.$$

Thus:

$$\left(\frac{x}{Z^2}\right)^\gamma \leq \lim_{t \to \infty} \frac{U(tx)}{U(t)} \leq (xZ^2)^\gamma,$$

and (7.27) follows for $x > 1$ by letting $Z \to 1$.

The final step to address $0 < x \leq 1$. The existence of the limit in (7.27) for $x > 1$ is sufficient to assure the existence of the limit in (7.26) as noted in the identity in (2). This identity then assures (7.27) for all $x > 0$. ∎

Definition 7.25 (Varying at infinity) *Given $\alpha \in \mathbb{R}$, a function f is called **regularly varying at infinity with index** α if for all $x \geq x_0 > 0$:*

$$\lim_{t \to \infty} \frac{f(tx)}{f(t)} = x^\alpha. \tag{7.29}$$

*When $\alpha = 0$, f is said to be **slowly varying at infinity**.*
*When f is **regularly varying at infinity with index** α, one often writes $f \in RV_\alpha$.*

Example 7.26 (Varying ay infinity: $F \in D(G_\gamma)$ with $\gamma > 0$) *If $F \in D(G_\gamma)$ with $\gamma > 0$, then both:*

- $U(y) \equiv \left(\frac{1}{1-F}\right)^* (y)$ *by (7.27), and,*

- $a(t)$ *by (7.28),*

  *are **regularly varying at infinity with index** γ.*
  Hence, if $F \in D(G_\gamma)$ with $\gamma > 0$, then both $a(t), U(y) \in RV_\gamma$:

Before continuing the current development we identify a corollary result to (7.27) which was promised in Remark II.9.36, regarding the normalizing sequences a_n and b_n in the statement of the **Fisher-Tippett-Gnedenko theorem**.

Corollary 7.27 (Fisher-Tippett-Gnedenko theorem, $\gamma > 0$) *If $F \in D(G_\gamma)$ with $\gamma > 0$, define $a_n \equiv U(n)$ and $b_n = 0$.*
Then as $n \to \infty$:

$$\lim_{n \to \infty} F^n(U(n)x) = \exp\left(-x^{-1/\gamma}\right), \qquad x \geq 0. \tag{7.30}$$

Thus for these normalizing sequences, $A = 1/\gamma$ and $B = -1/\gamma$ in Proposition 7.15, and:

$$F^n(a_n x) \Rightarrow G_\gamma\left((x-1)/\gamma\right),$$

where G_γ is defined in (7.14).
Proof. *The result in (7.27), when restricted to integer $t = n$, can be expressed in terms of left-continuous inverses as weak convergence:*

$$H_n^*(x) \Rightarrow K^*(x),$$

where:

$$H_n(x) = \frac{1}{n\left[1 - F(xU(n))\right]}, \qquad K(x) = x^{1/\gamma}.$$

Though not distribution functions, such weak convergence is defined in Remark II.8.5, as pointwise convergence at continuity points of $K^*(x)$.
    First:

$$K^*(x) = \inf\{y|y^{1/\gamma} \geq x\} = x^{\gamma}.$$

For $H_n(x)$:

$$
\begin{aligned}
H_n^*(x) &= \inf\{y|1/\{n[1 - F(U(n)y)]\} \geq x\} \\
&= \inf\{y|F(U(n)y) \geq 1 - 1/nx\}.
\end{aligned}
$$

Letting $z = yU(n)$, and recalling that $U(y) \equiv F^*(1 - 1/y)$ by (7.25):

$$H_n^*(x) = \frac{1}{U(n)} \inf\{z|F(z) \geq 1 - 1/nx\} = \frac{U(nx)}{U(n)}.$$

By Corollary II.8.28, the proof of which extends to increasing function sequences by Corollary II.3.27, $H_n^*(x) \Rightarrow K^*(x)$ implies that $H_n(x) \Rightarrow K(x)$. This obtains that for all $x \geq 0$:

$$\lim_{n \to \infty} n[1 - F(U(n)x)] = x^{-1/\gamma}. \tag{1}$$

Thus $1 - F(U(n)x) \to 0$ as $n \to \infty$, and so:

$$1 - F(U(n)x) = -\ln F(U(n)x) + O(n^{-2}).$$

The limit in (1) can thus be expressed:

$$\lim_{n \to \infty} \ln F^n(U(n)x) = -x^{-1/\gamma},$$

which is equivalent to (7.30). ∎

The next step in this development is to convert the limiting results for U to limiting results for the distribution function F. The following proposition states that if $F \in D(G_\gamma)$ with $\gamma > 0$, then the conditional distribution function $\Pr[X \leq tx|X > t]$ is asymptotically Pareto as $t \to \infty$ for all $x > 0$.

Proposition 7.28 ($F \in D(G_\gamma)$ is asymptotically Pareto for $\gamma > 0$) *A distribution function $F \in D(G_\gamma)$ with $\gamma > 0$ if and only if $x^* = \infty$, and for all $x > 0$:*

$$\lim_{t \to \infty} \frac{1 - F(tx)}{1 - F(t)} = x^{-1/\gamma}. \tag{7.31}$$

Equivalently, $F \in D(G_\gamma)$ with $\gamma > 0$ if and only if $x^* = \infty$, and for all $x > 0$, the conditional distribution function $\Pr[X \leq tx|X > t]$ is asymptotically Pareto as $t \to \infty$:

$$\lim_{t \to \infty} \Pr[X \leq tx|X > t] = 1 - x^{-1/\gamma}. \tag{7.32}$$

Proof. *The equivalence of the limits in (7.31) and (7.32) is Definition 1.12, recalling Example II.3.40:*

$$1 - \frac{1 - F(tx)}{1 - F(t)} = \frac{F(tx) - F(t)}{1 - F(t)} = \Pr[X \leq tx|X > t].$$

Assume that $F \in D(G_\gamma)$ with $\gamma > 0$. By Proposition 7.24 and (7.25), $U(t) = F^*(1 - 1/t) \to \infty$ as $t \to \infty$. This implies that $F(x) < 1$ for all x, and so $x^* = \infty$ by definition.
    By Proposition II.3.19:

$$F^*(F(t)) \leq t \leq F^*(F(t)^+),$$

and since $U\left(\frac{1}{1-F(t)}\right) = F^*\left(F(t)\right)$ *by (7.24), for any* $\epsilon > 0$:

$$U\left(\frac{1-\epsilon}{1-F(t)}\right) \le t \le U\left(\frac{1+\epsilon}{1-F(t)}\right),$$

and hence:

$$\frac{U\left(\frac{y}{1-F(t)}\right)}{U\left(\frac{1+\epsilon}{1-F(t)}\right)} \le \frac{1}{t} U\left(\frac{y}{1-F(t)}\right) \le \frac{U\left(\frac{y}{1-F(t)}\right)}{U\left(\frac{1-\epsilon}{1-F(t)}\right)}.$$

Since $F \in D(G_\gamma)$ with $\gamma > 0$, (7.27) applies. With $t' = \frac{1 \pm \epsilon}{1-F(t)}$ and $x = \frac{y}{1 \pm \epsilon}$, it follows that as $t' \to \infty$:

$$\frac{U\left(\frac{y}{1-F(t)}\right)}{U\left(\frac{1\pm\epsilon}{1-F(t)}\right)} = \frac{U\left(t'x\right)}{U\left(t'\right)} \to x^\gamma.$$

But $t' \to \infty$ *if and only* $F(t) \to 1$ *if and only if* $t \to \infty$, *and so for all* $\epsilon > 0$:

$$\left(\frac{y}{1+\epsilon}\right)^\gamma \le \lim_{t\to\infty} \frac{1}{t} U\left(\frac{y}{1-F(t)}\right) \le \left(\frac{y}{1-\epsilon}\right)^\gamma.$$

Hence:

$$\lim_{t\to\infty} \frac{1}{t} U\left(\frac{y}{1-F(t)}\right) = y^\gamma. \tag{1}$$

Defining $g_n(y) = \frac{1}{n} U\left(\frac{y}{1-F(n)}\right)$ *and* $g(y) = y^\gamma$, *then* $g_n(y) \to g(y)$ *for all* $y > 0$ *by (1). This implies that* $g_n^*(x) \to g^*(x)$ *for each continuity point of* g^* *by Proposition II.8.27. A calculation obtains that* $g^*(x) = x^{1/\gamma}$ *and* $g_n^*(x) = \frac{1-F(n)}{1-F(nx)}$, *and and so for* $x > 0$:

$$\frac{1-F(n)}{1-F(nx)} \to x^{1/\gamma}, \ n \to \infty. \tag{2}$$

For real t *define integer* $n(t)$ *so that* $n(t) \le t < n(t) + 1$. *Then for* $x > 0$, *by monotonicity of* F :

$$\frac{1-F([n(t)+1]x)}{1-F(n(t))} \le \frac{1-F(tx)}{1-F(t)} \le \frac{1-F(n(t)x)}{1-F(n(t)+1)}.$$

Thus (7.31) follows from (2) and an exercise that $x^* = \infty$ *implies that* $\frac{1-F(n(t))}{1-F(n(t)+1)} \to 1$. *Hint:* $F(t) \to 1$.

Conversely, if $x^* = \infty$ *and (7.31) is satisfied, then (2) follows, as does (1) by Corollary II.8.28, that* $g_n^*(x) \Rightarrow g^*(x)$ *implies that* $g_n(y) \Rightarrow g(y)$. *This corollary applies to this sequence of increasing functions* $g_n(y)$ *by the same proof, but replacing the reference to Proposition II.3.26 with one to Corollary II.3.27.*

Now let $s = \frac{1}{1-F(t)}$. *Since* $U(s) = F^*(1 - 1/s) = F^*(F(t))$:

$$\frac{U(sy)}{U(s)} = \frac{t}{F^*(F(t))} \frac{1}{t} U\left(\frac{y}{1-F(t)}\right). \tag{3}$$

The assumption that $x^* = \infty$ *assures that* $s \to \infty$ *as* $t \to \infty$, *and also* $\frac{t}{F^*(F(t))} \to 1$ *by Proposition II.3.19. Hence (1) and (3) obtain:*

$$\lim_{s\to\infty} \frac{U(sy)}{U(s)} = y^\gamma.$$

Defining normalizing sequences $a_n = \gamma U(n)$ and $b_n = U(n)$ obtains that for all $y > 0$:

$$\lim_{n\to\infty} \frac{U(ny) - b_n}{a_n} = \frac{y^\gamma - 1}{\gamma}. \tag{4}$$

We now recall that the first step of the proof of the Fisher-Tippett-Gnedenko theorem of Proposition 7.15 was to derive (4), and then proceed to obtain from this that $F \in D(G_\gamma)$. ∎

Example 7.29 (Varying ay infinity: $F \in D(G_\gamma)$ with $\gamma > 0$) *Adding to Example 7.26, the result in (7.31) states that if $F \in D(G_\gamma)$ with $\gamma > 0$, then:*

- *$1 - F$ is **regularly varying at infinity with index** $-1/\gamma$.*

   Hence if $F \in D(G_\gamma)$ with $\gamma > 0$, then $1 - F \in RV_{-1/\gamma}$.
   This can also be expressed in an even more descriptive way. If $F \in D(G_\gamma)$ for $\gamma > 0$, then as $x \to \infty$:

$$F(x) = 1 - L(x)x^{-1/\gamma}, \qquad L \in RV_0, \tag{7.33}$$

*which is to say that L is **slowly varying at infinity**. This result was derived by Gnedenko, and follows from (7.31) by considering $L(tx)/L(t)$.*
   *Thus, if $F \in D(G_\gamma)$ for $\gamma > 0$, then (7.33) states that F has a **fat tail** in the sense that $1 - F(x)$ effectively decays like a power function. This is a distributional observation often identified in finance applications.*

   Since the criterion that determines if $F \in D(G_\gamma)$ is based on a property of the distribution function $F^n(x)$ of the maximum of a sample of n variates, it seems logical to expect that this property is preserved in the various "tail" conditional distribution functions of $F(x)$. The answer is in the affirmative, as is the converse, and is proved with the criterion of Proposition 7.28.

Corollary 7.30 (Conditional tail distributions of $F \in D(G_\gamma)$ with $\gamma > 0$) *Given a distribution function $F \in D(G_\gamma)$ with $\gamma > 0$, define for $y > 0$:*

- *The **conditional tail distribution** function $F_y(x)$, defined on $x \geq y$ by:*

$$F_y(x) \equiv \frac{F(x) - F(y)}{1 - F(y)}. \tag{7.34}$$

- *The **relative conditional tail distribution** function $\tilde{F}_y(x)$, defined on $x \geq 1$ by:*

$$\tilde{F}_y(x) \equiv \frac{F(xy) - F(y)}{1 - F(y)}. \tag{7.35}$$

Then $F_y, \tilde{F}_y \in D(G_\gamma)$.

   Conversely, if either $F_y \in D(G_\gamma)$ or $\tilde{F}_y \in D(G_\gamma)$ for $\gamma > 0$ and some $y > 0$, then $F \in D(G_\gamma)$.

Proof. *First note that by definition, if $x^* = \infty$ for $F(x)$, then $x^* = \infty$ for F_y and $\tilde{F}_y$.*
   For $x > 0$ and $t \geq y/x$:

$$\frac{1 - F_y(tx)}{1 - F_y(t)} = \frac{1 - F(tx)}{1 - F(t)}. \tag{1}$$

Similarly, for $x > 0$ and $t \geq 1/x$:

$$\frac{1 - \tilde{F}_y(tx)}{1 - \tilde{F}_y(t)} = \frac{1 - F(tyx)}{1 - F(ty)}. \tag{2}$$

Thus F_y and $\tilde{F}_y$ satisfy (7.31) and $F_y, \tilde{F}_y \in D(G_\gamma)$ by Proposition 7.28.
   The converse follows from the identities in (1) and (2). ∎

Example 7.31 (Pareto conditional tail distributions) *If $F(x)$ is the Pareto distribution of (7.17), defined for $x \geq x_0$ by:*

$$F(x) = 1 - \left(\frac{x}{x_0}\right)^{-1/\gamma},$$

then for all t :

$$\frac{1 - F(tx)}{1 - F(t)} = x^{-1/\gamma}.$$

Given $y \geq x_0$:

$$F_y(x) \equiv 1 - \left(\frac{x}{y}\right)^{-1/\gamma},$$

so $F_y(x)$ is Pareto on $x \geq y$.
   Similarly:

$$\tilde{F}_y(x) = 1 - x^{-1/\gamma},$$

so $\tilde{F}_y(x)$ is Pareto on $x \geq 1$.

We present one last result and corollary that improves the above proposition regarding the asymptotic Pareto-like behavior of $F \in D(G_\gamma)$ with $\gamma > 0$. The proposition sharpens the result above by providing a uniform estimate of convergence. This proposition is a special case of **Karamata's Representation theorem**, named for **Jovan Karamata** (1902–1967). This result is also true with modifications for $\gamma < 0$ and $\gamma = 0$, but we do not develop this theory. See **de Hann and Ferreira** (2006).

Remark 7.32 (Pareto and Karamata's representation theorem) *In the special case of Karamata's result where $c(t) \equiv c$ and $g(t) \equiv t\gamma$, the result (7.37) states that for $t > t_0$:*

$$1 - F(t) \equiv c\left(\frac{t}{t_0}\right)^{-1/\gamma}.$$

Thus $F(t)$ is a Pareto distribution.

Proposition 7.33 (Karamata's representation theorem, $F \in D(G_\gamma)$, $\gamma > 0$) *A distribution function $F \in D(G_\gamma)$ with $\gamma > 0$ if and only if there are positive measurable functions $c(t)$ and $g(t)$ with:*

$$\lim_{t \to \infty} c(t) = c \in (0, \infty), \qquad \lim_{t \to \infty} \frac{g(t)}{t} = \gamma, \tag{7.36}$$

so that for all $t \in (t_0, \infty)$ with $t_0 > 0$:

$$F(t) = 1 - c(t) \exp\left[-\int_{t_0}^{t} \frac{ds}{g(s)}\right], \tag{7.37}$$

defined as a Lebesgue integral.
Proof. *If (7.37) is satisfied for a distribution function $F(x)$, then $x^* = \infty$ by (7.36). Further:*

$$\frac{1 - F(tx)}{1 - F(t)} = \frac{c(tx)}{c(t)} \exp\left[-\int_{t}^{tx} \frac{ds}{g(s)}\right].$$

By (7.36), for any $\epsilon > 0$ there is a T so that for $t \geq T$:

$$\gamma - \epsilon \leq \frac{g(t)}{t} \leq \gamma + \epsilon.$$

Hence for $x \geq 1$:

$$\exp\left[-(\gamma - \epsilon)^{-1} \int_t^{tx} \frac{ds}{s}\right] \leq \exp\left[-\int_t^{tx} \frac{ds}{g(s)}\right] \leq \exp\left[-\int_t^{tx} (\gamma + \epsilon)^{-1} \frac{ds}{s}\right].$$

Thus for $t \geq T$, $x \geq 1$:

$$x^{-1/(\gamma-\epsilon)} \leq \exp\left[-\int_t^{tx} \frac{ds}{g(s)}\right] \leq x^{-1/(\gamma+\epsilon)},$$

and since $\frac{c(tx)}{c(t)} \to 1$ for all $x \geq 1$, (7.37) obtains:

$$x^{-1/(\gamma-\epsilon)} \leq \lim_{t\to\infty} \frac{1 - F(tx)}{1 - F(t)} \leq x^{-1/(\gamma+\epsilon)}.$$

As ϵ is arbitrary, $F \in D(G_\gamma)$ with $\gamma > 0$ by Proposition 7.28.
Conversely, assume that $F \in D(G_\gamma)$ with $\gamma > 0$ and define:

$$r(t) = \frac{1 - F(t)}{\int_t^\infty (1 - F(x)) \frac{dx}{x}}. \tag{1}$$

In Proposition 7.35 below it will be proved that the Lebesgue integral in the definition of $r(t)$ is finite for all $t \geq T$ with T to be defined, and in (7.39) we prove that:

$$\lim_{t\to\infty} r(t) \to 1/\gamma, \tag{2}$$

a limit needed below.
Assuming these results, Proposition III.3.39 obtains for $t \geq 1$:

$$\frac{d}{dt}\left(\ln \int_t^\infty (1 - F(x)) \frac{dx}{x}\right) = \frac{r(t)}{t}, \quad a.e.,$$

meaning almost everywhere and defined as outside a set of Lebesgue measure 0. Thus if $t_0 \geq T$, it follows from Proposition III.3.62 and (1) that:

$$\begin{aligned}
-\int_{t_0}^t \frac{r(s)}{s} ds &= \ln \int_{t_0}^\infty (1 - F(x)) \frac{dx}{x} - \ln \int_t^\infty (1 - F(x)) \frac{dx}{x} \\
&= \ln \frac{1 - F(t)}{r(t)} - \ln \frac{1 - F(t_0)}{r(t_0)}.
\end{aligned}$$

Rewriting:

$$\exp\left[-\int_{t_0}^t \frac{r(s)}{s} ds\right] = \frac{1 - F(t)}{r(t)} \frac{r(t_0)}{1 - F(t_0)},$$

and so:

$$1 - F(t) = r(t)\frac{1 - F(t_0)}{d(t_0)} \exp\left[-\int_{t_0}^t \frac{r(s)}{s} ds\right].$$

Defining:

$$c(t) = r(t)\frac{1 - F(t_0)}{r(t_0)}, \quad \frac{1}{g(s)} = \frac{r(s)}{s},$$

yields (7.37).

Finally, the limits in (7.36) follow from Proposition 7.35:

$$c \equiv \lim_{t \to \infty} c(t) = \frac{1}{\gamma} \int_{t_0}^{\infty} (1 - F(x)) \frac{dx}{x} \in (0, \infty),$$

and:

$$\lim_{t \to \infty} \frac{g(t)}{t} = \gamma.$$

∎

The following corollary proves that for $F \in D(G_\gamma)$ with $\gamma > 0$, not only are the conditional distributions of $F(x)$ asymptotically Pareto as proved in Proposition 7.28, but these conditional distributions are bounded by Pareto distributions with arbitrarily close tail indexes of $\frac{1}{\gamma \pm \epsilon}$ if t is large enough.

Corollary 7.34 (Bounding $\frac{1-F(tx)}{1-F(t)}$ by Pareto) *Given a distribution function $F \in D(G_\gamma)$ with $\gamma > 0$ and c defined in (7.36), then for any $\epsilon > 0$ with $\epsilon < c/2$, there exists T' so that for $t \geq T'$ and all $x \geq 1$:*

$$(1 - \epsilon) x^{-1/(\gamma - \epsilon)} \leq \frac{1 - F(tx)}{1 - F(t)} \leq (1 + \epsilon) x^{-1/(\gamma + \epsilon)}. \qquad (7.38)$$

Proof. *Given (7.36), for any $\epsilon > 0$ with $\epsilon < 1$ there is a T' so that for $t \geq T'$:*

$$\gamma - \epsilon \leq \frac{g(t)}{t} \leq \gamma + \epsilon, \qquad c(1 - \epsilon/3) \leq c(t) \leq c(1 + \epsilon/3). \qquad (1)$$

As in the above proof:

$$\frac{1 - F(tx)}{1 - F(t)} = \frac{c(tx)}{c(t)} \exp\left[-\int_t^{tx} \frac{ds}{g(s)} \right],$$

and:

$$x^{-1/(\gamma - \epsilon)} \leq \exp\left[-\int_t^{tx} \frac{ds}{g(s)} \right] \leq x^{-1/(\gamma + \epsilon)}.$$

Thus for $t \geq T'$ and all $x \geq 1$:

$$\frac{1 - \epsilon/3}{1 + \epsilon/3} x^{-1/(\gamma - \epsilon)} \leq \frac{1 - F(tx)}{1 - F(t)} \leq \frac{1 + \epsilon/3}{1 - \epsilon/3} x^{-1/(\gamma + \epsilon)}.$$

The result now follows since $\frac{1+\epsilon/3}{1-\epsilon/3} \leq 1 + \epsilon$ and $\frac{1-\epsilon/3}{1+\epsilon/3} \geq 1 - \epsilon$. ∎

7.2.4 $F \in D(G_\gamma)$, $\gamma > 0$, then $\gamma_H \approx \gamma$

In Example 7.20 it was noted that if $F \in D(G_\gamma)$ is in fact a Pareto distribution with $\gamma > 0$, then the Hill estimator γ_H equals the conditional maximum likelihood estimate for the parameter γ. For general $F \in D(G_\gamma)$ with $\gamma > 0$, Corollary 7.34 proved that the conditional distribution defined for $x > 0$:

$$\Pr[X \leq tx | X > t] = 1 - \frac{1 - F(tx)}{1 - F(t)},$$

could be bounded arbitrarily closely by Pareto distributions for t large. Further, Proposition 7.28 proved that this conditional distribution was asymptotically Pareto as $t \to \infty$.

So it seems only logical to expect that the Hill estimator would:

1. Provide an approximation to the given γ for finite t, meaning when based on sample variates above a given order statistic or quantile;

2. Converge to γ exactly as $t \to \infty$, meaning when based on sample variates above an increasing sequence of quantiles.

We derive this first result in this section, and a version of the second result in the next section.

For the approximation result, the proposition below provides another representation for γ based on an integral of the distribution function F introduced in the proof of Proposition 7.33. This representation will then support the conclusion that the Hill estimator γ_H approximates the exact value of γ for $F \in D(G_\gamma)$ with $\gamma > 0$.

The formula for $\gamma > 0$ in (7.39) generalizes von Mises' condition in the sense that it is valid for all such F without differentiability conditions. On the other hand, von Mises' condition is valid for all γ.

While the proposition below is stated for $\gamma > 0$, it can also be formulated with modifications in the cases of $\gamma < 0$ and $\gamma = 0$. We do not develop this theory, but reference **de Haan and Ferreira** (2006).

Proposition 7.35 (A formula for γ for $F \in D(G_\gamma)$, $\gamma > 0$) *A distribution function $F \in D(G_\gamma)$ with $\gamma > 0$ if and only if $F(x) < 1$ for all $x > 0$, and there exists T so that defined as Lebesgue integrals:*

$$\int_t^\infty (1 - F(x)) \frac{dx}{x} < \infty, \quad t \geq T,$$

and:

$$\lim_{t \to \infty} \frac{\int_t^\infty (1 - F(x)) \frac{dx}{x}}{1 - F(t)} = \gamma. \tag{7.39}$$

Proof. *If $F \in D(G_\gamma)$ with $\gamma > 0$, then $F(x) < 1$ for all $x > 0$ since $x^* = \infty$ by Proposition 7.28.*

By (7.31), for any $\epsilon > 0$ there is a T so that for $t \geq T$:

$$\frac{1 - F(te)}{1 - F(t)} \leq (1 + \epsilon) e^{-1/\gamma} < e^{\epsilon - 1/\gamma}.$$

Since $te^k \geq T$ for $k \geq 0$:

$$\frac{1 - F(te^n)}{1 - F(t)} = \prod_{k=1}^n \frac{1 - F(te^k)}{1 - F(te^{k-1})} \leq e^{n(\epsilon - 1/\gamma)}.$$

Given $x > 1$, let $n \equiv \lceil \ln x \rceil$, the least integer greater than or equal to $\ln x$. Then $\lceil \ln x \rceil < \ln x + 1$, and substituting $x = e^{\ln x}$:

$$\frac{1 - F(tx)}{1 - F(t)} \leq e^{n(\epsilon - 1/\gamma)} \leq e^{\epsilon - 1/\gamma} x^{\epsilon - 1/\gamma}.$$

Consequently, for any $t \geq T$:

$$\frac{1 - F(tx)}{1 - F(t)} \frac{1}{x} \leq e^{\epsilon - 1/\gamma} x^{\epsilon - 1/\gamma - 1},$$

and since dominated by an integrable function when $\epsilon < 1/\gamma$:

$$\int_1^\infty \frac{1 - F(Tx)}{1 - F(T)} \frac{dx}{x} < \infty.$$

This proves by substitution that:

$$\int_T^\infty (1 - F(x)) \frac{dx}{x} < \infty.$$

By Lebesgue's dominated convergence theorem of Proposition III.2.52 and (7.31):

$$\lim_{t \to \infty} \frac{\int_t^\infty (1 - F(x)) \frac{dx}{x}}{1 - F(t)} = \lim_{t \to \infty} \int_1^\infty \frac{1 - F(tx)}{1 - F(t)} \frac{dx}{x}$$

$$= \int_1^\infty x^{-1/\gamma - 1} dx$$

$$= \gamma,$$

which is (7.39).

Conversely given (7.39), define as in the proof of Proposition 7.33:

$$r(t) = \frac{1 - F(t)}{\int_t^\infty (1 - F(y)) \frac{dy}{y}}, \quad t \geq T,$$

and note that with $b(t) = \int_t^\infty (1 - F(y)) \frac{dy}{y}$ for $t \geq T$, Proposition III.3.39 obtains:

$$\frac{r(t)}{t} = -\frac{b'(t)}{b(t)}, \quad a.e.$$

Now $b(t) > 0$ for all such t since $F(x) < 1$ for all $x > 0$ by assumption. Thus $\ln b(t)$ is well defined, monotonic, differentiable a.e. by Proposition III.3.12, and thus $[\ln b(t)]' = \frac{b'(t)}{b(t)}$ a.e. Consequently, Proposition III.3.62 obtains:

$$\int_T^t \frac{r(y)}{y} dy = -\ln \left[\int_t^\infty (1 - F(y)) \frac{dy}{y} \right] + \ln \left[\int_T^\infty (1 - F(y)) \frac{dy}{y} \right],$$

and so by definition of $r(t)$:

$$1 - F(t) = r(t) \int_t^\infty (1 - F(y)) \frac{dy}{y}$$

$$= r(t) \int_T^\infty (1 - F(y)) \frac{dy}{y} \exp\left[-\int_T^t \frac{r(y)}{y} dy \right].$$

With a similar expression for $1 - F(tx)$, and a change of variable:

$$\frac{1 - F(tx)}{1 - F(t)} = \frac{r(tx)}{r(t)} \exp\left[-\int_t^{tx} \frac{r(y)}{y} dy \right]$$

$$= \frac{r(tx)}{r(t)} \exp\left[-\int_1^x \frac{r(ty)}{y} dy \right].$$

Letting $t \to \infty$ obtains from (7.39) that $\frac{r(tx)}{r(t)} \to 1$ and $r(ty) \to 1/\gamma$, and so by Lebesgue's dominated convergence theorem:

$$\lim_{t \to \infty} \frac{1 - F(tx)}{1 - F(t)} = x^{-1/\gamma}.$$

Consequently $F \in D(G_\gamma)$ with $\gamma > 0$ by Proposition 7.28. ∎

With most of the hard work done and the integral formula for γ derived in (7.39), we will now demonstrate that the formula in (7.18) for the Hill estimator γ_H approximates the value of this integral. To this end, we begin with a transformation of the formula in (7.39) into a Riemann-Stieltjes integral of Chapter III.4.

The integral in (7.40) is well-defined over bounded intervals by Proposition III.4.17, since $F(x)$ is increasing and $\ln(x/t)$ is continuous, and this extends to the improper integral as will be addressed in the proof.

Proposition 7.36 (Another formula for γ for $F \in D(G_\gamma), \gamma > 0$) *Given a distribution function $F \in D(G_\gamma)$ with $\gamma > 0$:*

$$\gamma = \lim_{t \to \infty} \frac{\int_t^\infty \ln\left(\frac{x}{t}\right) dF(x)}{1 - F(t)}, \tag{7.40}$$

where the integral is defined as a Riemann-Stieltjes integral.

Proof. *Given $F \in D(G_\gamma)$ with $\gamma > 0$, define $h(x) \equiv 1 - F(x)$ and $k(x) \equiv \ln\left(\frac{x}{t}\right)$ for fixed $t \geq T$, where T is as in Proposition 7.35. Since $k(x)$ is increasing and continuous, define the Riemann-Stieltjes integral of bounded $h(x)$ relative to $k(x)$ over bounded intervals by Proposition III.4.17. Then by Proposition III.4.28, since $k(x)$ is continuously differentiable:*

$$\int_t^N h(x)dk = \int_t^N h(x)k'(x)dx.$$

The integral on the right is defined as a Riemann integral by Proposition III.1.22 since $k'(x)$ is continuous, and $h(x)$ is monotonic and differentiable almost everywhere by Proposition III.3.12, and so continuous almost everywhere.

Substituting for $h(x)$ and $k'(x) = 1/x$ obtains:

$$\int_t^N h(x)dk = \int_t^N (1 - F(x))\frac{dx}{x}.$$

Since the limit as $N \to \infty$ of the integral on the right exists by Proposition 7.35:

$$\lim_{N \to \infty} \int_t^N h(x)dk = \int_t^\infty (1 - F(x))\frac{dx}{x}. \tag{1}$$

Using the integration by parts formula for Riemann-Stieltjes integrals of Proposition III.4.14:

$$\int_t^N h(x)dk = h(N)k(N) - h(t)k(t) - \int_t^N k(x)dh. \tag{2}$$

Now $h(t)$ is finite and $k(t) = 0$ by definition. Also, by (7.38), if $F \in D(G_\gamma)$ with $\gamma > 0$, then for any $\epsilon > 0$ there is a $C_\epsilon = (1 + \epsilon)t^{1/(\gamma+\epsilon)}$ so that:

$$1 - F(N) \leq C_\epsilon N^{-1/(\gamma+\epsilon)},$$

and hence as $N \to \infty$:

$$h(N)k(N) = (1 - F(N))\ln\left(\frac{N}{t}\right) \to 0. \tag{3}$$

Combining (1) − (3) obtains:

$$\int_t^\infty (1 - F(x))\frac{dx}{x} = -\lim_{N \to \infty} \int_t^N k(x)dh = -\int_t^\infty \ln\left(\frac{x}{t}\right) d(1 - F),$$

where the integral on the right is again defined as a Riemann-Stieltjes integral.

By the construction of a Riemann-Stieltjes integral, using $-(1 - F)$ or F as an integrator produces the same result. This plus (7.39) proves (7.40). ∎

Remark 7.37 (γ and Conditional expectation) *The formula in (7.40) can be interpreted in the context of expectations as defined in (4.1).*

To this end, let X be a random variable defined on a probability space $(S, \mathcal{E}, \lambda)$ with distribution function $F \in D(G_\gamma)$ with $\gamma > 0$, and let $t \geq T$ be fixed. Given $F(x)$, this random variable and probability space exist by Proposition 1.4. Define the conditional distribution function $F_t(x)$ for $x \geq t$ by:

$$F_t(x) \equiv \frac{F(x) - F(t)}{1 - F(t)}.$$

Since $F(t)$ is a constant, items 5 and 6 of Proposition III.4.24 obtain:

$$\frac{\int_t^\infty \ln\left(\frac{x}{t}\right) dF}{1 - F(t)} = \int_t^\infty \ln\left(\frac{x}{t}\right) dF_t(x).$$

Then by (4.1):

$$\int_t^\infty \ln\left(\frac{x}{t}\right) dF_t(x) = E\left[\ln\left(\frac{X}{t}\right) \mid X \geq t\right],$$

where by notational convention $E\left[\ln\left(X/t\right)\right]$ of (4.1) is changed as above to emphasize the use of the conditional distribution function $F_t(x)$.

Then by (7.40):

$$\gamma = \lim_{t \to \infty} \int_t^\infty \ln\left(\frac{x}{t}\right) dF_t(x) = \lim_{t \to \infty} \frac{\int_t^\infty \ln\left(\frac{x}{t}\right) dF(x)}{1 - F(t)}, \tag{7.41}$$

and equivalently:

$$\gamma = \lim_{t \to \infty} E\left[\ln\left(\frac{X}{t}\right) \mid X \geq t\right].$$

With γ defined in (7.41) as a limit as $t \to \infty$, an approximation can be achieved for γ by evaluating this expression for large enough t.

Given a sample of random variates $\{X_i\}_{i=1}^n$ with distribution $F \in D(G_\gamma)$ with $\gamma > 0$, and associated order statistics $\{X_{(j)}\}_{j=1}^n$, it follows from Remark 7.37 that for $n - k$ large:

$$\gamma \approx \frac{\int_{X_{(n-k)}}^\infty \ln\left(x/X_{(n-k)}\right) dF}{1 - F(X_{(n-k)})}.$$

This is a nice formula, but for estimation of γ it is not yet useful because $F(x)$ depends on γ and is thus unknown.

The next result states that $\gamma_H \approx \gamma$ when $F \in D(G_\gamma)$ with $\gamma > 0$. The goal of the proof is to approximate the integral in (7.41), which is based on the unknown distribution function F, with an integral based on the empirical distribution function F_n implied by the given sample $\{X_i\}_{i=1}^n$. This empirical distribution function was introduced in (6.44) and assigns a probability of $\frac{1}{n}$ to each variate:

$$F_n(x) = \frac{1}{n} \sum_{j=1}^n \chi_{(-\infty, x]}(X_j),$$

where the characteristic function $\chi_{(-\infty, x]}(X_j)$ equals 1 if $X_j \leq x$ and is 0 otherwise.

By the Glivenko-Cantelli theorem of Proposition 6.60, $\sup_x |F_n(x) - F(x)| \to 0$ with probability 1 as $n \to \infty$. Thus when $X_{(n-k)}$ is large, it seems reasonable to approximate the last expression in (7.41) with $t = X_{(n-k)}$, and integral there becomes:

$$\int_{X_{(n-k)}}^\infty \ln\left(x/X_{(n-k)}\right) dF \approx \int_{X_{(n-k)}}^\infty \ln\left(x/X_{(n-k)}\right) dF_n.$$

The informal justification for this approximation is to choose n so that $\sup_x |F_n(x) - F(x)| < \epsilon$, since then $\sup_x |\Delta F_n - \Delta F| < 2\epsilon$ for any term in the defining Riemann-Stieltjes summations. However while intuitively plausible, to be made rigorous requires another approximation because $\ln\left(x/X_{(n-k)}\right)$ is unbounded.

In the next section it will be demonstrated that under more clearly defined conditions that $\gamma_H \to_1 \gamma$. That is, γ_H converges with probability 1 to γ, as defined in (6.31).

Proposition 7.38 ($\gamma_H \approx \gamma$ **for** $F \in D(G_\gamma)$, $\gamma > 0$) *Given a distribution function $F \in D(G_\gamma)$ with $\gamma > 0$, and random sample $\{X_i\}_{i=1}^n$ of X defined on $(S, \mathcal{E}, \lambda)$ with distribution F and order statistics $\{X_{(j)}\}_{j=1}^n$, then for n and $X_{(n-k)}$ large:*

$$\gamma_H^{(k,n)} \equiv \frac{1}{k}\sum_{j=0}^{k-1}\left[\ln\left(X_{(n-j)}\right) - \ln\left(X_{(n-k)}\right)\right] \approx \gamma. \tag{7.42}$$

Proof. *Given $\epsilon > 0$, by (7.40) choose T so that for $t \geq T$:*

$$\left|\frac{\int_t^\infty \ln\left(\frac{x}{t}\right) dF(x)}{1 - F(t)} - \gamma\right| < \epsilon.$$

For any such t choose $t'(t) \geq t$ so that:

$$\frac{\int_{t'(t)}^\infty \ln\left(\frac{x}{t}\right) dF(x)}{1 - F(t)} < \epsilon,$$

which is possible since the numerator converges to 0 as $t' \to \infty$. Thus for $t \geq T$:

$$\left|\frac{\int_t^{t'(t)} \ln\left(\frac{x}{t}\right) dF(x)}{1 - F(t)} - \gamma\right| < 2\epsilon. \tag{1}$$

For given ϵ, choose N by the Glivenko-Cantelli theorem so that with probability 1, for $n \geq N$:

$$\sup_x |F_n(x) - F(x)| < \epsilon. \tag{2}$$

Given $\{X_i\}_{i=1}^n$ with $n \geq N$ and associated $\{X_{(j)}\}_{j=1}^n$, assume that n and $n - k$ are large enough so that $X_{(n-k)} \geq T$ and $X_{(n)} \geq t'(X_{(n-k)})$. This is possible since $X_{(n)} \to \infty$ and $X_{(n-k_n)} \to \infty$ with probability 1 for $n \to \infty$ and $k_n/n \to 0$ by Proposition 6.52, since $x^* = \infty$ by Proposition 7.28. Thus by (1):

$$\left|\frac{\int_{X_{(n-k)}}^{X_{(n)}} \ln\left(\frac{x}{X_{(n-k)}}\right) dF(x)}{1 - F(X_{(n-k)})} - \gamma\right| < 2\epsilon. \tag{3}$$

Now by Definition III.4.3 of the Riemann-Stieltjes integral, given this ϵ there exists δ so that for any partition $\{x_i\}_{i=0}^m$ of $[X_{(n-k)}, X_{(n)}]$ of mesh size δ or smaller:

$$\left|\int_{X_{(n-k)}}^{X_{(n)}} \ln\left(\frac{x}{X_{(n-k)}}\right) dF(x) - \sum_{i=1}^m \ln\left(\frac{\widetilde{x}_i}{X_{(n-k)}}\right) \Delta F_i\right| < \epsilon, \tag{4}$$

where $\Delta F_i = F(x_i) - F(x_{i-1})$ and $\widetilde{x}_i \in [x_{i-1}, x_i]$ is arbitrary. By (2), we can replace ΔF_i with $\Delta F_{n,i}$:

$$\left|\sum_{i=1}^m \ln\left(\frac{\widetilde{x}_i}{X_{(n-k)}}\right) \Delta F_i - \sum_{i=1}^m \ln\left(\frac{\widetilde{x}_i}{X_{(n-k)}}\right) \Delta F_{n,i}\right| \leq 2\epsilon \sum_{i=1}^m \ln\left(\frac{\widetilde{x}_i}{X_{(n-k)}}\right)$$

$$\leq 2\epsilon m \ln\left(\frac{X_{(n)}}{X_{(n-k)}}\right). \tag{5}$$

Combining (4) and (5), and reflecting the definition of F_n in the associated Riemann-Stieltjes integral obtains:

$$\left| \int_{X_{(n-k)}}^{X_{(n)}} \ln\left(\frac{x}{X_{(n-k)}}\right) dF(x) - \frac{1}{n}\sum_{j=0}^{k-1}\ln\left(\frac{X_{(n-j)}}{X_{(n-k)}}\right) \right|$$

$$= \left| \int_{X_{(n-k)}}^{X_{(n)}} \ln\left(\frac{x}{X_{(n-k)}}\right) dF(x) - \frac{k}{n}\gamma_H^{(k,n)} \right|$$

$$\leq \epsilon\left[1 + 2m\ln\left(\frac{X_{(n)}}{X_{(n-k)}}\right)\right].$$

Dividing by $1 - F(X_{(n-k)})$ and noting that $1 - F_n(X_{(n-k)}) = \frac{k}{n}$ obtains:

$$\left| \frac{\int_{X_{(n-k)}}^{X_{(n)}} \ln\left(\frac{x}{X_{(n-k)}}\right) dF(x)}{1 - F(X_{(n-k)})} - \frac{k}{n}\frac{\gamma_H^{(k,n)}}{1 - F(X_{(n-k)})} \right|$$

$$= \left| \frac{\int_{X_{(n-k)}}^{X_{(n)}} \ln\left(\frac{x}{X_{(n-k)}}\right) dF(x)}{1 - F(X_{(n-k)})} - \gamma_H^{(k,n)}\frac{1 - F_n(X_{(n-k)})}{1 - F(X_{(n-k)})} \right|$$

$$\leq \epsilon\left[1 + 2m\ln\left(\frac{X_{(n)}}{X_{(n-k)}}\right)\right] \Big/ \left(1 - F(X_{(n-k)})\right). \tag{6}$$

Let:

$$I \equiv \frac{\int_{X_{(n-k)}}^{X_{(n)}} \ln\left(\frac{x}{X_{(n-k)}}\right) dF(x)}{1 - F(X_{(n-k)})}.$$

Comparing (3) and (6) obtains that for any $\epsilon > 0$, large samples will obtain that I is within 2ϵ of γ, and within $M\epsilon$ of a multiple of $\gamma_H^{(k,n)}$, where this multiple is approximately 1 by the Glivenko-Cantelli theorem. This supports the assertion of an approximation for large samples, but it is not clear here that M is uniformly bounded, and thus we cannot assert more on the quality of this approximation in the limit. ∎

7.2.5 $F \in D(G_\gamma), \gamma > 0$, then $\gamma_H \to_1 \gamma$

For the result of Proposition 7.38, that $\gamma_H \approx \gamma$ for $F \in D(G_\gamma)$ with $\gamma > 0$, it was seen that this approximation required the sample size n to be large, but also that the order statistics $X_{(n-k)}$ and $X_{(n)}$ used in γ_H be large. The requirements followed from the Glivenko-Cantelli theorem for n, and Proposition 6.52 because it was necessary for both $X_{(n)} \to \infty$ and $X_{(n-k)} \to \infty$. It then followed that $k = k_n$ was required to satisfy $k_n/n \to 0$ as $n \to \infty$. In other words, the lowest order statistic used in the Hill estimator must be based on a quantile $q_{k_n} \equiv 1 - k/n$ with the property that $q_{k_n} \to 1$, and thus $k \to \infty$ as $n \to \infty$.

The proposition below will formalize the result that if $k, n \to \infty$ and $k/n \to 0$, then $\gamma_H \to_1 \gamma$.

In order to prove this section's result, a more general version of **Karamata's representation theorem** is needed. In the special case of Proposition 7.33, this theorem stated that for $F \in D(G_\gamma)$ with $\gamma > 0$, that $1 - F(t)$ could be represented as an integral in (7.37), and with functions with limiting properties summarized in (7.36). As noted there, this result also applies to $F \in D(G_\gamma)$ with $\gamma < 0$ or $\gamma = 0$, but with appropriate modifications.

But far beyond this, Karamata's representation theorem applies not only to distribution functions, but to all functions f that are **regularly varying at infinity with index** $\alpha \in \mathbb{R}$. As noted in (7.29) of Definition 7.25, this terminology means that for all $x \geq x_0 > 0$:

$$\lim_{t \to \infty} \frac{f(tx)}{f(t)} = x^{\alpha},$$

and is denoted $f \in RV_{\alpha}$.

We again need this general result only for the case $\alpha > 0$, and it will be applied to the function $U(t) \equiv \left(\frac{1}{1-F}\right)^{*}(t)$, the left-continuous inverse of $1/(1 - F)$ which is not a distribution function. This representation theorem will then provide the needed uniform estimate of $U(tx)/U(t)$ for $t \in (t_0, \infty)$ and all $x \geq 1$, just as Proposition 7.33 supported such a uniform estimate of $(1 - F(tx))/(1 - F(t))$ in (7.38) of Corollary 7.34.

We state without proof the general result though again will only require the case $\alpha > 0$.

Proposition 7.39 (Karamata's Representation theorem) *A function f is regularly varying at infinity with index α if and only if there are positive measurable functions c and g, so that for all $t \in (t_0, \infty)$ with $t_0 > 0$:*

$$f(t) = c(t) \exp\left[\int_{t_0}^{t} \frac{h(s)}{s} ds\right],\tag{7.43}$$

where

$$\lim_{t \to \infty} c(t) = c \in (0, \infty), \qquad \lim_{t \to \infty} h(t) = \alpha.\tag{7.44}$$

Proof. *See de Haan and Ferreira, Theorem B.1.6.* ■

The proof of the needed corollary result now follows the proof of Corollary 7.34.

Corollary 7.40 (Bounding $\frac{f(tx)}{f(t)}$ by Pareto) *Given a function $f(x)$ that is regularly varying at infinity with index α, then for any $\epsilon > 0$ with $\epsilon < 1$ there is a T so that for $t \geq T$ and all $x \geq 1$:*

$$(1 - \epsilon) x^{(\alpha - \epsilon)} \leq \frac{f(tx)}{f(t)} \leq (1 + \epsilon) x^{(\alpha + \epsilon)}.\tag{7.45}$$

Proof. *If (7.43) is satisfied for $t > t_0$, then for $x > 1$:*

$$\frac{f(tx)}{f(t)} = \frac{c(tx)}{c(t)} \exp\left[\int_{t}^{tx} \frac{h(s)}{s} ds\right].$$

By the limits in (7.44), for any $\epsilon > 0$ with $\epsilon < 1$ there is a T so that for $t \geq T$:

$$\alpha - \epsilon \leq h(t) \leq \alpha + \epsilon, \qquad c(1 - \epsilon/3) \leq c(t) \leq c(1 + \epsilon/3).$$

Hence for $x > 1$:

$$\exp\left[(\alpha - \epsilon) \int_{t}^{tx} \frac{ds}{s}\right] \leq \exp\left[\int_{t}^{tx} \frac{h(s)}{s} ds\right] \leq \exp\left[\int_{t}^{tx} (\alpha + \epsilon) \frac{ds}{s}\right],\tag{1}$$

and equivalently:

$$x^{(\alpha - \epsilon)} \leq \exp\left[\int_{t}^{tx} \frac{h(s)}{s} ds\right] \leq x^{(\alpha + \epsilon)}.\tag{2}$$

The bounds in (1) and (2) then obtain:

$$\frac{1 - \epsilon/3}{1 + \epsilon/3} x^{(\alpha - \epsilon)} \leq \frac{f(tx)}{f(t)} \leq \frac{1 + \epsilon/3}{1 - \epsilon/3} x^{(\alpha + \epsilon)},$$

and the result in (7.45) then follows from $\frac{1+\epsilon/3}{1-\epsilon/3} \leq 1 + \epsilon$ and $\frac{1-\epsilon/3}{1+\epsilon/3} \geq 1 - \epsilon$. ■

We now turn to the result that if $F \in D(G_\gamma)$ with $\gamma > 0$, then recalling Definition 6.44, the Hill estimator γ_H converges to γ with probability 1 as $n \to \infty$. To achieve this convergence result we will also require that $k \to \infty$, and that the implied quantile $q_{k_n} \equiv 1 - k/n$ of the base order statistic $X_{(n-k)}$ converge to 1, or equivalently $k/n \to 0$. This assures that $n - k \to \infty$, and thus the base variate in the Hill calculation satisfies $X_{(n-k)} \to \infty$ by Proposition 6.52.

We identify the probability space on which these variates are defined to give meaning to the associated probability statements.

Proposition 7.41 (Hill Estimator: $\gamma_H^{(k,n)} \to_1 \gamma$.**)** *Let* $\{X_i\}_{i=1}^\infty$ *be independent and identically distributed random variables defined on a probability space* $(\mathcal{S}, \mathcal{E}, \lambda)$ *with distribution function* $F \in D(G_\gamma)$ *with* $\gamma > 0$, *and let* $\gamma_H^{(k,n)}$ *denote the Hill estimator defined in (7.18) with given* k, n.

Then as $k, n \to \infty$ *and* $k/n \to 0$, *the estimator* $\gamma_H^{(k,n)}$ *converges to* γ *with probability 1 :*

$$\gamma_H^{(k,n)} \to_1 \gamma. \tag{7.46}$$

Proof. *We first apply Corollary 7.40 to the left-continuous inverse function* $U(t) \equiv \left(\frac{1}{1-F}\right)^*(t)$, *which is regularly varying at infinity with index* γ *by (7.27). This obtains by (7.45) that for any* $\epsilon > 0$ *there is a* $T \equiv T(\epsilon)$ *so that for* $t \geq T$ *and all* $x \geq 1$:

$$(1 - \epsilon) x^{(\gamma - \epsilon)} \leq \frac{U(tx)}{U(t)} \leq (1 + \epsilon) x^{(\gamma + \epsilon)},$$

and so:

$$\ln(1 - \epsilon) + (\gamma - \epsilon) \ln x \leq \ln U(tx) - \ln U(t) \leq \ln(1 + \epsilon) + (\gamma + \epsilon) \ln x. \tag{1}$$

Now $U(t) \equiv F^*(1 - 1/t)$ *by (7.25), with* F^* *Borel measurable by Proposition II.3.16. Hence by Proposition II.4.9 in the Chapter 5 introduction,* $\{U(Y_j)\}_{i=1}^n$ *are independent and have distribution function* $F(x)$ *if* $\{1 - 1/Y_j\}_{i=1}^n$ *are independent and uniformly distributed on* $[0, 1]$. *By this same result, if* $\{Y_j\}_{i=1}^n$ *are independent random variables with continuous distribution function* $G(y) = 1 - 1/y$ *on* $y \geq 1$, *then* $\{G(Y_j)\}_{i=1}^n = \{1 - 1/Y_j\}_{i=1}^n$ *will be independent and uniformly distributed on* $(0, 1)$. *Hence given such* G-*distributed variates,* $\{U(Y_j)\}_{i=1}^n$ *will be independent and have distribution function* $F(x)$.

Thus γ_H *can be defined as in (7.19) in terms of the order statistics of a sample* $\{U(Y_j)\}_{i=1}^n$:

$$\gamma_H^{(k,n)} = \frac{1}{k} \sum_{j=0}^{k-1} \left[\ln U(Y_{(n-j)}) - \ln U(Y_{(n-k)})\right], \tag{2}$$

where $\{Y_j\}_{i=1}^n$ *is a sample with distribution function* $G(y) = 1 - 1/y$. *Note that for this representation that the order statistics of* $\{Y_j\}_{i=1}^n$ *determine those of* $\{U(Y_j)\}_{i=1}^n$ *since* $U(t)$ *is an increasing function.*

To use the bounds in (1) for $\ln U(tx) - \ln U(t)$, *let* $t = Y_{(n-k)}$. *Then by (6.40) of Proposition 6.52, if* $k/n \to 0$ *then* $Y_{(n-k)} \to \infty$ *with probability 1. So it follows that* $Y_{(n-k)} \geq T(\epsilon)$ *for* k, n *sufficiently large. With* $x \equiv Y_{(n-j)}/Y_{(n-k)}$, *(1) and (2) obtain that:*

$$\ln(1 - \epsilon) + (\gamma - \epsilon) Z_{k,n} \leq \gamma_H^{(k,n)} \leq \ln(1 + \epsilon) + (\gamma + \epsilon) Z_{k,n}, \tag{3}$$

where:

$$Z_{k,n} \equiv \frac{1}{k} \sum_{j=0}^{k-1} \ln \left[\frac{Y_{(n-j)}}{Y_{(n-k)}}\right].$$

We now show that as $k, n \to \infty$ and $k/n \to 0$, that $Z_{k,n}$ converges to 1 with probability 1:

$$Z_{k,n} \to_1 1. \tag{4}$$

As Y has distribution function $G(y) = 1 - 1/y$ on $y \geq 1$, the random variable $X \equiv \ln Y$ has a standard exponential distribution for $x \geq 0$:

$$F_X(x) = \Pr[\ln Y \leq x] = \Pr[Y \leq e^x] = 1 - e^{-x}.$$

This assures that $X_{(n-k)} \to \infty$ with probability 1, and that:

$$Z_{k,n} = \frac{1}{k} \sum_{j=0}^{k-1} \left[X_{(n-j)} - X_{(n-k)} \right].$$

Hence (4) follows by Proposition 6.22.

Summarizing, we have proved the following. Given $\epsilon > 0$ there exists T, so that given G-distributed variates $\{Y_j\}_{i=1}^n$ with order statistic $Y_{(n-k)} \geq T(\epsilon)$, then as $k, n \to \infty$ and $k/n \to 0$:

$$\ln(1 - \epsilon) + (\gamma - \epsilon) \leq \bar{\gamma}_H \leq \ln(1 + \epsilon) + (\gamma + \epsilon). \tag{5}$$

with probability 1. Here $\bar{\gamma}_H$ denotes either the limit supremum or limit infimum of all Hill estimators $\gamma_H^{(k,n)}$ with $X_{(n-k)} \equiv \ln Y_{(n-k)} \geq \ln T(\epsilon)$, and where $k, n \to \infty$ and $k/n \to 0$.

Given $\epsilon = 1/m$, let $A_m \subset \mathcal{S}$ be the set with $\lambda(A_m) = 1$ on which (5) is satisfied. For any m, there exists k, n so that $Y_{(n-k)} \geq T(1/m)$, and thus for all m there exists $A_m \subset \mathcal{S}$ with $\lambda(A_m) = 1$ on which:

$$\ln(1 - 1/m) + (\gamma - 1/m) \leq \bar{\gamma}_H \leq \ln(1 + 1/m) + (\gamma + 1/m). \tag{6}$$

Letting $A \equiv \bigcap_{m=1}^\infty A_m$ obtains a set with $\lambda(A) = 1$ for which (6) is satisfied for all m. Thus with probability 1:

$$\limsup \gamma_H^{(k,n)} = \liminf \gamma_H^{(k,n)} = \gamma,$$

and so $\gamma_H^{(k,n)} \to_1 \gamma$. ∎

By Proposition II.5.21, the above result:

$$\gamma_H^{(k,n)} \to_1 \gamma,$$

implies convergence in probability defined in (6.29):

$$\gamma_H^{(k,n)} \to_P \gamma.$$

This latter result has a converse, which we state without proof.

Proposition 7.42 $(\gamma_H^{(k,n)} \to_P \gamma, \gamma > 0 \Rightarrow F \in D(G_\gamma))$ *Let $\{X_i\}_{i=1}^\infty$ be independent and identically distributed random variables defined on a probability space $(\mathcal{S}, \mathcal{E}, \lambda)$ with distribution function F, and $\gamma_H^{(k,n)}$ the Hill estimator defined in (7.18) with given k, n. Assume that there exists a sequence $k_n \to \infty$ with $k_n/n \to 0$ and $k_{n+1}/k_n \to 1$ as $n \to \infty$, so that the estimator $\gamma_H^{(k,n)}$ converges in probability to a constant $\gamma > 0$:*

$$\gamma_H^{(k,n)} \to_P \gamma.$$

Then $F \in D(G_\gamma)$.
Proof. *See de Haan and Ferreira, Theorem 3.2.4.* ∎

7.2.6   Asymptotic Normality of the Hill Estimator

Although we do not develop this theory here, it is known that under additional assumptions on the distribution function $F \in D(G_\gamma)$ with $\gamma > 0$, that the Hill estimator is asymptotically normally distributed as was the case for the Hill estimator of the Pareto distribution in Proposition 7.23.

Recall that by Corollary II.9.35 that if $F \in D(G_\gamma)$ and $x > 0$, then as $t \to \infty$:

$$\frac{U(tx) - U(t)}{a(t)} \to \frac{x^\gamma - 1}{\gamma},$$

with $U(t) \equiv \left(\frac{1}{1-F}\right)^*(t)$, the left-continuous inverse of $1/(1-F)$. Here, the constant c in the earlier formula is integrated into the definition of $a(t)$, and so here $a(t)) \equiv a_c(t)$. The right hand limit is defined to be $\ln x$ when $\gamma = 0$, which equals $\lim_{\gamma \to 0} (x^\gamma - 1)/\gamma$.

In order to obtain the asymptotic normality result noted above, this distribution function F must satisfy an additional assumption known as a **second-order condition**, which provides information on the rate of convergence in the above limit.

Definition 7.43 (Second-order condition) *A distribution function $F \in D(G_\gamma)$ satisfies a second-order condition if:*

1. *There is a function $A(t)$ with:*

$$\lim_{t \to \infty} A(t) = 0,$$

   and there exists T so that $A(t)$ does not change sign for $t \geq T$;

2. *There is a function $H(x)$ so that as $t \to \infty$:*

$$\left[\frac{U(tx) - U(t)}{a(t)} - \frac{x^\gamma - 1}{\gamma}\right] \Big/ A(t) \to H(x). \qquad (7.47)$$

By (7.26), recalling as above that here $a(t) \equiv a_c(t)$, this is equivalent to:

$$\left[\frac{U(tx)}{U(t)} - x^\gamma\right] \Big/ A(t) \to H(x).$$

For well-definedness, it is required that $H(x)$ is not a multiple of $\frac{x^\gamma - 1}{\gamma}$.

Analogous to the development surrounding the Fisher-Tippett-Gnedenko theorem, it turns out that when such a function $H(x)$ exists, it must have a well defined structure. For example when $\gamma > 0$, the assumption underlying the above development of the Hill estimator, it is the case that any such $H(x)$ satisfies:

$$H(x) = x^\gamma \frac{x^\rho - 1}{\rho},$$

where $\rho \leq 0$. When $\rho = 0$:

$$H(x) \equiv x^\gamma \ln x,$$

which is equal to $x^\gamma \lim_{\rho \to 0} (x^\rho - 1)/\rho$.

The significance of the parameter ρ is that it is then the case that $A(t) \in RV_\rho$, meaning that $A(t)$ is regularly varying at infinity with index ρ.

We then have the following result.

Proposition 7.44 (Hill Estimator: Asymptotic normality) *Let $F \in D(G_\gamma)$ with $\gamma > 0$ and assume that F satisfies a second order condition where $\rho \leq 0$. Let $F_{\gamma'_H}$ denote the distribution function of the normalized Hill estimator:*

$$\gamma'_H = \frac{\gamma_H - \gamma}{1/\sqrt{k}}.$$

If:

$$\lambda \equiv \lim_{n \to \infty} \sqrt{k} A\left(\frac{n}{k}\right) < \infty,$$

as $k, n \to \infty$ and $k/n \to 0$, then:

$$F_{\gamma'_H} \Rightarrow N\left(\frac{\lambda}{1-\rho}, \gamma^2\right), \tag{7.48}$$

where $N\left(\lambda/(1-\rho), \gamma^2\right)$ denotes the normal distribution with mean $\frac{\lambda}{1-\rho}$ and variance γ^2.
Proof. *See **de Haan and Ferreira**, Theorem 3.2.5.* ∎

Remark 7.45 (On λ) *Since $n/k \to \infty$, it follows that $A(n/k) \to 0$ by the definition of second order condition. However, the assumption that $k/n \to 0$ implies that k/n may approach 0 at any rate as a function of n, and it is thus also the case that n/k may approach ∞ at any rate as a function of n.*

Consequently, the value of the parameter $\lambda = \lim_{n\to\infty} \sqrt{k_n} A(n/k_n)$ can depend on the actual sequence $\{k_n\}$ of parameters used, and need not be finite.

7.2.7 The Pickands-Balkema-de Haan Theorem: $\gamma > 0$

Both the **Fisher-Tippett-Gnedenko theorem** and the **Pickands-Balkema-de Haan theorem** address the question of the existence of a limiting distribution, defined relative to the "tail" of a distribution function $F(x)$ of a random variable X. The former theorem investigates $F^n(x)$, the distribution function of $M_n = \max_{m\leq n}\{X_m\}$ for independent $\{X_m\}_{m=1}^n$, and investigates a key question related to the limiting distribution of $F^n(x)$ as $n \to \infty$:

- If there exists sequences $\{a_n\}_{n=1}^\infty$ and $\{b_n\}_{n=1}^\infty$ where $a_n > 0$ for all n, and a nondegenerate distribution function $G(x)$ so that $F^n(a_n x + b_n) \Rightarrow G(x)$, what can be said about G?

The answer provided by this theorem is Proposition 7.15, which states that if such sequences and distribution exist, so $F \in D(G)$ in the notation of **domains of attraction**, then there are real constants $A > 0$, B, and γ so that $G(x) = G_\gamma(Ax + B)$ with $G_\gamma(x)$ defined for $\gamma \neq 0$ by:

$$G_\gamma(x) = \exp\left(-(1+\gamma x)^{-1/\gamma}\right), \qquad 1 + \gamma x \geq 0.$$

When $\gamma = 0$, $G_\gamma(x)$ is defined on $\mathbb{R}$:

$$G_0(x) \equiv \exp\left(-e^{-x}\right).$$

Proposition 7.28 then provided a characterization of such $F \in D(G_\gamma)$ for $\gamma > 0$ in (7.31), that for $x > 0$:

$$\lim_{t\to\infty} \frac{1 - F(tx)}{1 - F(t)} = x^{-1/\gamma}.$$

This is restated in an even more descriptive way in (7.33), that if $F \in D(G_\gamma)$ for $\gamma > 0$, then as $x \to \infty$:

$$F(x) = 1 - L(x)x^{-1/\gamma}, \qquad L \in RV_0,$$

where $L \in RV_0$ means that L is **slowly varying at infinity** as in Definition 7.25.

The **Pickands-Balkema-de Haan theorem** investigates another "tail" distribution, specifically the **conditional probability distribution of exceedances**, and the analysis underlying this result is often referred to as the **peaks over threshold** method. This investigation was initiated in Book II, where in Proposition II.9.38 was proved:

Proposition 7.46 (Pickands-Balkema-de Haan theorem I) *If $F \in D(G_\gamma)$ for any γ, then:*

- *For all x with $x > -1/\gamma$ when $\gamma \geq 0$, where $-1/0 \equiv -\infty$, or,*

- *For $0 \leq x < -1/\gamma$ when $\gamma < 0$:*

$$\lim_{t \to x^*} \frac{1 - F(t + xh_a(t))}{1 - F(t)} = \begin{cases} (1 + \gamma x)^{-1/\gamma}, & \gamma \neq 0, \\ \exp(-x), & \gamma = 0. \end{cases}$$

Here, x^* is defined in (6.39), and:

$$h_a(t) \equiv a\left(\frac{1}{1 - F(t)}\right),$$

with $a(t) \equiv a_c(t)$ is defined in terms of the normalizing function in (7.23).

Since:

$$1 - \frac{1 - F(t + xh_a(t))}{1 - F(t)} = \frac{F(t + xh_a(t)) - F(t)}{1 - F(t)},$$

this result can also be expressed as a conditional probability statement. For example with $\gamma \neq 0$:

$$\lim_{t \to x^*} \Pr\left[X \leq t + xh_a(t) | X > t\right] = 1 - (1 + \gamma x)^{-1/\gamma}.$$

Thus for fixed t "large" relative to x^* :

$$\Pr\left[X \leq t + y | X > t\right] \approx 1 - \left(1 + \frac{\gamma}{h_a(t)}y\right)^{-1/\gamma}. \tag{7.49}$$

These limiting and approximating distributions are examples of the **generalized Pareto distribution function** $H_{\gamma,0,\beta}(x)$, introduced in Definition II.9.39.

Definition 7.47 (Generalized Pareto distribution) *The distribution function $H_{\gamma,t,\beta}(x)$ defined for $\gamma \neq 0$ and $\beta > 0$ by:*

$$H_{\gamma,t,\beta}(x) \equiv 1 - \left(1 + \frac{\gamma}{\beta}(x - t)\right)^{-1/\gamma}, \tag{7.50}$$

*is called a **generalized Pareto distribution**, abbreviated **GPD**. When $\gamma = 0$, $H_{0,t,\beta}(x)$ is defined as the limit of $H_{\gamma,t,\beta}(x)$ as $\gamma \to 0$:*

$$H_{0,t,\beta}(x) \equiv 1 - \exp\left(-\frac{1}{\beta}(x - t)\right). \tag{7.51}$$

The distribution $H_{\gamma,t,\beta}(x)$ is defined for $x \geq t$ when $\gamma \geq 0$, and for $t \leq x \leq t - \beta/\gamma$ when $\gamma < 0$.

Remark 7.48 *For the result below, we will primarily be interested in* $H_{\gamma,0,\beta}(x)$ *with* $\gamma > 0$:

$$H_{\gamma,0,\beta}(x) \equiv 1 - \left(1 + \frac{\gamma}{\beta}x\right)^{-1/\gamma}, \tag{7.52}$$

but introduced the more general notation because it is commonly cited. Our interest in $H_{\gamma,0,\beta}(x)$ *with* $\gamma > 0$ *is motivated by (7.49).*

Note that for $\gamma > 0$:

$$H_{\gamma,t,\gamma t}(x) \equiv 1 - \left(\frac{x}{t}\right)^{-1/\gamma}, \qquad x \geq t,$$

and:

$$H_{\gamma,0,\gamma t}(x) \equiv 1 - \left(1 + \frac{x}{t}\right)^{-1/\gamma}, \qquad x \geq 0,$$

representing two common parametrizations for a standard **Pareto distribution.**

It is common to represent the exponential index by α, *and so* $\alpha = 1/\gamma > 0$.

The asymptotic result of Proposition II.9.38 for the conditional distribution function can be improved as stated in Proposition II.9.44, but there without proof. With the aid of Corollary 7.34 to Karamata's representation theorem, this earlier result can be proved in the case $\gamma > 0$ in which case $x^* = \infty$ by Proposition 7.28.

Proposition 7.49 (Pickands-Balkema-de Haan theorem II) *Assume that* $F(x)$ *is in the* **domain of attraction** *of* G_γ, $F \in \mathcal{D}(G_\gamma)$ *with* $\gamma > 0$. *Given* $t \geq 0$, *define the conditional distribution function* $F_t(y)$ *for* $y \geq 0$ *by:*

$$F_t(y) \equiv \Pr\left[X \leq t + y | X > t\right] = \frac{F(t+y) - F(t)}{1 - F(t)}.$$

Then the approximation in (7.49) is uniform in y *in the sense that there exists a positive function* $\beta(t)$, *so that:*

$$\lim_{t \to \infty} \sup_{0 \leq y < \infty} \left| F_t(y) - H_{\gamma,0,\beta(t)}(y) \right| = 0. \tag{7.53}$$

Further, (7.53) is true with $\beta(t) \equiv \gamma t$, *so* $H_{\gamma,0,\beta(t)}(y) \equiv H_{\gamma,0,\gamma t}(y)$, *and thus* $F_t(y)$ *is asymptotically Pareto:*

$$\Pr\left[X \leq t + y | X > t\right] \to 1 - \left(1 + \frac{y}{t}\right)^{-1/\gamma} \quad \text{as } t \to \infty. \tag{7.54}$$

Further, the error in this approximation converges to 0 uniformly in $y \geq 0$.

If $\beta(t)$ *is any other function which satisfies (7.53), then as* $t \to \infty$:

$$\frac{\beta(t)}{\gamma t} \to 1,$$

and thus $\beta(t) \equiv \gamma t$ *is asymptotically unique.*

Proof. *By Corollary 7.34, given a distribution function* $F \in \mathcal{D}(G_\gamma)$ *with* $\gamma > 0$ *and* c *defined in (7.36), then for any* $\epsilon > 0$ *with* $\epsilon < c/2$, *there exists* $T = T(\epsilon)$ *so that for* $t \geq T$ *and all* $x \geq 1$:

$$(1 - \epsilon)x^{-1/(\gamma-\epsilon)} \leq \frac{1 - F(tx)}{1 - F(t)} \leq (1 + \epsilon)x^{-1/(\gamma+\epsilon)}.$$

Writing $t + y = t(1 + y/t)$, $y \geq 0$, then for $t \geq T$:

$$(1 - \epsilon)\,(1 + y/t)^{-1/(\gamma - \epsilon)} \leq \frac{1 - F(t + y)}{1 - F(t)} \leq (1 + \epsilon)\,(1 + y/t)^{-1/(\gamma + \epsilon)}.$$

Hence bounds for the difference between $F_t(y)$ and $H_{\gamma,0,\beta(t)}(y)$ are:

$$(1 - \epsilon)\left(1 + \frac{y}{t}\right)^{-1/(\gamma - \epsilon)} - \left(1 + \frac{\gamma}{\beta(t)}y\right)^{-1/\gamma}$$

$$\leq \frac{1 - F(t + y)}{1 - F(t)} - \left(1 + \frac{\gamma}{\beta(t)}y\right)^{-1/\gamma} \tag{1}$$

$$\leq (1 + \epsilon)\left(1 + \frac{y}{t}\right)^{-1/(\gamma + \epsilon)} - \left(1 + \frac{\gamma}{\beta(t)}y\right)^{-1/\gamma}.$$

The proof of (7.53) will be completed by proving that for $\beta(t) \equiv \gamma t$, that the supremum in y of both these bounds converges to 0 as $t \to x^*$.

To this end we investigate the upper bound and leave the lower bound as an exercise. With $\beta(t) \equiv \gamma t$, the upper bound becomes:

$$M(y) \equiv (1 + \epsilon)\left(1 + \frac{y}{t}\right)^{-1/(\gamma + \epsilon)} - \left(1 + \frac{y}{t}\right)^{-1/\gamma}.$$

Letting $w = y/t$, $a \equiv 1/(\gamma + \epsilon)$ and $\delta \equiv \epsilon/\gamma$ obtains for $0 \leq w < \infty$:

$$M(wt) = (1 + \gamma\delta)(1 + w)^{-a} - (1 + w)^{-a(1+\delta)}.$$

For any t,

$$\sup_{0 \leq y < \infty} M(y) = \sup_{0 \leq w < \infty} M(wt).$$

Now $M(0) = \gamma\delta$, $M(\infty) = 0$, $M(wt) \geq 0$, and $M(wt)$ can be differentiated in w to reveal that $M'(w) \leq 0$ for all w, and this obtains:

$$0 \leq M(w) \leq \gamma\delta = \epsilon.$$

Thus, for all $t \geq T(\epsilon)$:

$$\sup_{0 \leq y < \infty} M(y) \leq \epsilon.$$

Since ϵ can be made arbitrarily small by choosing T large, the proof of (7.53) with $\beta(t) \equiv \gamma t$ is complete.

Finally, assume that (7.53) is true for given $\beta(t)$. Then

$$\left| 2^{-1/\gamma} - \left(1 + \frac{\gamma}{\beta(t)}t\right)^{-1/\gamma} \right| = \left| H_{\gamma,0,\gamma t}(t) - H_{\gamma,0,\beta(t)}(t) \right|$$

$$\leq \sup_{0 \leq y < \infty} \left| H_{\gamma,0,\gamma t}(y) - H_{\gamma,0,\beta(t)}(y) \right|$$

$$\leq \sup_{0 \leq y < \infty} \left| H_{\gamma,0,\gamma t}(y) - F_t(y) \right| + \sup_{0 \leq y < \infty} \left| F_t(y) - H_{\gamma,0,\beta(t)}(y) \right|.$$

As $t \to \infty$, the first supremum converges to 0 as proved above, while the second converges to 0 by assumption.

Thus as $t \to \infty$:

$$\left| 2^{-1/\gamma} - \left(1 + \frac{\gamma}{\beta(t)}t\right)^{-1/\gamma} \right| \to 0,$$

which obtains $\beta(t)/\gamma t \to 1$. ∎

Remark 7.50 ($H_{\gamma,0,\gamma t}(y)$ vs. $H_{\gamma,0,\beta}(y)$) *While the Pareto distribution $H_{\gamma,0,\gamma t}(y)$ is the exact asymptotic limit for the conditional distribution function $F_t(y)$ as $t \to \infty$, it is common in applications to assume the more general model of the generalized Pareto distribution, $H_{\gamma,0,\beta}(y)$. Given the chosen threshold t and data set, this approach provides two parameters to be determined by maximum likelihood or other estimation method rather than one. The desirability of two parameters is reinforced by the fact that the convergence to Pareto can be very slow indeed.*

Example 7.51 (Slow convergence in (7.54)) *As an illustration of the potential for very slow convergence in (7.54), we follow **Makarov (2007)** with:*

$$F(x) \equiv 1 - \frac{\ln x}{x}, \quad x \geq e.$$

Recalling (1.24), $F(x)$ is a mixed distribution function with both a continuous component on (e, ∞) and a saltus component at $x = e$.

Since for all $x > 0$:

$$\lim_{t \to \infty} \frac{1 - F(tx)}{1 - F(t)} = x^{-1},$$

$F \in \mathcal{D}(G_1)$ by (7.31).

For $t > e$:

$$F_t(y) = 1 - \frac{t}{t + y} \frac{\ln(t + y)}{\ln t},$$

while:

$$H_{1,0,t}(y) = 1 - \frac{t}{t + y}.$$

The value of

$$\sup_y \left[H_{1,0,t}(y) - F_t(y) \right],$$

is found by calculus to occur at $\hat{y} \equiv (e - 1)t$.

Thus:

$$\sup_y \left[H_{1,0,t}(y) - F_t(y) \right] = \frac{1}{e \ln t},$$

which converges to zero very slowly as $t \to \infty$. For example, to halve this supremum one must square the threshold t.

Bibliography

I have listed below a number of textbook references for the mathematics and finance presented in this series of books. All provide both theoretical and applied materials in their respective areas that are beyond those developed here and are worth pursuing by those interested in gaining a greater depth or breadth of knowledge. This list is by no means complete and is intended only as a guide to further study. In addition, various published research papers have been identified in some chapters where these results were discussed.

The reader will no doubt observe that the mathematics references are somewhat older than the finance references and upon web searching will find that some older texts have been updated to newer editions, sometimes with additional authors. Since I own and use the editions below, I decided to present these editions rather than reference the newer editions which I have not reviewed. As many of these older texts are considered "classics," they are also likely to be found in university and other libraries.

That said, there are undoubtedly many very good new texts by both new and established authors with similar titles that are also worth investigating. One that I will at the risk of immodesty recommend for more introductory materials on mathematics, probability theory and finance is:

[1] Reitano, Robert, R. *Introduction to Quantitative Finance: A Math Tool Kit.* Cambridge, MA: The MIT Press, 2010.

Topology, Measure, and Integration

[2] Doob, J. L. *Measure Theory.* New York, NY: Springer-Verlag, 1994.

[3] Dugundji, James. *Topology.* Boston, MA: Allyn and Bacon, 1970.

[4] Edwards, Jr., C. H. *Advanced Calculus of Several Variables.* New York, NY: Academic Press, 1973.

[5] Gemignani, M. C. *Elementary Topology.* Reading, MA: Addison-Wesley Publishing, 1967.

[6] Halmos, Paul R. *Measure Theory.* New York, NY: D. Van Nostrand, 1950.

[7] Hewitt, Edwin, and Karl Stromberg. *Real and Abstract Analysis.* New York, NY: Springer-Verlag, 1965.

[8] Royden, H. L. *Real Analysis,* 2nd Edition. New York, NY: The MacMillan Company, 1971.

[9] Rudin, Walter. *Principals of Mathematical Analysis,* 3rd Edition. New York, NY: McGraw-Hill, 1976.

[10] Rudin, Walter. *Real and Complex Analysis,* 2nd Edition. New York, NY: McGraw-Hill, 1974.

[11] Shilov, G. E., and B. L. Gurevich. *Integral, Measure & Derivative: A Unified Approach.* New York, NY: Dover Publications, 1977.

[12] Strang, Gilbert. *Introduction to Linear Algebra,* 4th Edition. Wellesley, MA: Cambridge Press, 2009.

Probability Theory & Stochastic Processes

[13] Billingsley, Patrick. *Probability and Measure,* 3rd Edition. New York, NY: John Wiley & Sons, 1995.

[14] Chung, K. L., and R. J. Williams. *Introduction to Stochastic Integration.* Boston, MA: Birkhäuser, 1983.

[15] Davidson, James. *Stochastic Limit Theory.* New York, NY: Oxford University Press, 1997.

[16] de Haan, Laurens, and Ana Ferreira. *Extreme Value Theory, An Introduction.* New York, NY: Springer Science, 2006.

[17] Durrett, Richard. *Probability: Theory and Examples,* 2nd Edition. Belmont, CA: Wadsworth Publishing, 1996.

[18] Durrett, Richard. *Stochastic Calculus, A Practical Introduction.* Boca Raton, FL: CRC Press, 1996.

[19] Feller, William. *An Introduction to Probability Theory and Its Applications,* Volume I. New York, NY: John Wiley & Sons, 1968.

[20] Feller, William. *An Introduction to Probability Theory and Its Applications,* Volume II, 2nd Edition. New York, NY: John Wiley & Sons, 1971.

[21] Friedman, Avner. *Stochastic Differential Equations and Applications, Volume 1 and 2.* New York, NY: Academic Press, 1975.

[22] Ikeda, Nobuyuki, and Shinzo Watanabe. *Stochastic Differential Equations and Diffusion Processes.* Tokyo: Kodansha Scientific, 1981.

[23] Karatzas, Ioannis, and Steven E. Shreve. *Brownian Motion and Stochastic Calculus.* New York, NY: Springer-Verlag, 1988.

[24] Kloeden, Peter E., and Eckhard Platen. *Numerical Solution of Stochastic Differential Equations.* New York, NY: Springer-Verlag, 1992.

[25] Lowther, George, *Almost Sure, A Maths Blog on Stochastic Calculus,* https://almostsure. wordpress.com/stochastic-calculus/

[26] Lukacs, Eugene. *Characteristic Functions.* New York, NY: Hafner Publishing, 1960.

[27] Nelson, Roger B. *An Introduction to Copulas,* 2nd Edition. New York, NY: Springer Science, 2006.

[28] Øksendal, Bernt. *Stochastic Differential Equations, An Introduction with Applications,* 5th Edition. New York, NY: Springer-Verlag, 1998.

[29] Protter, Phillip. *Stochastic Integration and Differential Equations, A New Approach.* New York, NY: Springer-Verlag, 1992.

[30] Revuz, Daniel, and Marc Yor. *Continuous Martingales and Brownian Motion,* 3rd Edition. New York, NY: Springer-Verlag, 1991.

[31] Rogers, L. C. G., and D. Williams. *Diffusions, Markov Processes and Martingales*, Volume 1, Foundations, 2nd Edition. Cambridge, UK: Cambridge University Press, 2000.

[32] Rogers, L. C. G., and D. Williams. *Diffusions, Markov Processes and Martingales*, Volume 2, Itô Calculus, 2nd Edition. Cambridge, UK: Cambridge University Press, 2000.

[33] Sato, Ken-Iti. *Lévy Processes and Infinitely Divisible Distributions.* Cambridge, UK: Cambridge University Press, 1999.

[34] Schilling, René L. and Lothar Partzsch. *Brownian Motion: An Introduction to Stochastic Processes,* 2nd Edition. Berlin/Boston: Walter de Gruyter GmbH, 2014.

[35] Schuss, Zeev, *Theory and Applications of Stochastic Differential Equations.* New York, NY: John Wiley and Sons, 1980.

Finance Applications

[36] Etheridge, Alison. *A Course in Financial Calculus.* Cambridge, UK: Cambridge University Press, 2002.

[37] Embrechts, Paul, Claudia Klüppelberg, and Thomas Mikosch. *Modelling Extremal Events for Insurance and Finance.* New York, NY: Springer-Verlag, 1997.

[38] Hunt, P. J., and J. E. Kennedy. *Financial Derivatives in Theory and Practice,* Revised Edition. Chichester, UK: John Wiley & Sons, 2004.

[39] McLeish, Don L. *Monte Carlo Simulation and Finance.* New York, NY: John Wiley, 2005.

[40] McNeil, Alexander J., Rüdiger Frey, and Paul Embrechts. *Quantitative Risk Management: Concepts, Techniques, and Tools.* Princeton, NJ.: Princeton University Press, 2005.

Research Papers for Book IV

[41] Bailey, R. W. "Polar generation of random variates with the t-distribution." Mathematics of Computation, 62(206), 779–781, 1994.

[42] Balkema, A., de Haan, L. "Residual life time at great age." Annals of Probability, 2, 792–804, 1974.

[43] Box, G. E. P., Muller, Mervin E. "A Note on the Generation of Random Normal Deviates." Annals of Mathematical Statistics, 29(2), 610–611, 1958.

[44] Cantelli, F. P. "Sulla determinazione empirica delle leggi di probabilita." Giornale dell Istituto Italiano degli Attuari, 4, 221–424, 1933.

[45] Dvoretzky, A., Kiefer, J., Wolfowitz, J. "Asymptotic minimax character of the sample distribution function and of the classical multinomial estimator." Annals of Mathematical Statistics, 27(3), 642–669, 1956.

[46] Fisher, R. A., Tippett, L. H. C. "Limiting forms of the frequency distribution of the largest or smallest member of a sample." Mathematical Proceedings of the Cambridge Philosophical Society, 24, 180–190, 1928.

[47] Glivenko, V. "Sulla determinazione empirica della legge di probabilita." Giornale dell Istituto Italiano degli Attuari, 4, 92–99, 1933.

[48] Gnedenko, B. "Sur la distribuion limite du terme maximum d'une série aléatoire." Annals of Mathematics, 44, 423–453, 1943.

[49] Heyde, C. C. "On a Property of the Lognormal Distribution." In: Maller R., Basawa I., Hall P., Seneta E. (eds) Selected Works of C.C. Heyde. Selected Works in Probability and Statistics. New York, NY: Springer, 2010.

[50] Hill, B. "A simple general approach to inference about the tail of a distribution." The Annals of Statistics, 3(5):1163–1174, 1975.

[51] Kolmogorov, A. "Sulla determinazione empirica di una legge di distribuzione." Giornale dell Istituto Italiano degli Attuari, 4, 83–91, 1933.

[52] Makarov, Mikhail. "Applications of exact extreme value theorem." Journal of Operational Risk, 2(1), 115–120, 2007.

[53] Massart, P. "The tight constant in the Dvoretzky–Kiefer–Wolfowitz inequality." The Annals of Probability, 18(3), 1269–1283, 1990.

[54] Pickands, J. "Statistical inference using extreme order statistics." Annals of Statistics, 3, 119–131, 1975.

[55] Rényi, Alfréd. "On the theory of order statistics." Acta Mathematica Academiae Scientiarum Hungarica, 4, 191–231, 1953.

[56] Scheffé, H. "A Useful Convergence Theorem for Probability Distributions," Annals of Mathematical Statistics, 18, 434–438, 1947.

[57] Smirnov, N. V. "Sur les écarts de la courbe de distribution empirique." (Russian, French summary). Matematicheskii Sbornik, 6, 3–26, 1939.

[58] Smirnov, N. V. "On the estimation of the discrepancy between empirical curves of distribution for two independent samples." Moscow University Mathematics Bulletin, 2(2), 3–11 1939.

[59] Smirnov, N. V. "Table for estimating the goodness of fit of empirical distributions." Annals of Mathematical Statistics, 19, 279–281, 1948.

Index

absolute monents, 89
absolutely continuous, 6
    measures, 12
alternating series theorem, 108

Balkema, A. A.
    Pickands-Balkema-de Haan theorem,
        211
Bernoulli, Jakob
    Bernoilli trial, 17
beta distribution, 23, 50, 54
    moments, 105
    simulating samples, 140
binomial coefficient, 17
binomial distribution
    Bernoulli distribution, 37
    general, 17
    moments, 103
    simulating samples, 138
    standard, 16
binomial theorem, 17
Borel-Cantelli theorem, 184
Box, George E. P.
    "All models are wrong...," 190
    Box–Muller transform, 142

canonical decomposition
    distribution function, 6
Cantelli, Francesco Paolo
    Glivenko-Cantelli theorem, 195
Cantor, Georg
    Cantor function, 7
Carathéodory, Constantin
    Carathéodory measurable, 2
Cauchy distribution, 25, 42, 52
    moments, 106
    simulating samples, 141
Cauchy, Augustin-Louis
    Cauchy convergence criterion, 177
    Cauchy's functional equation, 76
    Cauchy-Schwarz inequality, 114
Cauchy-Schwarz inequality, 114
ceiling function, 138

central limit theorem, 165
central moments, 89
    of a sum, 96
Chebyshev, Pafnuty
    Chebyshev's inequality, 109
Chernoff, Herman
    Chernoff bound, 202
    Cramér-Chernoff theorem, 208
chi-squared distribution, 22, 31, 51
    moments, 105
concave function
    strictly concave, 111
conditional distribution function
    of a random vector, 4, 73
conditional value at risk, 147
conjugate indexes, 117
continuous from above, 3
continuous probability theory, 19
convergence in probability
    random variable sequence, 178
convergence set
    sequence of random variables,
        177
convergence with probability 1
    random variable sequence, 183
convex function
    strictly convex, 111
convolution
    of functions, 37
copula, 9
correlation
    between two random variables, 97
covariance
    of two random variables, 97
Cramér, Harald
    Cramér-Chernoff theorem, 208
cumulant generating function, 102
cumulants, 102
cumulative distribution function (c.d.f.)
    of a random variable, 1
    of a random vector, 3
cylinder set
    infinite dimensional product space, 154

de Haan, Laurens
  Pickands-Balkema-de Haan theorem,
    211
de Moivre, Abraham
  central limit theorem, 161
  De Moivre-Laplace theorem, 24, 162
density function, 10, 13
  empirical, 191
density functions
  on $\mathbb{R}$, 10
  on $\mathbb{R}^n$, 67
discrete probability theory, 16
distribution function
  conditional distribution function,
    4
  continuous probability theory, 20
  discrete probability theory, 16
  empirical, 192
  left-continuous inverse, 20
  marginal distribution function, 4
  of a random variable, 1, 2
  of a random vector, 3
distribution functions
  beta, 23, 50, 54
  binomial, 16, 17, 37
  Cauchy, 25, 42, 52
  Chi-squared, 31
  chi-squared, 22, 51, 125
  continuous uniform, 20, 55, 60, 61
  discrete uniform, 16
  exponential, 21, 38, 55, 60
  extreme value, 60, 211
  F-distribution, 51
  gamma, 22, 38, 50, 54, 124
  generalized extreme value, 211
  generalized Pareto, 238
  geometric, 17, 124
  kth order statistic, 59
  lognormal, 25, 119, 124
  negative binomial, 18, 124
  normal, 23, 42, 51, 52, 124, 125
  Pareto, 212, 239
  Poisson, 19, 38
  Student T, 52, 125
domain of attraction
  extreme value theory, 211
Dvoretzky-Kiefer-Massart-Wolfowitz
    theorem, 199

equal in distribution
  random variables, 76

expectation
  of a function, 83
  of $g(X)$, 81
expected shortfall, 147
exponential distribution, 21, 38, 55, 60
  moments, 105
  simulating samples, 140
extreme value distribution, 60, 211
  index, 211
  simulating samples, 141

F-distribution, 51
fat tail, 223
Fisher, R.A.
  Fisher-Snedecor distribution, 51
Fisher, Ronald
  Fisher-Tippett-Gnedenko theorem,
    210
Fisher-Tippett theorem
  extreme value theory, 210
floor function
  greatest integer function, 174
Fourier, Jean-Baptiste Joseph
  Fourier transform, 90

gamma distribution, 22, 38, 50, 54
  moments, 105
  simulating samples, 140
gamma function, 22
generalized extreme value (GEV), 211
generalized Pareto distribution, 238
geometric distribution, 17
  generalized, 18
  moments, 103
  simulating samples, 138
GEV distribution
  generalized extreme value , 211
Glivenko, Valery
  Glivenko-Cantelli theorem, 195
Gnedenko, Boris
  Fisher-Tippett-Gnedenko theorem,
    210
Gosset, William Sealy
  Student's T distribution, 52, 194
greatest integer function
  floor function, 173, 174

Heyde, C. C.
  lognormal moments example, 119
Hill, Bruce M.
  Hill estimator, 212

histogram, 191
    bins, 192
Hölder, Otto
    Hölder's Inequality in $\mathbf{R}^n$ or $\mathbb{C}^n$, 117

i.i.d., 214
i.i.d.-X, 57
independent
    random variables, 5
    random vectors, 5

Jacobi, Carl Gustav Jacob , 143
Jacobian matrix
    determinant, 143
Jensen's inequality, 112

Karamata, Jovan
    Karamata's Representation theorem,
      224
Kolmogorov, Andrey
    Kolmogorov's inequality, 113
    Kolmogorov's theorem, 197
    Kolmogorov's zero-one law, 177
    Kolmogorov-Smirnov statistic, 195
kth order statistic, 57, 59

Laplace, Pierre-Simon
    De Moivre-Laplace theorem, 24, 162
    Laplace transform, 90
Lebesgue, Henri
    Lebesgue's decomposition theorem, 13
    Lebesgue-Stieltjes measure, 86
left-continuous inverse
    of a distribution function, 20
likelihood function
    maximum likelihood estimate, 213
limit
    of a sequence of sets, 183
limit inferior
    of a sequence of sets, 183
limit superior
    of a sequence of sets, 183
lognormal distribution, 25
    moments, 107
    simulating samples, 141
Lyapunov, Aleksandr
    Lyapunov's inequality, 118

marginal distribution function
    of a random vector, 4, 70
Markov, Andrey
    Markov's inequality, 110

mean
    of a distribution, 88
    of a sum, 95
method of moments, 126
moment generating function, 89
    of a sum, 97
moments
    absolute moments, 89
    mean, 88
    moment generating function,
      89
    nth central moment, 89
    nth moment, 88
    of a sum, 95
    of distributions, 88
    standard deviation, 89
    variance, 89
Muller, Mervin E.
    Box-Muller transform, 142
multinomial theorem, 95
mutually singular
    measures, 12

n-increasing, 3
negative binomial distribution, 18
    moments, 103
    simulating samples, 139
Nikodým, Otto
    Radon-Nikodým theorem, 12
normal density function
    approximation to binomial
      half integer adjustment, 165
normal distribution, 23, 31, 42, 51, 52
    moments, 106
    simulating samples, 141
normalized random variable, 162

odd function, 43, 120
ordered samples
    order statistic, 57

Pareto, Vilfredo
    Pareto distribution, 212, 238
peaks over threshold, 238
permutation, 62
Pickands III, James
    Pickands-Balkema-de Haan theorem,
      211
Poisson, Siméon-Denis
    Poisson distribution, 19, 38
      simulating samples, 139

Poisson Limit theorem, 158
Poisson moments, 104
probability density function
    continuous probability theory, 20
    discrete probability theory, 16
product space
    infinite dimensional, 154

Rényi, Alfréd
    order statistics, 75, 149
Radon, Johann
    Radon-Nikodým theorem, 12
random sample, 58
random variable (r.v.), 1
    independent, 5
random variable sequence
    convergence in probability, 178
    convergence with probability
        1, 183
random variables
    triangular array, 159
random vector, 3, 91
    independent, 5
ratio test
    for a series, 108
rectangular distribution
    discrete , 16, 20
    moments, 102
    simulating samples, 138
regularly varying function
    at infinity, 220, 223

saltus function, 6
sample
    M-sample, 57
Scheffé, Henry
    Scheffé's theorem, 171
Schwarz, Hermann
    Cauchy-Schwarz inequality, 114
second order condition
    extreme value theory, 236
sigma algebra
    generated by a random variable, 176
singular function, 6
Sklar, Abe
    Sklar's theorem, 9
slowly varying function
    at infinity, 220
Smirnov, N. V. (Nikolai Vasil'evich)
    Kolmogorov-Smirnov statistic, 195

Smirnov's limit theorem on order
    statistics, 168
Smirnov's theorem, 199
Snedecor, George W.
    Snedecor's F-distribution, 51
standard deviation, 89
Stieltjes, Thomas
    Lebesgue-Stieltjes measure, 86
Stirling's formula
    Stirling's approximation, 124
strong convergence set
    i.i.d. random variables, 183
Student T distribution, 52, 144, 194

tail conditional distribution, 223
    relative, 223
tail event
    tail sigma algebra, 177
tight
    sequence of distribution functions, 126
    sequence of probability measures, 126
tilted distribution, 207
Tippett, L. H. C.
    Fisher-Tippett-Gnedenko theorem, 210
triangular array
    random variables, 159
truncation
    of a random variable, 179
twisted density, 209
twisted distribution, 207

uniform distribution
    continuous, 20, 55, 60, 61
        simulating samples, 140
    discrete , 16
    moments, 104
uniform integrablility
    random variables, 129

value at risk, 147
variance, 89
    general sum, 97
    independent sum, 96
Von Mises, Richard
    von Mises' condition, 212

weak convergence
    distributions, 156

Young, W. H.
    Young's inequality, 117

Printed in the United States
by Baker & Taylor Publisher Services